Quantitative Risk Assessment of *Listeria monocytogenes* in Foods

Quantitative Risk Assessment of *Listeria monocytogenes* in Foods

Guest Editor

Michel Federighi

MDPI

Basel • Beijing • Wuhan • Barcelona • Belgrade • Novi Sad • Cluj • Manchester

Guest Editor
Michel Federighi
HQSA
EnvA
Maisons-Alfort
France

Editorial Office
MDPI AG
Grosspeteranlage 5
4052 Basel, Switzerland

This is a reprint of the Special Issue, published open access by the journal *Foods* (ISSN 2304-8158), freely accessible at: https://www.mdpi.com/journal/foods/special_issues/83GCU364KY.

For citation purposes, cite each article independently as indicated on the article page online and as indicated below:

Lastname, A.A.; Lastname, B.B. Article Title. *Journal Name* **Year**, *Volume Number*, Page Range.

ISBN 978-3-7258-5467-7 (Hbk)
ISBN 978-3-7258-5468-4 (PDF)
https://doi.org/10.3390/books978-3-7258-5468-4

Contents

foods

MDPI

Editorial

Quantitative Risk Assessment of *Listeria monocytogenes* in Foods

Juliana De Oliveira Mota [1] and Michel Federighi [2,3,*]

1 Department of Nutrition and Food Safety, World Health Organization, 1211 Geneva, Switzerland
2 Laboratoire de Sécurité des Aliments, Anses, 94700 Maisons-Alfort, France
3 HQSA Unit, Ecole Nationale Vétérinaire d'Alfort, 94700 Maisons-Alfort, France
* Correspondence: michel.federighi@vet-alfort.fr

1. Introduction

Listeriosis is a rare foodborne infection caused by *Listeria monocytogenes*, with 0.1 to 10 cases per 1 million people per year according to the World Health Organization [1]. It is a 1–2 µm long bacillus that can multiply at low temperatures and in the presence of high salt concentrations. This bacterium is ubiquitous: it can be found in water, soil, plants, and many animals. In addition, this bacterium has the ability to contaminate ready-to-eat foods and food production sites. It is sensitive to heat but can multiply slowly at low temperatures.

Often occurring sporadically, their widespread dissemination in a wide range of foods can cause epidemics. It is important to remember that vertical transmission (from mother to fetus) through the placenta also occurs. The main clinical forms of listeriosis include septicemia and infection of the central nervous system or neuro-listeriosis, associated with a mortality rate of 30 to 45%, respectively, and maternal–neonatal forms. For example, in France, there is a ratio of 10 between neonatal forms (40/year) and other forms (400 to 500/year).

The scope of this Special Issue 'Quantitative Risk Assessment of *Listeria monocytogenes* in Foods' is twofold: (1) it aims to present a critical analysis of the existing human listeriosis quantitative risk assessment models conducted on milk and dairy products, produce, seafood, and meat products, highlighting common results from intervention strategies despite the unavoidable differences in model scopes and assumptions; (2) it also aims to present new quantitative risk assessment models of a longer food chain scope, in the light of new data and approaches, to control listeriosis linked to specific foods such as frozen vegetables, cantaloupe, and ready-to-eat seafood. In response to a request by the Codex Committee on Food Hygiene (CCFH) during its 52nd session, formal risk assessment models were developed by the Joint FAO/WHO Expert Meeting on Microbiological Risk Assessment of *Listeria monocytogenes* in Foods: Part 1 [2], taking into account the effects of agrifood practices, climate change, and the latent possibility of cross-contamination along the food chain. Programmed models were subsequently tested and revised at the Joint FAO/WHO Expert Meeting on Microbiological Risk Assessment of *Listeria monocytogenes* in Foods, Part 2: Risk Assessment Models [3].

 https://doi.org/10.3390/foods14193353

2. Overview of the Published Articles

2.1. Critical Reviews of Risk Assessment Models for Listeria monocytogenes in Foods

This Special Issue provided the opportunity to critically review existing risk assessment models for four food commodities groups, namely dairy products [4], meat and meat products [5], seafood [6], and produce [7]. The reviews aimed to identify and evaluate the relative effectiveness of the control measures and intervention strategies along the food chain.

The first review focused on dairy products [4]. Eighteen quantitative risk assessment (QRA) models were identified with a focus on cheeses, especially those made from raw and pasteurized milk, often within long supply chains. A key finding was the important role of the on-farm contamination in overall listeriosis risk, particularly in the context of raw milk cheeses. Consumer handling practices had greater influence on listeriosis risk than retail conditions, while storage temperature was found to have a higher impact than the storage time.

The review of the existing QRA for *L. monocytogenes* in meat and meat products identified 23 models with a main focus on ready-to-eat (RTE) meat products and short supply chains [5]. Key factors influencing risk estimations included lag time, growth rate, and maximum microbial density, especially under the influence of growth inhibitors and lactic acid bacteria. Retail conditions emerged as an important contributing factor to risk, with deli meat sliced during retail presenting higher potential risks than pre-packed products. As for dairy, storage temperature was shown to have more impact on risk than storage duration. However, reducing *L. monocytogenes* prevalence and levels at the end of processing was found to be more effective than relying on storage conditions later in the chain, highlighting the need to incorporate processing modules into risk models.

For seafood [6], most of the 13 models focused on smoked or graved fish, all with short scopes, from the end of processing (packed products) to consumption, thus excluding cross-contamination modeling. As with previous food commodities, controlling storage temperature was more effective than acting on the storage duration. Again, prevalence and levels at the end of processing were found to be more effective than relying on storage conditions later in the chain. Therefore, this highlighted the importance of incorporating processing modules in order to better control the levels of microbial contamination at an earlier stage.

Finally, a review of produce identified 13 relevant articles, most of which focused on fresh or RTE leafy greens [7]. None of the models accounted for key factors or sources of contamination during primary production, such as the type of cultivation, water, fertilizers, or irrigation method/practices. Cross-contamination during processing was modeled using transfer rates, which moderately influenced the final risk, emphasizing the importance of accurately representing the transfer coefficient parameters. Temperature fluctuations at retail and temperature misuse during home storage were consistently identified as major contributors to listeriosis risk. The review advocated for extending the scope to include a primary module to assess current on-farm practices and controls, as well as to incorporate a refined sanitation module to test their effectiveness.

Across all food categories, the review concluded that the enhanced integration of cross-contamination pathways and emerging technologies, the incorporation of detailed processing modules, and the use of well-informed kinetic parameters, such as the exponential growth rate, minimum growth temperature, and maximum population density, are essential steps toward more robust and realistic QRA models for *L. monocytogenes*.

2.2. Updated Dose–Response Models

Pouillot et al. [8] developed the updated dose–response (DR) model incorporating both the virulence class of the *L. monocytogenes* strain and demographic factors such as age and sex. This model improves the accuracy of risk estimation across population groups and strain types, providing the necessary data are available. An open-source R package was developed to facilitate the practical application of these DR models.

2.3. Quantitative Risk Assessment Models for Listeria monocytogenes

Building on the previous critical reviews and expert inputs, QRA models for *L. monocytogenes* were developed in order to estimate the risk of invasive listeriosis associated with the consumption of three different food commodities, namely non-RTE frozen vegetables [9], RTE smoked and graved fish [10], and RTE cantaloupe melon [11].

Gonzales-Barron et al. [9] developed the QRA model for non-RTE frozen vegetables, beginning at the freezing facility, with the assumption that incoming vegetables were already preconditioned (e.g., trimmed, peeled, and washed). The model incorporates the effect of blanching, potential cross-contamination during freezing and packaging, and consumer handling activities such as defrosting and cooking.

Gonzales-Barron et al. [10] described a QRA model for RTE cold-smoked salmon produced by brining. The model covered other production steps such as smoking, slicing, and packing. It also accounted for cold-chain storage and consumer handling.

The last model developed by Guillier et al. [11] aimed to estimate the risk of listeriosis from the consumption of RTE cantaloupe. The model includes stages from pre-harvesting, harvesting, cleaning, and washing to processing, cold-chain storage, and consumer handling.

Stochastic modeling was applied to simulate the risk of listeriosis, accounting for both within-lot variability (differences between individual units in a single production lot) and between-lot variability (differences across production lots). The models also simulated within-lot end-product testing scenarios.

Risk is first estimated for each unit and then averaged across the units to produce a lot-level risk. Subsequently, the average or median risk per serving is calculated across all lots, providing an overall estimate used for comparison and scenario analysis.

The QRA model was made freely available as an open-source R package with full documentation and can be used as a tool to inform the consideration of strengthened risk management measures in view of the current changes in consumer behavior and new diet trends.

3. Conclusions

The models developed in this Special Issue are valuable tools for health authorities and food business operators to make the necessary decisions regarding this important public health issue. These tools are particularly useful and, moreover, are scalable. Nevertheless, they face certain biological limitations inherent to the specific microorganism *Listeria monocytogenes*.

In particular, they do not take into account this microorganism's ability to colonize various abiotic surfaces by forming biofilms, particularly in industrial food environments. The genomic diversity of *L. monocytogenes* is significant and is linked to this ability to form biofilms for certain strains but also to different levels of virulence (hyper- or hypovirulent strains) [12]. There is no doubt that incorporating this data into models will improve the accuracy of quantitative risk assessments. In addition, it is known that this microorganism is capable of adopting viable but non-culturable forms, which are undetectable by standard research methods and can lead to risk underestimation [13].

This Special Issue highlighted the potential variations in risks associated with *L. monocytogenes* under different agricultural and processing conditions. The main results also highlight the importance of maintaining good agricultural practices in the field, adhering to strict hygiene practices during harvesting and processing, and ensuring appropriate storage conditions and handling practices by consumers in order to mitigate the risk of listeriosis.

At the same time, phylodynamic studies are needed to better understand the plasticity of the *Listeria* genome, which leads to the coexistence of several subtypes of *Listeria monocytogenes* in the same reservoir [14]. We also need to better understand the relationship between the improved fitness of *L. monocytogenes* and its ability to spread in many biotopes and/or cause more disease [15]. This "old" pathogen certainly continues to surprise us, and there is still much to learn about it.

Author Contributions: Conceptualization, writing—review and editing, J.D.O.M. and M.F. All authors have read and agreed to the published version of the manuscript.

Funding: This research received no external funding.

Data Availability Statement: Not applicable.

Conflicts of Interest: The authors declare no conflicts of interest.

References

1. WHO Listeriosis. Available online: https://www.who.int/news-room/fact-sheets/detail/listeriosis (accessed on 9 September 2025).
2. FAO & WHO Listeria monocytogenes in Ready-to-Eat (RTE) Food: Attribution, Characterization and Monitoring: Meeting Report. Available online: https://www.who.int/publications/i/item/9789240034969 (accessed on 1 July 2025).
3. FAO & WHO Joint. FAO/WHO Expert. In Proceedings of the Microbiological Risk Assessment of Viruses in Foods Part 2: Prevention and Intervention Measures, Geneva, Switzerland, 12–16 February 2024.
4. Gonzales-Barron, U.; Cadavez, V.; Guillier, L.; Sanaa, M. A Critical Review of Risk Assessment Models for *Listeria monocytogenes* in Dairy Products. *Foods* **2023**, *12*, 4436. [CrossRef] [PubMed]
5. Gonzales-Barron, U.; Cadavez, V.; De Oliveira Mota, J.; Guillier, L.; Sanaa, M. A Critical Review of Risk Assessment Models for *Listeria monocytogenes* in Meat and Meat Products. *Foods* **2024**, *13*, 359. [CrossRef] [PubMed]
6. Gonzales-Barron, U.; Cadavez, V.; De Oliveira Mota, J.; Guillier, L.; Sanaa, M. A Critical Review of Risk Assessment Models for *Listeria monocytogenes* in Seafood. *Foods* **2024**, *13*, 716. [CrossRef] [PubMed]
7. Gonzales-Barron, U.; Cadavez, V.; De Oliveira Mota, J.; Guillier, L.; Sanaa, M. A Critical Review of Risk Assessment Models for *Listeria monocytogenes* in Produce. *Foods* **2024**, *13*, 1111. [CrossRef] [PubMed]
8. Pouillot, R.; Kiermeier, A.; Guillier, L.; Cadavez, V.; Sanaa, M. Updated Parameters for *Listeria monocytogenes* Dose–Response Model Considering Pathogen Virulence and Age and Sex of Consumer. *Foods* **2024**, *13*, 751. [CrossRef] [PubMed]
9. Gonzales-Barron, U.; Pouillot, R.; De Oliveira Mota, J.; Hasegawa, A.; Allende, A.; Dong, Q.; Stasiewicz, M.J.; Kovacevic, J.; Cadavez, V.; Guillier, L.; et al. A Quantitative Risk Assessment Model for *Listeria monocytogenes* in Non-Ready-to-Eat Frozen Vegetables. *Foods* **2024**, *13*, 3610. [CrossRef] [PubMed]
10. Gonzales-Barron, U.; Pouillot, R.; Skjerdal, T.; Carrasco, E.; Teixeira, P.; Stasiewicz, M.J.; Hasegawa, A.; De Oliveira Mota, J.; Guillier, L.; Cadavez, V.; et al. A Quantitative Risk Assessment Model for *Listeria monocytogenes* in Ready-to-Eat Smoked and Gravad Fish. *Foods* **2024**, *13*, 3831. [CrossRef] [PubMed]
11. Guillier, L.; Gonzales-Barron, U.; Pouillot, R.; De Oliveira Mota, J.; Allende, A.; Kovacevic, J.; Guldimann, C.; Fazil, A.; Al-Qadiri, H.; Dong, Q.; et al. A Quantitative Risk Assessment Model for *Listeria monocytogenes* in Ready-to-Eat Cantaloupe. *Foods* **2025**, *14*, 2212. [CrossRef] [PubMed]
12. Félix, B.; Sevellec, Y.; Palma, F.; Douarre, P.E.; Felten, A.; Radomski, N.; Mallet, L.; Blanchard, Y.; Leroux, A.; Soumet, C.; et al. A European-wide dataset to uncover adaptive traits of *Listeria monocytogenes* to diverse ecological niches. *Sci. Data* **2022**, *9*, 190. [CrossRef] [PubMed]
13. Carvalho, F.; Carreaux, A.; Sartori-Rupp, A.; Tachon, S.; Gazi, A.D.; Courtin, P.; Nicolas, P.; Dubois-Brissonnet, F.; Barbotin, A.; Desgranges, E.; et al. Aquatic environment drives the emergence of cell wall-deficient dormant forms in *Listeria*. *Nat. Commun.* **2024**, *15*, 8499. [CrossRef] [PubMed]

14. Félix, B.; Capitaine, K.; Te, S.; Felten, A.; Gillot, G.; Feurer, C.; van den Bosch, T.; Torresi, M.; Lancz, Z.S.; Delannoy, S.; et al. Identification by High-Throughput Real-Time PCR of 30 Major Circulating *Listeria monocytogenes* Clonal Complexes in Europe. *Microbiol. Spectr.* **2023**, *11*, e0395422. [CrossRef] [PubMed]
15. Palma, F.; Radomski, N.; Guérin, A.; Sévellec, Y.; Félix, B.; Bridier, A.; Soumet, C.; Roussel, S.; Guillier, L. Genomic elements located in the accessory repertoire drive the adaptation to biocides in *Listeria monocytogenes* strains from different ecological niches. *Food. Microbiol.* **2022**, *106*, 103757. [CrossRef] [PubMed]

foods

MDPI

Article

A Quantitative Risk Assessment Model for *Listeria monocytogenes* in Ready-to-Eat Cantaloupe

Laurent Guillier [1,*], Ursula Gonzales-Barron [2], Régis Pouillot [3], Juliana De Oliveira Mota [4], Ana Allende [5], Jovana Kovacevic [6], Claudia Guldimann [7], Aamir Fazil [8], Hamzah Al-Qadiri [9], Qingli Dong [10], Akio Hasegawa [4], Vasco Cadavez [2] and Moez Sanaa [4,*]

1 Risk Assessment Department, French Agency for Food, Environmental and Occupational Health & Safety (Anses), 14 rue Pierre et Marie Curie, 94700 Maisons-Alfort, France
2 CIMO, LA SusTEC, Instituto Politécnico de Bragança, Campus de Santa Apolónia, 5300-253 Bragança, Portugal; ubarron@ipb.pt (U.G.-B.); vcadavez@ipb.pt (V.C.)
3 Consultant, 18 rue Mohamed Al Ghazi, Rabat 10170, Morocco; rpouillot.work@gmail.com
4 Department of Nutrition and Food Safety, World Health Organization, 1211 Geneva, Switzerland; deju@who.int (J.D.O.M.); hasegawaa@who.int (A.H.)
5 CEBAS-CSIC, Microbiology and Quality of Fruits and Vegetables Group, Campus de Espinardo, 25, 30100 Murcia, Spain; aallende@cebas.csic.es
6 Food Innovation Center, Oregon State University, 1207 NW Naito Parkway, Portland, OR 97209, USA; jovana.kovacevic@oregonstate.edu
7 Competence Center for Food Safety, Chair for Food Safety and Analytics, Faculty of Veterinary Medicine, Ludwig-Maximilians-Universität München (LMU), Schönleutnerstrasse 8, 85764 Oberschleissheim, Germany; c.guldimann@lmu.de
8 Public Health Risk Sciences Division (PHRS), Public Health Agency of Canada, 370 Speedvale Avenue West, Guelph, ON N1H 7M7, Canada; aamir.fazil@phac-aspc.gc.ca
9 Department of Nutrition and Food Technology, School of Agriculture, The University of Jordan, Amman 11942, Jordan; h.qadiri@ju.edu.jo
10 School of Health Science and Engineering, University of Shanghai for Science and Technology, Shanghai 200093, China; qdong@usst.edu.cn
* Correspondence: laurent.guillier@anses.fr (L.G.); sanaam@who.int (M.S.); Tel.: +33-01-4977-1300 (L.G.)

Abstract: This study introduces a farm-to-fork quantitative risk assessment (QRA) model for invasive listeriosis from ready-to-eat diced cantaloupe. The modular model comprises seven stages—preharvest (soil and irrigation contamination), harvest (cross-contamination and survival), pre-processing (brushing), processing (flume tank washing, dicing and equipment cross-contamination), lot testing, cold-chain transport and retail growth, and consumer storage/handling. Each stage employs stochastic functions to simulate microbial prevalence and concentration changes (growth, inactivation, removal, partitioning, cross-contamination) using published data. In a reference scenario—good agricultural practices (soil barriers, no preharvest irrigation), hygienic processing and proper cold storage—the model predicts low lot- and pack-level contamination, with few packs >10 CFU/g and most servings below detection; the mean risk per serving is very low. "What-if" analyses highlight critical control points: the absence of soil barriers with preharvest irrigation can increase the risk by 10,000-fold; flume tank water contamination has a greater impact than harvest-stage cross-contamination; and poor consumer storage can raise the risk by up to 500-fold. This flexible QRA framework enables regulators and industry to evaluate and optimize interventions—from improved agricultural measures to targeted sampling plans and consumer guidance—to mitigate listeriosis risk from RTE diced cantaloupe.

Keywords: melon; primary production; listeriosis; exposure assessment; simulation

1. Introduction

Historical data on foodborne outbreaks demonstrate that cantaloupes are frequently implicated in incidents of disease caused by foodborne pathogens, posing a substantial public health concern. Common implicated pathogens include Shiga toxin-producing *Escherichia coli* (STEC), *Salmonella* [1–4], norovirus [5], and *Listeria monocytogenes* [6,7]. Given that listeriosis cases mostly occur sporadically [8], findings from a meta-analysis of case–control studies focusing on sporadic listeriosis [9] highlight the role of the consumption of produce as a significant risk factor among various categories of ready-to-eat (RTE) foods. Notably, within the produce category, fruit consumption—specifically melons, cantaloupes, strawberries, and RTE fruit salads—was significantly associated with an increased risk of listeriosis among susceptible populations (pooled odds ratio = 1.538; 95% CI: 1.1431–2.070).

Given the ubiquitousness of *L. monocytogenes* in the environment, multiple routes of contamination for cantaloupes can be identified. *Listeria* species thrive in diverse habitats, predominantly in various soil types where they exist at low levels, below 10^4 CFU·g^{-1} [10,11]. Additionally, *L. monocytogenes* colonizes the root systems of plants [12,13] and is hosted by various soil-dwelling organisms, including unicellular eukaryotes, nematodes, isopods, dipterans, and beetles [14,15]. Agricultural soils can further be contaminated through the application of organic fertilizers [16,17]. Beyond soil, irrigation water serves as another contamination route, with studies identifying it as a risk factor for *L. monocytogenes* presence in agricultural fields [16,18]. The inherent growth pattern of cantaloupes, in direct contact with the ground, further increases their vulnerability to contamination from soil and irrigation water. Additionally, the intricate net-like structure of the rind provides an ideal surface for pathogenic bacteria to adhere and persist on the fruit's surface [19,20].

An additional route of contamination is throughout the different processing steps after harvesting. Despite the implementation of good hygiene practices and environmental controls in production settings, it has been observed that some strains of *L. monocytogenes* can establish themselves within (any) food processing environment and remain there for several months, or even years [21,22]. The investigation of the 2011 outbreak linked to cantaloupes that resulted in 147 listeriosis cases in the United States (US) allowed identifying several potential factors that may have contributed to the contamination. These factors were particularly associated with the food contact surfaces in a cantaloupe packing facility [23]. In 2013, 17 cantaloupe packinghouses in US were surveyed by the FDA, examining both food-contact and non-food-contact surfaces for *L. monocytogenes*. The survey reported *L. monocytogenes* at one facility and nonpathogenic *Listeria* species at eight others [24]. A research study involving Florida cantaloupe packinghouses, which investigated the prevalence of *L. monocytogenes* and indicator microorganisms, found a low prevalence of *L. monocytogenes* in 2013 (2/374 samples), and no detected *L. monocytogenes* in subsequent testing in 2014 [25]. Notably, total plate counts, coliforms and *E. coli* were found to be poor indicators of *L. monocytogenes* presence in cantaloupe packinghouse environments [25]. The role of water use during washing and rinsing has also been highlighted in the cross-contamination of fruit and vegetables [26]. Cross-contamination can also occur at the cutting stage [27,28]. During slicing, either in processing facilities or in household kitchens, these bacteria can be transferred from the rind to fresh-cut pieces [29]. The internal environment of the cantaloupe, characterized by high sugar content and relatively low acidity (pH~6.7) [30], provides ideal conditions for the growth of *L. monocytogenes* and other foodborne pathogens [31]. If these contaminated fruits are not properly refrigerated, *L. monocytogenes* can proliferate to dangerous levels before any visible spoilage occurs [32]. Given the potential for contamination at any stage—from initial field production through to postharvest processing and final consumer handling—an expert committee of the Joint FAO/WHO Expert Meetings on Microbiological Risk Assessment (JEMRA) has underscored the need for a comprehensive production-to-consumption risk assessment to track and manage the introduction and propagation of *L. monocytogenes* in cantaloupes.

A recent review was conducted on existing quantitative risk assessment (QRA) models for *L. monocytogenes* in produce [33]. Most (9/13) QRA models examined focused on fresh or RTE leafy greens. Only one QRA model addressed cantaloupes [34]. This model considered the contamination of RTE cantaloupes in processing plants and aimed to assess the impact of improved hygiene and cold storage of the products. Yet, none of these QRA models incorporated critical contamination factors during primary production, including cropping systems, water, fertilizers, or irrigation methods and practices.

The JEMRA expert committee has thus recommended the development of a new QRA model that comprehensively encompasses the journey of cantaloupe from primary production to consumption, utilizing a modular approach supported by existing data. The JEMRA expert committee analyzed and provided recommendations on existing methods, data and models on the basis of the review carried out previously [33], and informed when gaps existed. Key stages to be incorporated would include growing, harvesting, cooling, washing, sanitizing, and dicing, alongside transportation, retail display, and consumer handling practices. The model must offer the flexibility to incorporate variables such as farming techniques, extreme weather conditions, climate change impacts, and varying market practices. It would consider potential cross-contamination at multiple points—ranging from preharvest environments (e.g., irrigation water and soil) to processing activities (e.g., equipment surfaces and pooling of fruits) and consumer-level interactions (e.g., contamination transfer during cutting).

Within this context, the main objective of this work was to construct a QRA model of longer scope for RTE cantaloupe than previously published, able to represent the survival, re- and cross-contamination, and potential growth of *L. monocytogenes* on/in cantaloupes from production to consumption. Widely based on published data, the model was structured to be able to assess the effectiveness of interventions and control measures, in line with the recent JEMRA recommendations [35,36]. The functionality of the QRA model proposed was illustrated by a reference scenario and by the development of what-if scenario analysis.

2. Materials and Methods

2.1. Exposure Assessment

The exposure assessment model for RTE cantaloupes comprises seven modules (Figure 1). The "Preharvest" module covers practices related to crop growth before harvest. It incorporates the use of barriers (mulch, foil) against soil contaminants by minimizing direct contact between crops, and the soil growing of cantaloupe was with or without soil barriers (i.e., mulch); also incorporated are the type of irrigation system (i.e., drip or overhead) and the use and type of soil amendments (e.g., non-composted organic fertilizers). These choices can impact the microbial environment around the crops and influence the potential risk of contamination. The "Harvest" module focuses on the hygiene management of food contact surfaces to minimize contamination during harvest. The "Pre-Processing" module addresses the management of brushing and washing, which is essential for removing contaminants from the surface of the cantaloupes. However, poor maintenance of the microbiological quality of the washing water can become a source of contamination. The "Processing module" emphasizes the importance of food contact surface hygiene and the potential transfer of contamination from the rind to flesh. The "Cold Chain Storage" module refers to the control of storage time and temperature during transport to retail to inhibit the growth of *L. monocytogenes*. Finally, the "Consumer Handling" module covers how the RTE cantaloupe is handled by consumers, with an emphasis mainly on transport from retail to consumers' homes and storage at home (i.e., time and temperature). Each stage was modeled as a stochastic function estimating microbial prevalence and quantity based on fundamental processes [37] such as microbial growth, death, partitioning, removal or cross-contamination, according to the stage's purpose. Additionally, a seventh module, "Microbiological Lot Testing," is included to assess the effectiveness of microbial testing strategies in identifying and rejecting contaminated batches before distribution, thereby reducing the overall risk.

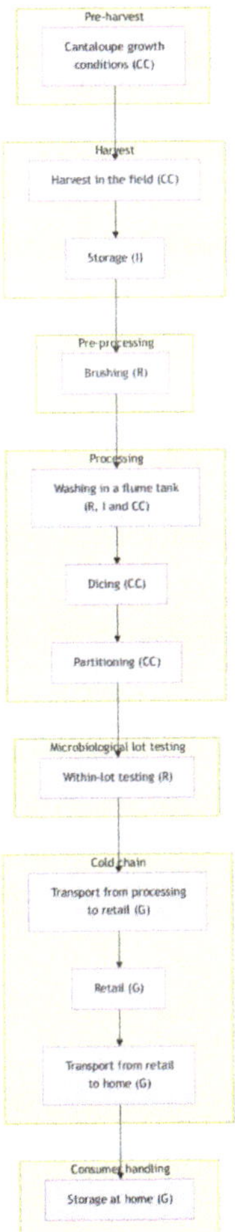

Figure 1. Flowchart for the RTE diced cantaloupe production process with indication of the modeled processes: CC: cross-contamination; I: inactivation; R: removal; G: growth.

Table 1 summarizes the modules, the sequence of stages and processes they consisted of, the assumption and data sources employed, and the corresponding functions programmed in the R software (version 4.4.1).

2.1.1. Preharvest Module

Data and assumptions:

The preharvest module for cantaloupe addresses agricultural practices and the potential contamination of *L. monocytogenes* from environmental sources, such as soil and irrigation water.

Factors such as growing cantaloupes using mulch as a soil barrier and the likelihood of soil contamination due to its ground-level growth are incorporated into the model. The prevalence of *L. monocytogenes* in agricultural soils varies widely (0.0 to 19.0%; see Table 2) and is influenced by factors such as the use of animal manure, soil conditions (e.g., pH, temperature), and irrigation methods [10,18,38–45]. Table 3 lists the effect of various risk factors on the soil prevalence of *L. monocytogenes*, as identified in the scientific literature. The survival of *L. monocytogenes* in soil is not significantly affected by manure type but varies depending on soil contamination methods [43,45]. Water sources such as municipal, rain, ground, and especially surface water also present variable contamination risks [46,47]. Additionally, fields located near pastures or water bodies face higher contamination risks, with surrounding land use, topography, and climate further affecting the risk [36,48]. Lastly, the model considers the natural decline in *L. monocytogenes* on cantaloupe rind before harvest [20,49].

Table 4 presents data on the occurrence of *L. monocytogenes* in irrigation water used for crop production. Three studies have quantified *L. monocytogenes* concentrations in positive water samples. Sharma et al. [50] and Acheamfour et al. [51] reported very low levels of *L. monocytogenes* in surface water sources, including creeks, ponds, tidal brackish river, and non-tidal fresh rivers, with MPN values ranging from <0.03 to 11 MPN/L. In contrast, Iwu et al. [52] reported higher mean concentrations of *L. monocytogenes* in irrigation water samples from 19 agricultural sites in South Africa (mean: 1196 CFU/100 mL; range: 0–5667 CFU/100 mL).

Table 1. Sequence of stages, microbial processes represented, data sources, assumptions and corresponding functions coded in R for the construction of the exposure assessment model of *Listeria monocytogenes* (LM) in RTE cantaloupe.

Module	Stage	Microbial Process	Assumptions	Sources	Function in R
Preharvest	Soil and irrigation water contamination	Contamination	The function assumes cantaloupe contamination through soil and irrigation water, with soil and water contamination characteristics such as the prevalence (pSoil and pIrrig) and distributions of *L. monocytogenes* concentration in soil and water as inputs. Risk factors such as irrigation prior to harvest, the use of organic fertilizer, or the use of a soil barrier (mulch) are included. It makes use of the outputs of caIrrig2rind and caSoil2rind sub-functions (see below).	[38,53–55]	caPrimaryProduction
	Irrigation water to rind contamination	Contamination	The function evaluates the contamination of cantaloupes through irrigation water only. It considers water contamination characteristics: prevalence (pIrrig) and concentration (cIrrig), which have to be chosen by the user based on existing data. cIrrig is conditional to water sources contaminated with *L. monocytogenes*.	[38,53–55]	caIrrig2rind
	Soil-to-rind contamination	Contamination	The function evaluates the contamination of cantaloupes through soil only. It considers soil contamination characteristics such as prevalence (pSoil) and concentration (cSoil) and the quantity of soil deposited on the cantaloupe rind. pSoil is conditional to risk factors such as irrigation before harvesting and the use of organic fertilizer, affecting pSoil with associated odds ratios (F_irrig_rain and fManure). It also takes into account the proportion of fields grown with a barrier (pFoil) and the reduction fraction of the quantity of soil transferred to the rind when a barrier is used (rFoil).	[18,56]	caSoil2rind

Table 1. *Cont.*

Module	Stage	Microbial Process	Assumptions	Sources	Function in R
Harvest	Contamination during harvest	Cross-contamination	The function simulates cross-contamination that might occur at the moment of harvesting from elements such as conveyors, crates, or plastic surfaces. It can be used for cantaloupes harvested in farms (intended for RTE and sold as whole cantaloupe in formal retail). Parameters such as the probability of cross-contamination (probCCH) and the variability in the transfer coefficient (trMean, trSd) help to assess the transfer.		caHarvestCC
	Holding time post-harvest	Survival	The function simulates the survival of *L. monocytogenes* on cantaloupe rind during post-harvest holding time or during any short storage before cantaloupes are washed in the packinghouse. It calculates the decline in *L. monocytogenes*, assuming no growth on surfaces because any injury to the rind that would promote growth would be recent and the holding time too short for significant growth. The function allows defining a probability that the lot is kept at cold temperatures (4–10 °C).		caHoldingTime
Pre-processing	Brushing cantaloupes	Removal	The function caBrush() models the removal of bacteria during the brushing or scrubbing step of cleaning cantaloupes. It requires the mean $\log_{10}$ reduction due to brushing as an input, which quantifies how effectively bacteria are removed from the rind during this cleaning step.		caBrush
Processing	Flume tank cross-contamination	Cross-contamination	The caFlumeTankCC function simulates the potential contamination of cantaloupe when in direct contact with contaminated water in a flume tank. It accounts for four possible scenarios: cross-contamination in lots already contaminated, re-contamination in lots not previously contaminated, and scenarios where no cross-contamination occurs, regardless of initial lot status.		caFlumeTankCC
	Dicing of cantaloupe	Cross-contamination	The caDicing function simulates the transfer of *L. monocytogenes* from rind to flesh during the dicing of cantaloupe in a processing environment. It is assumed that each cantaloupe is separately diced, and if contaminated on the rind, a fraction of *L. monocytogenes* cells is transferred to the diced pieces. The function does not consider cross-contamination from dicing machines or knives.	[29,57]	caDicing
	Partitioning	Cross-contamination	The caPartitioningCC() function simulates the potential cross-contamination of cantaloupes during dicing and partitioning into packed units. It accounts for four possible scenarios of contamination involving sublots already contaminated or not. The algorithm also models the random distribution of *L. monocytogenes* from a contaminated sublot of diced cantaloupe into packed units using a dispersion factor, indicating the heterogeneity in the distribution of cells among pack units.	[58,59]	caPartitioningCC
Microbiological lot testing	Microbiological testing of RTE cantaloupe	Removal	The caTesting() function simulates the microbiological testing of RTE cantaloupe samples from a lot or sublot. It models sampling and testing based on a defined sampling plan (two-class or three-class). The algorithm uses bootstrapping to estimate the probability of detecting contaminated lots and returns the output matrix either in the original lot or sublot arrangement, depending on the user's choice.		caTesting
Cold Chain Storage	All stages after processing	Growth	Bacterial growth is estimated using the primary growth model of with a lag phase Baranyi and Roberts [60], taking into account temperature conditions and the initial physiological state of cells (q0).	A range of research and studies, including [31,61], and several others, provide the data and parameters used to calibrate the growth model under temperatures below 30 °C.	caGrowthBaranyi

Table 1. *Cont.*

Module	Stage	Microbial Process	Assumptions	Sources	Function in R
Cold Chain Storage	Transport from processing to retail	Growth	The caTrans2RetRTE() function simulates the growth of *L. monocytogenes* in RTE diced cantaloupe during cold transport to retail, utilizing the caGrowthBaranyi() function. It assumes uniform initial conditions for all RTE diced cantaloupe packs from each lot, including the same initial q0, transport temperature, and time.	[62,63]	caTrans2RetRTE
	Display of RTE diced cantaloupe packs at retail	Growth	The caRetRTE() function simulates the growth of *L. monocytogenes* in RTE diced cantaloupe during display at retail, using the caGrowthBaranyi() function. It assumes uniform retail conditions for all RTE diced cantaloupe packs, including the same lnQt (from the previous logistic stage), retail temperature, and sampled retail time. The Pert distributions represent the variability in retail time and temperature.	[64]	caRetRTE
	Transport of RTE diced cantaloupe packs from retail to home	Growth	The caRet2HomeRTE() function simulates the growth of *L. monocytogenes* in RTE diced cantaloupe during transport from retail to home, using the caGrowthBaranyi() function. The transportation time and temperature are sampled at the unit level to reflect the variability depending on the consumer. The algorithm uses a gamma distribution for the variability in transport time and a Pert distribution for transport temperature.	-	caRet2HomeRTE
Consumer handling	Storage of RTE diced cantaloupe packs at home	Growth	The caHomeRTE() function simulates the growth of *L. monocytogenes* in RTE diced cantaloupe during home storage. It samples home storage time and temperature at the unit level, depending on consumer practices. The input data includes lot-specific values of EGR5 and unit-specific values of lnQt from the previous stage. Pert distributions are used to represent the variability in home storage time and temperature.	[62,65,66]	caHomeRTE

Table 2. *L. monocytogenes* prevalence in soils of produce fields.

Country	Characteristics	Positive/Total (%Prevalence)	Source
Canada	Cultivated fields, 7 fields fertilized with animal manure in addition to inorganic fertilizer	1/13 (7.7)	[38]
Malaysia	Vegetable fields in traditional farming	4/21 (19.0)	[67]
USA	Organic/Irrigate/Manure/Compost Farm 1: no/no/yes/yes Farm 2: yes/yes/yes/yes Farm 3: no/yes/yes/no Farm 4: no/yes/no/no Farm 5: no/no/no/no (data broken down by farm not available)	16/178 (8.9)	[55]
France	Cultivated soils from France	9/53 (17.0)	[39]
Poland	Lands fertilized with manure Lands fertilized with artificial fertilizers Garden plots intensively fertilized with manure Wastelands	2/173 (1.2) 0/173 (0.0) 5/47 (10.6) 0/120 (0.0)	[40]
Austria	Soil types (humus, sand, and clay)	28/467 (6.0)	[68]
USA	Soil samples from spinach fields Low-risk fields High-risk fields	24/546 (4.4) 62/546 (11.4)	[18]

Table 3. Data on likelihood of *L. monocytogenes* being detected in soils of produce fields, expressed as an odds ratio (OR) by risk factor.

Source	Risk Factor	Description	OR	95% CI	*p*-Value
[55]	Manure	Last time manure was applied			
		Within 365 days	7.0	[3.1–15.4]	<0.001
		Over 365 days	0.6	[0.2–1.7]	0.381
		Not applied	1.0		
	Irrigation	Last time field was irrigated			
		Within 3 days	6.0	[2.0–18.1]	0.010
		4–7 days	1.2	[0.3–4.5]	0.793
		8–14 days	0.4	[0.1–2.0]	0.288
		Over 14 days/not irrigated	1.0		
	Soil cultivation	Last time soil was cultivated			
		Within 7 days	2.9	[1.1–8.6]	0.050
		8–14 days	1.4	[0.4–5.1]	0.660
		15–30 days	0.4	[0.1–1.7]	0.224
		Over 30 days	1.0		
[18]	Irrigation/rain	Time since irrigation/rain occurred			
		24 h	25	[5.7–99]	0.010
		48 h	2.5	[0.49–12]	0.27
		72 h	3.4	[0.74–15]	0.11
		144–192 h	1.0		
	Amount of irrigation water (mm) applied to field 2 days before sample collection [1]		1.2	[1.1–1.3]	0.010
[48]	Areas within 37.5 m of surface water		3.0	[2.0–4.6]	<0.001
	Areas within 62.5 m of pasture		2.9	[1.4–6.0]	0.005

[1] Effect of a 1-mm increase in irrigation water application on the odds of isolating *L. monocytogenes*.

Table 4. Data *on L. monocytogenes* in water environments ordered by prevalence.

Country	Type of Water	Positive/Total (% Prevalence)	Source
Austria	River and pond	0/68 (0.0)	[68]
USA	Engineered water	0/28 (0.0)	[56]
USA	Engineered water	0/14 (0.0)	[55]
Malaysia	Irrigation water of vegetable farms	0/15 (0.0)	[67]
India	River water	8/100 (8.0)	[69]
Switzerland	River, stream, inland canal	25/191 (13.1)	[53]
Canada	Rural and urban watersheds	56/329 (17.0)	[70]
USA	Pond, river used for irrigation	2/9 (22.2)	[56]
South Africa	Roof-harvested rain water	72/297 (22.0)	[71]
Canada	Surface (river)	32/134 (23.9)	[72]
USA	Surface water	48/146 (33.0)	[55]
USA	Lake, stream, river, pond	605/1405 (43.1)	[73]
South Africa	Irrigation canal and river	19/36 (52.8)	[74]
USA	Surface water for irrigation	33/52 (63.5)	[18]

The R functions:

The function caPrimaryProduction() assesses the contamination of cantaloupes through soil and irrigation water. This function calls two auxiliary functions, caSoil2rind() and caIrrig2rind(), specifically built for each source of contamination.

The auxiliary function caSoil2rind() evaluates the contamination of cantaloupes through soil. It considers initial soil contamination characteristics of prevalence p_{Soil} (based on Table 2, the default value of 0.089 was selected). The initial prevalence of *L. monocytogenes* in soil (p_{Soil}) is replicated for each lot (n_{Lots}). This replication serves as the base probability of contamination for each field without considering any specific agricultural practices.

The probability of contamination is then adjusted for the first risk factor: usage of organic manure. A binary array (b_{manure} in the R script) is created where each element is either 0 or 1, determined randomly based on the proportion of fields that receive organic manure (p_{Manure}). For fields indicated to use organic manure, the base soil contamination probability (p_{Soil}) is adjusted using an odds ratio (f_{Manure}). This adjustment reflects the increased risk of contamination due to manure use. The adjusted prevalence ($p_{SoilManure}$) accounts for the base prevalence (p_{Soil}) increased by the odds ratio if manure is used. The odds ratio has been estimated at 7.0 (f_{Manure}) when manure is applied within 1 year [55].

$$p_{SoilManure} = \frac{f_{Manure} \cdot p_{Soil}}{1 - p_{Soil} + f_{Manure} \cdot p_{Soil}}$$

A second adjustment is performed for irrigation/rain events. Similarly, another binary array ($b_{irrig\_raining}$) is created to indicate which fields experience irrigation or rain up to two days before harvest, based on a given probability ($p_{IrrigRaining}$). For those fields, the previously adjusted prevalence ($p_{SoilManure}$) is further modified by another odds ratio ($f_{IrrigRaining}$). This odds ratio has been estimated at 25 ($f_{IrrigRaining}$) based on [18]. This step reflects the additional risk of contamination from irrigation or rain events. The final soil prevalence ($p_{SoilIrrigRaining}$) thus incorporates both manure and irrigation/rain factors.

$$p_{SoilIrrigRaining} = \frac{f_{IrrigRaining} \cdot p_{SoilManure}}{1 - p_{SoilManure} + f_{IrrigRaining} \cdot p_{Soil}}$$

The final step aggregates the adjusted prevalences across all fields to compute the simulated overall prevalence of soil contamination ($p_{SoilSim}$). This value represents the average prevalence of contamination across all simulated lots, taking into account variations due to manure use and weather conditions.

Together with the global prevalence of fields with contaminated soil, the function caSoil2rind() returns a matrix (N_0) for quantities of *L. monocytogenes* coming from contaminated soils per cantaloupe. The function first calculates the quantity of soil that adheres to each cantaloupe ($soil_{gain}$). This is modeled as a triangular distribution to account for variability from one cantaloupe to another within each lot. The distribution parameters—minimum ($q_{SoilMin}$ = 0.05 g), mode ($q_{SoilMode}$ = 0.5 g), and maximum ($q_{SoilMax}$ = 5 g) values—are established as initial values by experts to represent realistic variations, though they do not directly rely on existing empirical evidence.

$$soil_{gain} \sim Triangular(q_{SoilMin}, q_{SoilMode}, q_{SoilMax})$$

To model the effect of using soil barriers (such as foil, plastic mulch) to protect cantaloupes from soil contact, a binary decision for each lot ($soil_{Foil}$) is generated based on the probability of using barrier (p_{Foil}). For lots with barrier protection, the quantity of soil adhering to the cantaloupes is reduced by a factor (r_{Foil}), which represents the effectiveness of the barrier in preventing soil contact.

$$soil_{Foil} \sim Binom(n_{Lots}, p_{Foil})$$

$$soil_{gain} = soil_{gain} \cdot (1 - soil_{Foil} \cdot r_{Foil})$$

The concentration of L. *monocytogenes* in the soil (c_{Soil}) is also modeled with a triangular distribution, reflecting variability from field to field. The parameters of this distribution are given by $c_{SoilLogMin}$, $c_{SoilLogMode}$, and $c_{SoilLogMax}$, and represent the $\log_{10}$ CFU/g of soil using data from Dowe et al. [38], who reported a mean of 4.0 MPN *L. monocytogenes*/g soil, with a 95% confidence interval of <1.0–28 MPN/g.

$$c_{Soil} \sim Triangular\left(c_{SoilLogMin}, c_{SoilLogMode}, c_{SoilLogMax}\right)$$

Finally, for caSoil2rind() function, the amount of *L. monocytogenes* transferred to the cantaloupe rind (N_{0s}) is calculated by multiplying the quantity of soil attached to the rind by the pathogen concentration in the soil. This product gives the raw number of pathogens per cantaloupe. To introduce the possibility of stochastic variation and account for the Poisson distribution of pathogens, a Poisson process is applied to simulate the actual number of *L. monocytogenes* cells on each cantaloupe (the number of cantaloupes per lot, *sizeLot*, can be defined by the users).

$$N_{0s} \sim Poisson\left(soil_{gain} \cdot c_{soil}\right)$$

The auxiliary function caIrrig2rind() evaluates the contamination of cantaloupes through water.

The proportion of lots contaminated by irrigation water is given by the prevalence of contamination in the water used (p_{Irrig}). Table 4 lists many estimates of prevalence in water environments. The estimate of 0.131 provided by Raschle et al. [53] has been chosen as default value for p_{Irrig}. The function assumes that the amount of water deposited on the cantaloupe rind after the last irrigation ($water_{gain}$) is expressed as a proportion of the cantaloupe weight ($canta_{weigtht}$), and is sampled from a uniform distribution between a minimum value of zero ($p_{WaterGainMin}$) and a maximum value of 0.004 ($p_{WaterGainMax}$), taken from Richards and Beuchat [54].

$$water_{gain} = canta_{weight} \cdot Uniform\left(p_{WaterGainMin}, p_{WaterGainMax}\right)$$

The distribution of the concentration of *L. monocytogenes* in irrigation water (c_{Irrig}) is represented as a uniform distribution, using the minimum and maximum values from Sharma et al. [50], who reported <0.03 to 11 MPN *L. monocytogenes*/L of water.

$$c_{Irrig} \sim 10^{Uniform\left(c_{IrrigLogMin}, c_{IrrigLogMax}\right)}$$

Finally, for caIrrig2rind() function, the amount of *L. monocytogenes* transferred to the cantaloupe rind (N_{0i}) is calculated by multiplying the quantity of water remaining on the rind by the pathogen concentration in the irrigation water. A Poisson process is applied to simulate the actual number of *L. monocytogenes* cells on each cantaloupe.

$$N_{0i} \sim Poisson\left(water_{gain} \cdot c_{Irrig}\right)$$

Then, the caPrimaryProduction() function computes the prevalence of contamination from the prevalence in the two sources (soil and irrigation water) considered as independent. The results are then combined to estimate the overall prevalence of contamination (P_0) across the entire production process.

$$P_0 = p_{fromsoil} + p_{fromirrig} - \left(p_{fromsoil} \cdot p_{fromirrig}\right)$$

The main function caPrimaryProduction() also returns the contamination matrix (N_0) by computing and combining the pathogen loads from two contamination sources (soil and irrigation water) on cantaloupes, in a probabilistic and combinatorial manner.

It first calculates exclusive and joint probabilities of the three possible situations. The probability of observing only soil contamination (p_{OnlySoil}) is calculated as the probability of contamination coming from soil while not coming from irrigation water. The probability of observing only water contamination ($p_{\mathrm{OnlyIrrig}}$) is the inverse scenario, where contamination comes exclusively from irrigation water. The third scenario corresponds to the probability of observing both soil and water contamination simultaneously ($p_{\mathrm{SoilandIrrig}}$).

$$p_{OnlySoil} = p_{FromSoil} \cdot (1 - p_{FromIrrig})$$

$$p_{OnlyIrrig} = p_{FromIrrig} \cdot (1 - p_{FromSoil})$$

$$p_{SoilIrrig} = p_{FromIrrig} \cdot p_{FromSoil}$$

Then the contamination matrix (N0) is populated with contamination data based on the above probabilities. When soil is the only source of contamination (N_{OnlySoil}), a proportion (p_{OnlySoil}) of the lots received the contamination outputs from caSoil2rind(). When irrigation water is the only source of contamination ($N_{\mathrm{OnlyIrrig}}$), a proportion of the lots ($p_{\mathrm{OnlyIrrig}}$) received the contamination outputs from caIrrig2rind(). For lots contaminated by the two sources the sum of soil and irrigation matrices were used to represent the combined contamination levels.

2.1.2. Harvest of Cantaloupes

Data and assumptions:

Cantaloupes are hand-harvested and loaded onto transport vehicles via mobile conveyor belts, with loading either automated or assisted manually. Rough handling, such as throwing or dropping, can damage the rind, potentially allowing *L. monocytogenes*, present on surfaces, to penetrate and grow. After harvest, cantaloupes are typically transported to a packinghouse, where they undergo cleaning, washing and cooling. Given that transport times are usually under a few hours, the potential for growth of *L. monocytogenes* during this time was not considered relevant for this model.

During transport, there is also a risk of cross-contamination between cantaloupes and food contact surfaces. In packinghouses, fruits from different farms or fields may be mixed, but there was no data to support the inclusion of a transportation module in the model. Additionally, stem scars from harvesting and poor worker hygiene could contribute to contamination. However, based on expert opinions and the lack of quantitative data, these factors were not considered significant for the model.

Data were collected to prepare the two functions needed to describe the fate of *L. monocytogenes* on cantaloupes, the decline in *L. monocytogenes* on intact rind and the cross-contamination.

For the decline from harvest to processing, data from Ukuku and Fett [57] and Nyarko et al. [49] were used (Table 5). The bacteria are more prevalent on cantaloupes stored at 25 °C than at 4 °C, with studies confirming die-off on the rind surface under low temperatures.

The R functions:

The function caHarvestCC() simulates cross-contamination that might occur at the moment of harvesting from elements such as conveyors, crates or plastic surfaces.

It first considers the contamination of already contaminated lots ($P_{\text{preharvest}}$). The function generates a matrix of probabilities (pcc_{flag}) for each cantaloupe ($n_{\text{Lots}} \times size_{\text{Lot}}$), based on the binomial probability that cross-contamination occurs ($prob_{CCH}$).

$$pcc_{flag} \sim Binomial(n_{Lots} \cdot size_{Lot}, prob_{CCH})$$

Table 5. Data on the survival of *L. monocytogenes* on intact cantaloupe rind.

Source	Inoculation	Temp. (°C)	Time (Day)	Counts
	Spot-inoculated (6 $\log_{10}$) on drawn circles (11 cm^2) of cantaloupe, then allowed to dry for 1 h.			$\log_{10}$ CFU/cantaloupe (Athena/Rocky Ford Cultivars)
		4	0	5.77/5.77
			1	4.96/5.57
			3	3.65/4.35
			5	2.78/3.48
			7	2.87/3.39
			9	0.78/2.00
			15	1.75/2.96
		10	0	5.80/5.80
			1	5.59/5.52
			3	4.83/4.35
			5	4.14/4.62
			7	3.73/4.28
			9	2.76/4.00
		25	0	5.85/5.85
			1	5.10/5.10
			3	5.10/2.03
			5	3.30/2.10
			7	3.23/3.23
	Inoculated by immersion in 3 L suspension at 8 $\log_{10}$ CFU/mL, then allowed to dry for 1 h.			$\log_{10}$ CFU/cm^2
		4	0	3.47
			1	3.47
			3	3.08
			6	2.93
			9	2.77
			15	2.46
		20	0	3.47
			1	3.47
			3	3.08
			6	2.70
			9	2.31
			15	1.50

The transfer from conveyors, crates or plastic surfaces depends on the transfer rate, which is assumed to vary from lot to lot (TR_i). The parameters characterizing the variability in the transfer rate were taken from Hoelzer et al. [58].

$$TR_i \sim Normal(-1.42, 0.55), TR_i \leq 0$$

If a recontamination event occurs at the cantaloupe level (as determined by pcc_{flag}), a fraction (TR_i) of the bacterial load present on the contaminated equipment surface(s) (N_{plas}) is transferred to the rind ($N_{from\text{-}equip}$).

$$N_{from-equip} \sim Binomial\left(TR_i, N_{plas}\right) \cdot pcc_{flag}$$

The total number of L. *monocytogenes* ($N_{post\text{harvest}}$) is then simply the sum of cells already present before harvest ($N_{\text{prehavest}}$) and the ones transferred from equipment.

$$N_{postharvest} = N_{preharvest} + N_{from-equip}$$

The function also accounts for previously uncontaminated lots (probability 1 − *Ppre-harvest*) that may become contaminated through contact with contaminated equipment. For each of these lots, the potential bacterial transfer ($N_{from\text{-}equip}$) is evaluated. If $N_{\text{from-equip}}$ is > 0, the load of the lots is simply equal to $N_{\text{from-equip}}$ (as $N_{\text{preharvest}}$ is null for this type of lot). The proportion of the batch with no cells transferred is thereafter noted $prob_{\text{clean-batch}}$.

The overall prevalence of contamination in the batch is then reassessed:

$$p_{harvest} = p_{preharvest} + \left(1 - p_{preharvest}\right) \cdot \left(1 - prob_{clean-batch}\right)$$

and, along with $N_{postharvest}$, $p_{harvest}$ becomes a function's output.

The caHoldingTime() function models the survival of L. *monocytogenes* on cantaloupe rind during post-harvest holding or short pre-wash storage in packinghouses. The algorithm focuses on the decline in L. *monocytogenes*, assuming that any rind injury from harvesting would prevent significant bacterial growth due to the brief holding period. Users can specify the probability p_{Cooled} that a batch of newly harvested cantaloupes is stored at cold temperatures (4–10 °C). If the batch is not stored at a cold temperature, the model assesses survival at room temperature (25 °C).

Survival data for L. *monocytogenes* on intact rinds of two cantaloupe cultivars at 4, 10, and 25 °C, sourced from Nyarko et al. (2016) [49] (see Table 5), were used. A non-linear mixed model using the two-parameter Weibull decay equation, with random effects by cantaloupe variety, revealed that decay rates at 25 °C differ significantly from those at 4 and 10 °C, which showed no substantial difference. This model segments temperatures into two categories: cool (4/10 °C) and ambient (25 °C), with a fixed *shape* parameter for each. Random effects allow for cultivar-specific variability in decay times (*D*), generating estimates for $Dmean_{410}$, Dsd_{410}, $Dmean_{25}$, and Dsd_{25}, which represent average decay and standard deviations for cool and ambient temperatures, respectively. To represent variability among cantaloupe cultivars, *D* values are sampled for each lot.

$$D_T \sim Norm(D_{meanT}, D_{SDT}), D_T > 0$$

$$p_{survive} = 10^{-\left(\frac{t}{24 \cdot D_T}\right)^{shape}}$$

The number of L. *monocytogenes* on rinds of cantaloupes after the holding period (*t*) is then assessed according to the initial number present (N_{harvest}) and the probability of survival of one bacterial cell (p_{survive}),

$$N_{endharvest} \sim Binomial(N_{harvest}, p_{survive})$$

It is assumed that even after the holding period, no initially contaminated lot will be entirely free of L. *monocytogenes*, and thus, a minimum of one bacterial cell will be allocated to one cantaloupe from any contaminated batch. The prevalence of contamination in the batch is thus unchanged by the decline in L. *monocytogenes*. The proportion of contaminated batches remains unchanged, even if the concentration of L. *monocytogenes* decreases.

2.1.3. Pre-Processing: Cleaning and Washing

Data and assumptions:

The cleaning and washing module addresses hygiene concerns associated with brushes used for dry cleaning and with washing equipment used for the spray or immersion washing of freshly harvested cantaloupe.

The first treatment in the processing facility was the cleaning of cantaloupes with a dry-brushing step. Abrasive brushing was found to reduce biofilms and to reduce the *L. monocytogenes* load by 1.4 $\log_{10}$ [75].

The second step involved washing cantaloupes through spray or immersion washing, as well as hydrocooling to remove field heat. Washing with water alone typically achieves reductions in L. monocytogenes on cantaloupe surfaces below 1.0 $\log_{10}$ CFU/cm^2 [29,76,77]. The reduction in *L. monocytogenes* on the cantaloupe rind due to washing with water and sanitizers can be modeled using the data compiled in Table 6. Table 6 gathers literature data on the effectiveness of chemical sanitizers, including the mean reduction effect of washing and determinant features of the washing treatment, such as exposure time, temperature and sanitizer concentration. Nonetheless, in the function, only washing with water was considered.

Table 6. Reduction in *L. monocytogenes* (and standard error) on cantaloupe surface rind after treatment with water alone or with sanitizing agent.

Treatment	Study	Concentration (%)	Exposure Time (min)	Temperature (°C)	Reduction ($\log_{10}$ CFU/cm^2)	St. Error (Reduction)	n [1]
Tap water	[76]	-	3	25	0.18	0.053	8
	[29]		5	25	0.20	-	3
	[77] [2]						
	Stored at 5 °C x 0 day		5	20	0.20	-	3
	Stored at 5 °C x 7 days		5	20	0.15		3
	[57] [2]						
	Stored at 4 °C x 1 day	-	2	25	0.30	-	3
	Stored at 4 °C x 5 days	-	2	25	0.22	-	3
	Stored at 4 °C x 15 days	-	2	25	0.22	-	3
ClO_2 gas	[19]	0.00005	2	25	1.2	-	6
		0.0001	2	25	1.8	-	6
		0.00015	2	25	2.1	-	6
		0.0003	2	25	2.1	-	6
		0.0005	2	25	2.2	-	6
		0.00005	10	25	3.3	-	6
		0.0001	10	25	3.2	-	6
		0.00015	10	25	3.7		6
		0.0003	10	25	3.8	-	6
		0.0005	10	25	4.3	-	6
SH [3]		0.020	5	25	0.57	0.250	8
	Stored at 4 °C x 1 day	0.100	2	25	>3.0	-	3
	Stored at 4 °C x 5 days	0.100	2	25	>3.0	-	3
	Stored at 4 °C x 15 days	0.100	2	25	>3.0	-	3
H_2O_2	[29]	2.5	5	25	2.8		3
	[77] [2]						
	Stored at 5 °C x 0 day	2.5	5	20	2.3	-	3
	Stored at 5 °C x 7 days	2.5	5	20	2.8		3
	[57] [2]						
	Stored at 4 °C x 1 day	5.0	2	25	>3.0	-	3
	Stored at 4 °C x 5 days	5.0	2	25	>3.0	-	3
	Stored at 4 °C x 15 days	5.0	2	25	>3.0	-	3
HPLNC [4]	[77] [2]						
	Stored at 5 °C x 0 day	-	5	20	>4.0	-	3
	Stored at 5 °C x 7 days	-	5	20	>4.0	-	3

[1] Number of samples in inoculation experiments. [2] Studies where inoculated cantaloupes were stored at different temperatures and times before the washing treatment. [3] Sodium hypochlorite. [4] Solution of 1% H_2O_2, 25 µg/mL nisin, 1% sodium lactate and 0.5% citric acid.

The reductions reported in Table 6 are quite large, most probably due to the use of optimal lab conditions. In industrial settings, maintaining proper application conditions is challenging, significantly reducing the effectiveness of sanitizers in preventing cross-contamination [26,78]. In the best situation, the washing would not contribute to an increase in contamination; in the worst situation, flume tank washing could increase contamination through contaminated water.

The R functions:

Three functions were prepared to account for the brushing effect and the two alternate washing steps (ideal situation with potential population decrease, and one with potential cross-contamination).

The function caBrush() models the removal of bacteria during the brushing or scrubbing step. This function takes into account the effectiveness of the cleaning step, quantified as a mean $\log_{10}$ reduction in bacteria, that users need to provide as an input parameter. The function processes a list containing an initial matrix of bacterial counts on the cantaloupe rinds and the prevalence of contamination in harvested lots. This reduction can vary at lot level or at cantaloupe level, depending on how *logDecBrush* is defined (either as a single value, a vector per lot, or a matrix per cantaloupe).

$$N_{afterbrush} \sim Binomial\left(N_{endharvest}, 10^{-LogDecBrush}\right)$$

The function caFlumeTank() simulates three events taking place during flume tank washing: the washing of bacteria off the cantaloupe rind at a washing efficiency of *logDecWash*, the elimination of bacteria in the water containing sanitizers at a sanitizing efficiency of *log10SaniWash*, and the redistribution of survivors on the cantaloupe rind (characterized by the *b* parameter).

For the first event (washing of the bacteria from the rind to water), the number of bacteria remaining on the rind is calculated according to the parameter *logDecWash*, the reduction in $\log_{10}$ attained by washing.

$$N_{NotWash} \sim Binomial\left(N_{afterbrush}, 10^{-LogDecWash}\right)$$

The quantity of *L. monocytogenes* in the water (N_{wash}) is given by

$$N_{wash} = N_{afterbrush} - N_{NotWash}$$

Some of the *L. monocytogenes* in water will be inactivated by the presence of a sanitizer. The $\log_{10}$ reduction associated with the sanitizer (*logDecSani*) helps determine the number of *L. monocytogenes* that will be redistributed onto the cantaloupe rinds ($N_{\text{toDistribute}}$)

$$N_{toDistribute} \sim Binomial\left(N_{wash}, 10^{-LogDecSani}\right)$$

Finally, the redistribution of survivors on the cantaloupe rind was modeled according to a Dirichlet–multinomial process. The parameter to assess the dispersion factor representing the clustering of surviving cells during redistribution is *b*. A value of *b* higher than 1 should be used to assume random homogeneous distribution. The redistribution is handled computationally by first drawing proportions (*p*) from a Dirichlet distribution parameterized by *b*, which gives the fraction of surviving bacteria that each cantaloupe receives. Then, a multinomial distribution is used to allocate the actual counts of bacteria to each cantaloupe (N_{out}) based on these proportions.

$$p \sim Dirichlet(b)$$

where $\boldsymbol{b}$ is a vector of $b/size_{Lot}$ repeated $size_{Lot}$ time and $N_{out} \sim Multinomial(N_{toDistribute}, p)$.

Finally, the total number of *L. monocytogenes* per cantaloupe rind ($N_{\text{endwashing}}$) is the sum of the cells that remained on the rind plus the redistributed ones.

$$N_{endwashing} = N_{NotWash} + N_{out}$$

A second function was prepared for modeling the effect of washing cantaloupes. The caFlumeTankCC() function simulates potential contamination scenarios for cantaloupes when they come into direct contact with contaminated water in a flume tank. The cross-contamination algorithm considers four different scenarios [79]: cross-contamination in lots that are already contaminated, contamination in lots that were initially not contaminated, the absence of cross-contamination in lots that were already contaminated, and no cross-contamination occurring in lots that were not initially contaminated. The probabilities of each scenario are calculated according to the initial prevalence (P_{harvest}) and the probability that the flume tank water is contaminated ($prob_{\text{CCW}}$):

$$P(CantaLot_{Pos\_}CC_{Negi}) = P_{harvest} \times (1 - Prob_{CCW})$$

$$P(CantaLot_{Pos\_}CC_{Posi}) = P_{harvest} \times Prob_{CCW}$$

$$P(CantaLot_{Neg\_}CC_{Posi}) = (1 - P_{harvest}) \times Prob_{CCW}$$

The status of every lot is then randomly sampled from the probabilities $[P(CantaLot_{Pos\_}CC_{Negi}), P(CantaLot_{Pos\_}CC_{Posi}), P(CantaLot_{Neg\_}CC_{Posi})]$. This strategy is adopted to maintain the dimension [r, c] of the input contamination matrix $N_{harvest}$. The number of *L. monocytogenes* transferred (N_{added}) depends on the concentration in the flume tank water (C_{water}) and the value of the fraction of water gain (mL) ($p_{\text{WaterGain}}$) relative to the cantaloupe weight in g,

$$C_{water} \sim Pert(logWaterMin, logWaterMode, logWaterMax)$$

$$N_{added} = C_{water} \cdot p_{WaterGain} \cdot canta_{Weight}$$

Then, the values of N_{added} and the total contamination on rinds before washing ($N_{tot\ i}$,) are summed in order to determine the numbers of *L. monocytogenes* in the lot *i* after the potential recontamination event ($N_{tot\ postWaterCC\ i}$), according to the following:

$$N_{tot\ postWaterCC\ i} = \begin{cases} N_{tot\ i} & i \in CantaLot_{Pos\_}CC_{Neg} \\ N_{tot\ i} + N_{added\ i} & i \in CantaLot_{Pos\_}CC_{Pos} \\ N_{added\ i} & i \in CantaLot_{Neg\_}CC_{Pos} \end{cases}$$

Next, the partitioning algorithm randomly distributes the total numbers of cells $N_{tot\ postWaterCC\ i}$ from a contaminated lot onto all cantaloupe rinds present in the lot, following a Dirichlet–multinomial process according to a dispersion factor (b_w) representing the clustering of cells during flume tank washing:

$$p_w \sim Dirichlet(b_w, n_{Lots})$$

where $\boldsymbol{b_w}$ is a vector of $b_w/size_{Lot}$ repeated $size_{Lot}$ times.

$$N_{washedrindi} \sim Multinomial(N_{totpostCC}, p_w)$$

2.1.4. Processing

Data and assumptions:

This module addresses the contamination of the RTE diced cantaloupe, considering two main sources. The first represents contamination from the cantaloupe rind to the flesh during slicing and dicing. Previous studies mimicking an industrial processing line and home preparation procedures for fresh-cut cantaloupe have shown that *L. monocytogenes* readily transfer from the rind to the edible cantaloupe flesh during slicing and cutting, with a positive correlation between the contamination level on the rind and the numbers of cells transferred [29,80,81]. Patil [81] conducted an experiment to measure the penetration of *L. monocytogenes* into cantaloupe flesh from contaminated rinds. Cantaloupes with an initial rind contamination of 5 $\log_{10}$ CFU/cm^2 were stored at 4 °C and 30 °C for 24 h, then sampled using a cork borer to create 25 mm long cylinders. The bacteria's penetration was measured at various depths up to 25 mm. The results showed that the transfer of bacteria from the rind to the first 5 mm of flesh ranged between 2.8 and 3.6 $\log_{10}$ CFU/cm^2 at 4 °C and 1.9 and 4.0 $\log_{10}$ CFU/cm^2 at 30 °C. No significant differences were observed in the bacterial counts transferred at different depths or between the two storage temperatures, indicating that all depths of the edible flesh were equally susceptible to contamination regardless of the storage condition.

Other studies aimed to quantify the transfer of *L. monocytogenes* from cantaloupe rind to fresh-cut pieces [20,29]. Inoculated whole cantaloupes stored at 4 °C were cut into four sections, with each section further diced into 3 cm cubes before rind removal. The overall transfer rate data are summarized in Table 7. As few as 150 bacterial cells per cm^2 on the cantaloupe rind were sufficient to contaminate the edible flesh during cutting [81]. Transfer rates were generally between 0.08 and 1.18%.

Table 7. Available data to build a transfer equation of *L. monocytogenes* load from cantaloupe rind to flesh (diced pieces in g or surface slices in cm^2).

Source		*L. monocytogenes* on the Rind ($\log_{10}$ CFU/cm^2)	*L. monocytogenes* in Cantaloupe Flesh	Transfer Rate (%) ($10^{flesh}/10^{rind}$) × 100
To fresh-cut (diced pieces)			($\log_{10}$ CFU/g)	
[20]		2.16	0.23	1.175
		3.26	0.54	0.191
		3.98	1.31	0.214
		4.52	1.46	0.087
[29]		4.60	2.60	1.000
		4.40	2.20	0.631
To slices (flesh surface)			($\log_{10}$ CFU/cm^2)	
[81]	4 °C	5.94	2.45	0.032
		5.44	1.42	0.010
	30 °C	5.22	1.64	0.026
		5.44	1.17	0.005

The second relevant source of contamination is the cross-contamination of cantaloupes with *L. monocytogenes* from slicing and dicing equipment as well as other food contact surfaces [81]. The modeling approach used for this process was similar to the one used for cross-contamination during harvesting. The transfer rate was taken from [58]. Based on environmental data established for *Listeria* spp. [27], the initial concentration on food contact surfaces was estimated at 0.25 CFU per site (25 cm^2). Given this surface area and considering the estimated contact area of a cantaloupe with the contaminated surface (5 cm^2), the expected *L. monocytogenes* load transferred to the cantaloupe is approximately 9 CFU.

The R function:

The function caDicing() models the transmission of *L. monocytogenes* from the rind to the flesh when cantaloupes are diced in a processing environment. It operates under the assumption that each cantaloupe is diced individually, and if the rind is contaminated, a certain percentage of *L. monocytogenes* cells are transferred to the diced pieces. It is important to note that this function does not account for any potential cross-contamination that might occur from the dicing equipment or knives. The function returns the matrix N_{dicing} organized according to sublot configurations, where each row represents a sublot and each column corresponds to the number of cantaloupes processed within that sublot. This structure is used because the size of production lots may differ from the size of processing lots. Thus, the numbers and prevalence of contaminants are reported on a sublot basis (processing lot size). The variability in the transfer rate of *L. monocytogenes* from the rind to the flesh (CFU/g of diced cantaloupe per CFU/cm^2 of rind) is modeled using a Pert distribution. This modeling utilizes data digitized from studies by Ukuku and Fett [57] and Ukuku et al. [29]. The transfer rate ($TR_{\text{rind2flesh}}$) is expressed as a percentage (CFU/g of diced pieces per CFU/cm^2 of rind) × 100 and is described using by the parameters min_{TR}, $mode_{\text{TR}}$, and max_{TR}.

$$TR_{rind2flesh} \sim PERT(min_{TR}, mode_{TR}, max_{TR})$$

For calculating the number of *L. monocytogenes* bacteria transferred from the rind to the flesh of cantaloupes during the dicing process (N_{trans}), the density of bacteria per square centimeter on the rind of each cantaloupe within the sublots is assessed. This is achieved by dividing the total bacterial count on the rind ($N_{sublots}$) by the surface area of the cantaloupe (*cantaSurface*). Following this, the transfer rate ($TR_{rind2flesh}$) is applied to this density. Defined as a percentage, this rate quantifies the effectiveness of the bacterial transfer from the rind to the flesh during dicing. For calculation purposes, it is adjusted by dividing by 100. After applying the transfer rate, the resulting value, which indicates the number of bacteria transferred per gram of diced cantaloupe, is then scaled up by the total weight of the diced pieces produced from one rind-free cantaloupe (*cantaRindFree*).

$$N_{trans} = cantaRindFree \cdot TR_{rind2flesh} \cdot \frac{N_{sublots}}{cantaSurface} \cdot \frac{1}{100}$$

The overall prevalence is adjusted taking into account that, in some lots, no *L. monocytogenes* are transferred (P_{i0}):

$$P_{diced} = P_{beforedicing} \cdot (1 - P_{i0})$$

The function caPartitioningCC()models the possible cross-contamination of cantaloupes in scenarios where they come into direct contact with dicing machines or knives. This function also handles the distribution of diced cantaloupes from a processing lot (sublot) into packed units. The cross-contamination algorithm evaluates four scenarios: contamination and no contamination occurring in both already contaminated and initially uncontaminated sublots, with probabilities calculated for each event. Following contamination assessment, the partitioning algorithm allocates the total number of bacterial cells from a contaminated sublot into individual packed units. This distribution considers the dispersion factor b_i, a parameter of the beta distribution, which measures the extent of cell clustering within the bulk of diced cantaloupes in the sublot and reflects the variability in cell numbers among the packed units.

If the recontamination event takes place, at a probability $Prob_{CCDice}$, a fraction (TR_{Dicer}) of the numbers of cells on the contaminated equipment surface(s) (N_{Dicer}) is transferred to the diced pieces of cantaloupe ($N_{from\text{-}dicer}$). The algorithm considers that, if the lot of RTE diced cantaloupe is not contaminated with *L. monocytogenes* (actually not represented in the contamination matrix), it may become contaminated from equipment; if the lot is contaminated, its load will increase or remain the same depending on the recontamination event happening or not, respectively. This raises three possible scenarios of contaminated lots: (1) no recontamination occurring in lots already contaminated ($Dice_{Pos}\_CC_{Neg}$); (2) recontamination occurring in lots already contaminated ($Dice_{Pos}\_CC_{Pos}$); and (3) recontamination occurring in lots that were not contaminated ($Dice_{Neg}\_CC_{Pos}$). The probability of each scenario taking place for every lot i is calculated as

$$P\left(Dice_{Pos}\_CC_{Neg\ i}\right) = P_{diced} \cdot (1 - Prob_{CCDice})$$

$$P\left(Dice_{Pos}\_CC_{Pos\ i}\right) = P_{diced} \times Prob_{CCDice}$$

$$P\left(Dice_{Neg}\_CC_{Pos\ i}\right) = (1 - P_{diced}) \times Prob_{CCDice}$$

The status of every lot is then randomly sampled from the probabilities $[P(Dice_{Pos\_}CC_{Neg\ i}),\ P(Dice_{Pos\_}CC_{Pos\ i}),\ P(BDice_{Neg\_}CC_{Pos\ i})]$. The extent of recontamination in number of cells per lot i is computed as

$$TR_{Dicer\ i} \sim Normal\ (-1.42,\ 0.52),\quad TR_i \leq 0$$

$$N_{from-dicer\ i} \sim Binomial\left(N_{Dicer},\ 10^{TR_{Dicer\ i}}\right),\quad N_{Dicer\ i} > 0$$

where N_{Dicer} can be understood as the total load of *L. monocytogenes* cells on the surface of the blades of the dicing machine and/or packaging machine in contact with the diced pieces of cantaloupe. Next, the value of $N_{from\text{-}Dicer\ i}$ is added to $N_{tot\ i}$, in order to determine the numbers of *L. monocytogenes* in the lot i after the potential recontamination event ($N_{tot\ postCC\ i}$), according to the following:

$$N_{tot\ postCC\ i} = \begin{cases} N_{tot\ i} & i \in Dice_{Pos\_}CC_{Neg} \\ N_{tot\ i} + N_{from-equip\ i} & i \in Dice_{Pos\_}CC_{Pos} \\ N_{from-equip\ i} & i \in Dice_{Neg\_}CC_{Pos} \end{cases}$$

Next, the partitioning algorithm randomly distributes the total number of cells $N_{tot\ postCC\ i}$ from a contaminated lot into c packed units, following a Dirichlet–multinomial process:

$$p \sim Dirichlet(b_i, n_{Lots})$$

where $\boldsymbol{b_i}$ is a vector of $b_i/size_{Lot}$ repeated $size_{Lot}$ times.

$$N_{pack\_c\ i} \sim Multinomial(N_{tot\ postCC\ i}, p)$$

The resulting matrix N_{pack_c} contains the numbers of *L. monocytogenes* cells in the packs of diced cantaloupe in contaminated lots. Finally, the mean prevalence of contaminated lots after packaging P_{Pack} is updated.

$$P_{Pack} = 1 - (1 - P_{diced}) \times (1 - Prob_{CCDice})$$

2.1.5. Microbiological Lot Testing

Data and assumptions:

The QRA model enables the microbiological testing of food unit samples taken from a lot, according to a two-class or a three-class mixed sampling plan [82]. In the commonly employed two-class plan for *L. monocytogenes*, a predetermined number n of units is randomly selected from each lot for testing. Each selected unit undergoes an enrichment assay using g grams of the sample, and the lot is considered unacceptable if more than c units test positive. In contrast, the three-class mixed sampling plan not only tests but also enumerates the sampled units. Here, a lot is rejected if more than c units are positive or if any unit has a microbial concentration exceeding a set threshold M. Enumeration tests are typically performed only on samples that test positive in the initial detection, involving the direct plating of a specific amount $g_{TestedEnum}$ from the same sample.

The R function:

The function caTesting() receives the outputs of the function caPartitioningCC(): n, the number of tested units; g, the sub-sample weight in grams used for detection; c, the number of positive samples accepted (two-class or three-class mixed plan); M, the maximum limit concentration; $p_{lot\ tested}$, the proportion of tested lots; Se, the probability of the test (enumeration or detection) to detect, independently, each bacteria present in a sample; and $g_{TestedEnum}$, the sub-sample weight in grams used for the enumeration assay.

Assuming perfect homogenization in the sample, each of the bacteria present in each of the $r \times c$ units of $Unit_{size}$ weight has a probability of being present in the g grams of the sub-sample and detected with probability equal to $Se \times g/Unit_{size}$. The number of bacteria detected in the detection assay is then as follows:

$$N_{detected\ ij} \sim Binomial\left(N_{Pack\ ij}, Se \times g/Unit_{size}\right)$$

and the detection test is positive if $N_{detected\ ij} > 0$.

If the sampling plan is a 3-class plan, an enumeration test (direct plating) is performed. The number of bacteria enumerated in the sample is as follows:

$$N_{enumerated\ ij} \sim Binomial\left(N_{Pack\ ij}, Se \times g_{TestedEnum}/Unit_{size}\right)$$

and the estimated concentration is $N_{enumerated\ ij}/g_{TestedEnum}$ CFU/g. The algorithm assumes that the enumeration is performed only on samples detected as positive.

In order to evaluate the probability of each of the lots to be rejected, 1000 (by default) Monte Carlo samples of n samples are generated for each lot and, for each of these Monte Carlo samples, the microbiological criteria are applied (i.e., in a 2-class plan, the test is positive if $>c$ samples among n are detected, whereas in a 3-class plan, the test is positive if $>c$ samples among n are detected or if at least one sample has an estimated concentration $>M$ CFU/g). The mean number of times the lot is rejected among the Monte Carlo samples multiplied by the probability for the lot to be tested is an estimate of $P_{pos\ i}$, the probability of lot i to be rejected.

Contaminated lots detected after testing are not removed from the matrix, and therefore, the input matrix N_{pack} is returned unchanged. The function caTesting() only updates the prevalence outputs.

2.1.6. Cold Chain During Transport to Retail

RTE diced cantaloupe is typically refrigerated at retail. However, time–temperature measurements reveal that cold storage in retail display cabinets is often insufficient, frequently exceeding recommended limits [83]. Derens-Bertheau et al. [64] observed that about 31% of temperature measurements for various food products in France exceeded 6 °C, with retail display temperatures ranging from 1.4 °C to 9.8 °C, averaging 5.6 °C. Similarly, another study [84] recorded average temperatures of 4.1 °C and 3.5 °C for fresh-cut lettuce in winter and summer, respectively.

Herein, the impact of temperature on *L. monocytogenes* growth in cantaloupe was assessed using predictive microbiology models. The aim was to develop secondary models for lag phase duration and growth rate based on the published data presented below. Scolforo et al. [85] determined lag phase duration of *L. monocytogenes* serovar 1/2b inoculated in the pulp of Canary melons stored at different temperatures. A Q0 parameter characterizing the physiological state of the cells in cantaloupe was derived from these experiments.

For the growth rate, we used the challenge study of *L. monocytogenes* 4b, 1/2a, and 1/2b carried out by Fang et al. [61], who assessed the effect of serotypes on growth rates across all temperature conditions. They did not find any serotype effect or interaction with temperature, and subsequently, they determined the cardinal parameters for temperature of *L. monocytogenes* in fresh-cut cantaloupe (T_{min} = 1.90 °C, T_{opt} = 38.3 °C, T_{max} = 45.7 °C, μ_{opt} = 0.975 h^{-1}). Table 8 compiles the growth rate estimates for cantaloupe from various studies, including the results from squash (a fruit also from the *Cucurbitaceae* family; Farber et al. [86]), since its estimates are comparable to those of cantaloupe.

Table 8. Growth kinetics data of *L. monocytogenes* in cantaloupe flesh or similar from published articles.

Study	Medium	Strain	Stressed	Temperature (°C)	Specific GR (h^{-1})
[61]	Cantaloupe	F2365, H7858, ATCC19115	Stressed (rifampicin-resistant and cold-resistant)	4.0	0.0120
				8.0	0.0470
				12	0.1260
				16	0.1860
				20	0.2930
				25	0.5250
				30	0.7300
				33	0.8150
				40	0.9160
				37	0.8600
				43	0.6920
		F4260	Stressed (rifampicin-resistant and cold-resistant)	4.0	0.0110
				8.0	0.0580
				12	0.1230
				16	0.1940
				20	0.3210
				25	0.5300
				30	0.7470
				33	0.9000
				37	0.9900
				40	0.9730
				43	0.7590
				38	0.9750
		V7	Stressed (rifampicin-resistant and cold-resistant)	43	0.7540
				40	0.9980
				33	0.9010
				37	0.9640
				30	0.7860
				25	0.5430
				20	0.3220
				16	0.1970
				12	0.1350
				8.0	0.0640

Table 8. *Cont.*

Study	Medium	Strain	Stressed	Temperature (°C)	Specific GR (h^{-1})
[87]	Cantaloupe	ATCC BAA839, ATCC BAA839, ATCC 19111, ATCC 13932	Stressed (rifampicin-resistant and cold-resistant)	10	0.0852
				15	0.1983
				20	0.4030
				25	0.5803
				10	0.0852
				15	0.2118
				20	0.4030
				25	0.5112
	Diced cantaloupe	Scott A, H7778, ATCC-15313, CCR1LG	Not stressed	20	0.3720
	Fresh-cut cantaloupe	LCDC 81-861, Scott A, 101M, V7	Not stressed	5.0	0.0850
				10	0.2000
				15	0.2300
				20	0.3900
				25	0.4745
	Fresh-cut cantaloupe	NRCC B33076	Not stressed	5.0	0.0368
				10	0.0898
				25	0.6240
				30	0.7161
				36	0.9233
[81]	Diced cantaloupe	J22F, J29H, M3	Not stressed	4.0	0.0520
				7.0	0.0670
				10	0.1840
[88]	Fresh-cut watermelon	LCDC 81-861, V7, 101M, Scott A	Not stressed	4.0	0.0318
				13	0.1213
				13	0.1438
[86]	Squash	Not stated	Not stressed	4.0	0.0370
				10	0.0910

The differential form of the Baranyi and Roberts model [60] was employed to predict the growth of *L. monocytogenes* in cantaloupe, providing a value for Q_0 to assume the presence of lag phase.

$$\frac{dN}{dt} = \mu_{max}\frac{Q}{1+Q}\left(1 - \frac{N}{10^{MPD}}\right)N$$

$$\frac{dQ}{dt} = \mu_{max}Q$$

Figure 2 plots the specific growth rate estimates against storage temperature. The values obtained for temperatures below 30 °C were used to prepare a secondary model for temperature.

For temperature, data were collected for the three stages of the logistic chain: transport from processing to retail, storage at retail, and transport from retail to the consumer home.

Fresh-cut cantaloupe is usually transported in refrigerated trucks, whose temperature should be lower than 7 °C [89]. A Pert distribution is used to account for the temperature variability. Considering the values encountered for the storage of produce, the temperature is assumed to be between 3 °C and 10.3 °C with a mean of 5 °C [62,63].

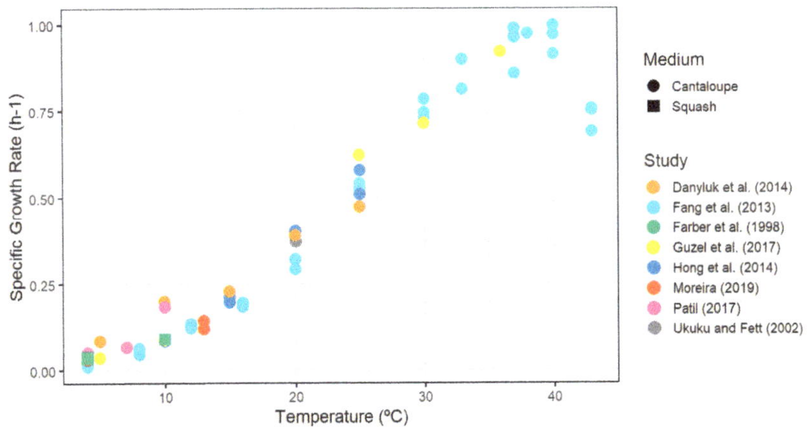

Figure 2. Specific growth rate estimates of *L. monocytogenes* in cantaloupe flesh against temperature, extracted from the literature. Refer to Table 8 for data sources [31,57,61,76,81,86–88].

From retail to home, the temperature typically rises, as the consumer usually does not take measures to prevent an increase in surface temperature. Risk assessments for *L. monocytogenes* usually account for this stage. For example, Pouillot et al. [90] considered a mean temperature of cold-smoked salmon of 10.4 °C during transport. Additionally, several risk assessments have reported an average transport time from retail to home of 43 min (with a 19 min standard deviation) [63,66].

As the shelf life of RTE diced cantaloupe is quite short, at usually less than 5 days [91], the packs are expected to stay a day on retail shelves.

The R functions:

Three functions were developed for modeling the growth of *L. monocytogenes* from processing to the consumer home. They all rely on the auxiliary function caGrowthBaranyi(). As inputs, this function takes time and temperature for a logistic stage, the levels of initial contamination at the beginning of this stage, and its physiological state (Q_0). It returns the level of contamination at the end of the stage and its new physiological state (updated Q_0).

From user inputs, this function first provides an estimation of the growth rate according to temperature

$$\mu_{max} = EGR_5 \cdot \left(\frac{(T - T_{min})}{5 - T_{min}}\right)^2$$

where EGR_5 is the growth rate (in h^{-1}) of *L. monocytogenes* in cantaloupe flesh at 5 °C and T_{min} is the nominal minimum temperature for the growth of *L. monocytogenes* in cantaloupe flesh.

Then the function uses μ_{max} and the time and duration of the logistic stage to assess the final level. The maximum population in the RTE cantaloupe pack depends on the *unitSize* and the maximum population density (MPD) of *L. monocytogenes* (set by default to 8 $\log_{10}$ cfu/g). The primary model of Baranyi and Roberts [60] in static environmental conditions is given by

$$A(t) = t + \frac{1}{\mu_{max}} \cdot ln\left(\frac{exp(-\mu_{max} \cdot t) + Q_0}{1 + Q_0}\right)$$

$$ln(N(t)) = ln(N(t=0)) + \mu_{max} \cdot A - ln\left(1 + \frac{exp(-\mu_{max} \cdot A) - 1}{exp\left(ln\left(10^{MPD \cdot unitSize}\right) - ln(N(t=0))\right)}\right)$$

The parameters used for the three functions thereafter were fitted according to the data presented in Figure 2 (below 30 °C): T_{min} = −2.02 °C (standard error = 0.576 °C) and EGR_5 = 0.036 h^{-1} (standard error = 0.004 h^{-1}).

Finally the caGrowthBaranyi() returns the physiological state value at the end of the logistic state:

$$ln(Q_t) = ln(Q_0 + \mu_{max} \cdot t)$$

The first function of the logistic chain is the caTrans2RetRTE() function. It models the growth of *L. monocytogenes* in RTE diced cantaloupe during cold transportation to retail outlets, utilizing the auxiliary caGrowthBaranyi() function for growth dynamics. This function assumes that all cantaloupe packs in a lot experience the same initial physiological state of cells (Q_0), consistent transportation temperature (Temp), and transportation duration. The variability in the physiological state parameter, ln(Q0), was represented by a normal distribution with a mean (lnQ0Mean) and standard deviation (lnQ0Sd) of −0.097 and 0.0640, respectively, according to [85].

Similarly, the variability in the specific growth rate at 5 °C (EGR_5) is modeled using a normal distribution with parameters meanEGR_5 = 0.03557288 h^{-1} and sdEGR_5 = 0.004 h^{-1}.

Additionally, the function uses Pert distributions to account for lot-specific differences in transport time ($time_{2Ret}$) and temperature ($temp_{2Ret}$) for this logistic step.

$$time_{trans2Ret} \sim Pert(time_{trans2RetMin}, time_{trans2RetMode}, time_{trans2RetMax})$$

$$temp_{trans2Ret} \sim Pert(temp_{trans2RetMin}, temp_{trans2RetMode}, temp_{trans2RetMax})$$

The caRetRTE() function models the growth of *L. monocytogenes* in RTE diced cantaloupe during retail display, utilizing the caGrowthBaranyi() function to determine growth patterns. This algorithm assumes that all packs of diced cantaloupe from each lot contain bacterial cells with the same physiological state, ln(Qt) passed from the prior logistic phase. All packs from a lot are considered to have the same retail temperature ($temp_{Ret}$). However, the variation in retail temperatures across different lots is described by a Pert distribution:

$$temp_{Ret} \sim Pert(temp_{RetMin}, temp_{RetMode}, temp_{RetMax})$$

The duration of retail display, however, is determined individually at the unit level. Pert distributions are employed to account for variability in duration for packs on shelves:

$$time_{Ret} \sim Pert(time_{RetMin}, time_{RetMode}, time_{RetMax})$$

The caRet2HomeRTE() function simulates the growth of *L. monocytogenes* in RTE diced cantaloupe during the transportation from retail to home, based on the caGrowthBaranyi() auxiliary function. This algorithm determines transportation time and temperature at the unit level, reflecting variations based on individual consumer behaviors. It uses input data to integrate lot-specific growth rates at 5 °C (EGR_5) and unit-specific initial quantities of bacteria (ln(Qt)) that are carried over from the preceding logistics stage. The variability in transportation time is modeled using a gamma distribution of parameters defined by [66], while a Pert distribution captures the variability in transportation temperature:

$$time_{Ret2Home} \sim Gamma(scale, time_{Ret2HomeMode}, time_{Ret2HomeMax})$$

Parameter values for the caTrans2RetRTE(), caRetRTE() and caRet2HomeRTE() functions are provided in the Supplementary Materials.

2.1.7. Consumer Handling

Data and assumptions:

It is assumed that growth can occur in RTE cantaloupes at home. The variability in temperatures at home may be important and generally further from the target temperature of 4 °C [92]. Rocatto et al. [93] showed that the overall variability in European domestic refrigerators is described by a normal distribution, with a mean of 7 °C for Southern European countries and 6.1 °C for the northern countries. Nauta et al. [66] noted that the average temperature inside European domestic refrigerators was 6.64 °C. A similar value was measured in Brazil [94], with an average refrigerator temperature of 6 °C and with fluctuations ranging from 3.1 to 10.8 °C. Ding et al. [62] used an average refrigerator temperature ranging from 4 to 8.3 °C in a Korean risk assessment model for lettuce. Carrasco et al. [65] estimated the maximum refrigerator temperature at home to be 11.3 °C. RTE diced cantaloupe, a highly perishable food, is expected to be in a consumer's home for no longer than 10 days [95,96].

The R function:

The caHomeRTE() function models the growth of *L. monocytogenes* in RTE diced cantaloupe in home settings, utilizing the caGrowthBaranyi() for underlying growth calculations. This function tailors the simulation of home storage conditions—time and temperature—specifically to each unit, reflecting variations that depend on individual consumer behaviors. It incorporates lot-specific growth rates at 5 °C (EGR_5) and unit-specific initial bacterial quantities (ln(Qt)) that are derived from earlier stages in the logistics chain. To accurately capture the variability in how consumers store the product at home, Pert distributions are used to model the fluctuations in both storage time and temperature.

$$time_{Home} \sim Pert(time_{HomeMin}, time_{HomeMode}, time_{HomeMax})$$

$$temp_{Home} \sim Pert(temp_{HomeMin}, temp_{HomeMode}, temp_{HomeMax})$$

Parameter values for time and temperature are provided in the Supplementary Materials.

2.2. Risk Characterization

Several dose–response relationships are available for *L. monocytogenes* [97]. To estimate the risk of listeriosis per serving, the dose response in the susceptible population of FAO-WHO [98] was chosen. According to this exponential model, each ingested *L. monocytogenes* cell has an independent probability r of causing invasive listeriosis, which is assumed to be constant within a given population. The FAO-WHO [98] inferred a median value for r of 1.06×10^{-12} for the "population with increased susceptibility". The function DRQuick() from the dose response model R package [97] was employed to calculate the marginal probabilities of invasive listeriosis in the susceptible population $RiskServing_{ij}$ from the input matrix $N_{Cooked\ ij}$, containing the *L. monocytogenes* doses (CFU) in servings j

from contaminated lots i. The mean risk for every lot i ($RiskLot_i$) was calculated as a risk averaged across servings, j:

$$RiskLot_i = \frac{\sum_{j=1}^{c} RiskServing_{ij} \times Prob_{UnitPos\ i}}{c}$$

2.3. QRA Model's Ouputs

The model's outputs were summarized at four stages: after harvesting, the end of processing, the point of consumption, and risk characterization. After harvesting, the probability of contamination of lots of cantaloupe and the median contamination are provided. At the end of processing, the descriptors are generated based on the prevalence and the contamination matrix outputs from the function caTesting(). These include the following: (1) descriptive statistics (mean, median, and 95% confidence interval) of the mean concentration of *L. monocytogenes* (CFU/g) in the fraction of contaminated lots; (2) the prevalence of contaminated packs; and (3) the probability that a contaminated pack contains more than 10 CFU *L. monocytogenes* per g of RTE cantaloupe. At the point of consumption, the model's descriptors are estimated from the outputs of the function caHomeRTE(), which simulates home storage, and include the following: (1) descriptive statistics (mean, median, and 95% confidence interval) of the concentration of *L. monocytogenes* in any serving; (2) the prevalence of contaminated servings; (3) the probability that a contaminated serving contains more than 10 CFU *L. monocytogenes* per g of RTE cantaloupe; and (4) the probability that a contaminated serving contains more than 100 CFU *L. monocytogenes* per g of RTE cantaloupe. For risk characterization, the descriptors include summary statistics such as the mean, median, and 2.5, 97.5, and 99.5 percentiles of the lot-level mean risk per serving, RiskLot.

2.4. QRA Model's Implementation

All the functions described in Section 2.1 were programmed in R version 4.4.1 [99], and compiled in the package qraLm, which can be installed from the Github repository: https://github.com/WorldHealthOrganization/qraLm, accessed on 8 June 2025. The reference manual can be found at https://WorldHealthOrganization.github.io/qraLm/reference/, accessed on 8 June 2025.

3. Results and Discussion

3.1. Use of the Model for Risk Management Scenarios

The developed model was used to explore three types of control measures. The first type of intervention concerns the preharvest conditions of cantaloupes. Two parameters were explored: the presence of water on the cantaloupes before harvest (pIrrigRaining) and the use of protective barriers (i.e., barriers (mulch) to prevent contact between the cantaloupes and the soil (pFoil)). Figure 3 shows that these conditions have a strong impact on the risk of listeriosis. Compared to the reference condition (pIrrigRaining = 0, pFoil = 1), not using a barrier and applying overhead irrigation just before harvest across all plots (cantaloupe lots) increases the risk by a factor of 10,000. This increase is not explained by a higher level of contamination at the time of consumption but by a higher prevalence.

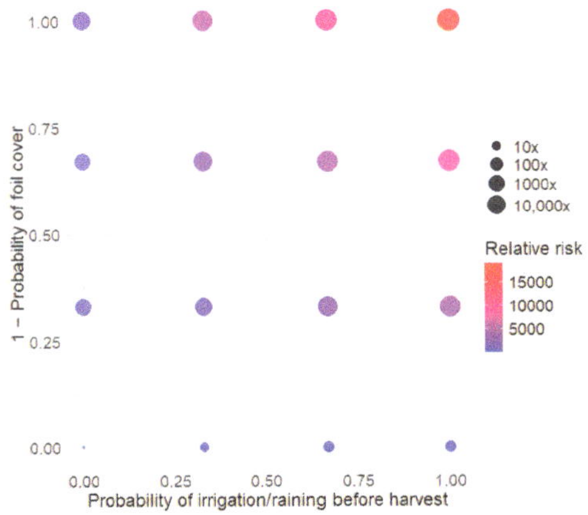

Figure 3. Assessment of the impact of different conditions in the fields on relative listeriosis risk for RTE diced cantaloupe. The reference risk value is defined for all fields using foil (pFoil = 1) and those with an absence of rain or irrigation just before harvesting (pIrrigRaining = 0).

The hygiene conditions during the harvest and washing of cantaloupes were also explored (Figure 4). The probCCH parameter, which represents hygiene control during harvest, ranged from 0.25 (reference condition) to 0.75, while the probCCH parameter, related to the contamination probability in the washing water, varied from 0 (reference condition) to 0.25. Figure 4 shows that the probability of contamination in the flume tank has a higher impact on the risk of listeriosis than cross-contamination during harvest.

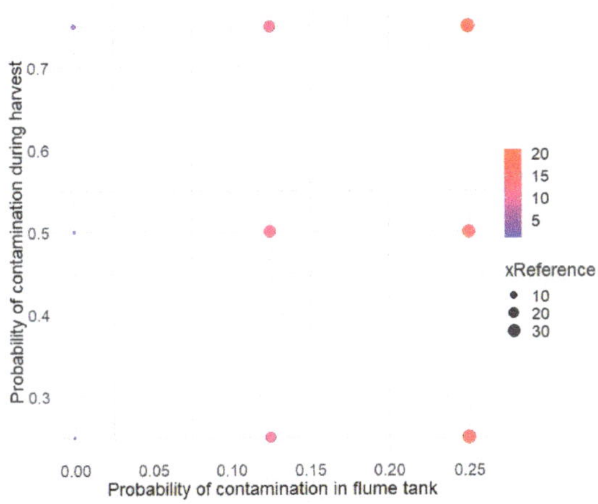

Figure 4. Assessment of the impact of cross-contamination events on relative listeriosis risk for RTE diced cantaloupe. The reference risk value is defined for a probability of cross-contamination during harvest (probCCH) of 0.25 and probability of contamination during washing of cantaloupes (probCCW) of 0.

Finally, as in most risk assessments for listeriosis, the storage conditions at consumers' homes were explored by varying the maximum values of the triangular distribution for the maximum temperature (Temp_max_home going from 11.1 °C (reference condition) to 13.1 °C) and the maximum storage duration (t_max_home going from 120 h (reference condition) to 168 h). The combination covering the riskiest practices leads to an increase in risk by a factor of 500 (Figure 5).

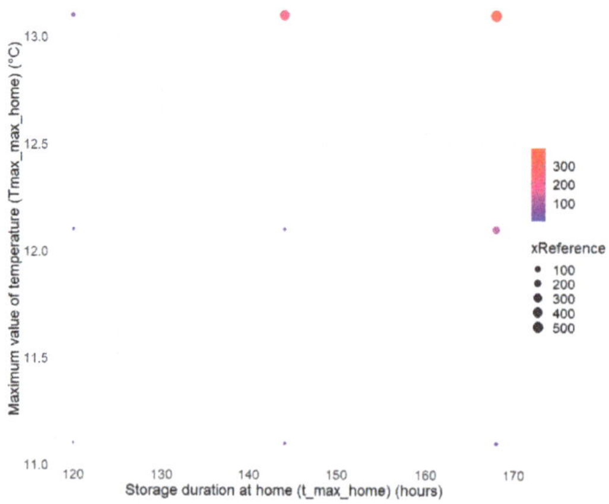

Figure 5. Assessment of the impact of home storage conditions on relative listeriosis risk for RTE diced cantaloupe. The reference risk value is defined for a maximum storage duration at home (*t_max_home*) of 120 h and maximum temperature (*Temp_max_home*) at 11.1 °C.

All parameters used for the reference condition and these different scenarios are listed in the Supplementary Materials presented in the Supplementary Materials.

3.2. Validation and Sensitivity Analysis

Johnston et al. [100] investigated the presence of *L. monocytogenes* on melons sampled from 36 fields. None of the samples were positive for the pathogen. Similarly, in our model, under the reference scenario (representing good agricultural practices where all fields use soil barriers and no irrigation occurs just before harvest), the predicted *L. monocytogenes* concentrations per cantaloupe remain below 100 CFU/cantaloupe. However, under the worst-case scenario, where poor agricultural practices are applied, the model predicts that *L. monocytogenes* is present in 73% of cantaloupe batches at the field level. Despite this, given a detection limit of 100 CFU per surface, only 0.05% of individual cantaloupes would test positive. These results highlight the low probability of detecting L. monocytogenes in cantaloupes at the field exit, even when contamination is widespread at the batch level.

Zhang et al. [101] detected *L. monocytogenes* in 0.72% (n = 699) of sampled cut cantaloupes. The model with reference parameters (good agricultural practices and no contamination at the time of washing the cantaloupes) predicts a prevalence of 0.08%. If the contamination frequency is increased to 0.25, then the predicted prevalence is 0.26%.

Townsend et al. [16], through a meta-analysis, found that the prevalence of *Listeria* spp. and *L. monocytogenes* on surfaces at the level of harvest or production environment could be very variable, ranging from none to over 50% of samples tested. Although the overall risk may appear low for the model in the reference situation (the mean risk of listeriosis per lot of RTE diced cantaloupe is 3.10^{-18}), the risk could be considerably increased by poor hygiene practices at primary production and cantaloupe handling conditions. However, it should be noted that some parameters are poorly supported scientifically and the values used are based on expert elicitation. Figure 6 shows the impact of uncertainty on parameters that appear to have the least evidence. The contamination levels of surface in contact with cantaloupes and present on the dicer appear to have a large effect on risk estimate.

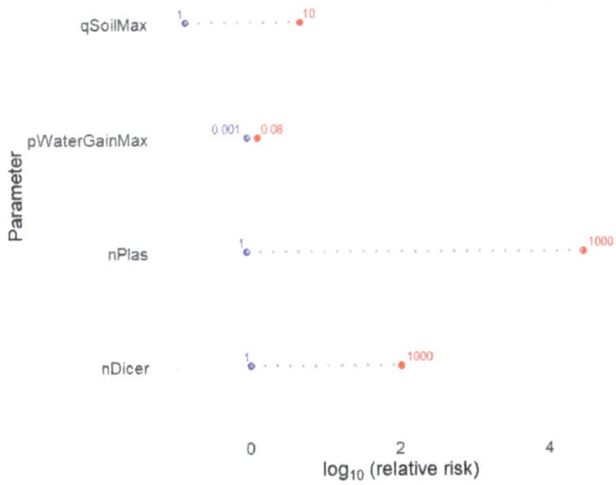

Figure 6. Sensitivity analysis for four parameters of QRA for RTE diced cantaloupe. The values used in the model that provide the reference risk are, respectively, 5g for the max value of the triangular distribution describing variability in the quantity of soil deposited on cantaloupe (qSoilMax), 0.004 for the max value of the % of weight gain (pWaterGainMax), and 9 cfu for the number of *L. monocytogenes* on food contact surfaces (conveyors or crates at harvest) touching the cantaloupes (nPlas) and for the number the surface of the dicing machine (nDicer).

3.3. Perspectives and Conclusions

In this study, we evaluated the effectiveness of several relevant "what-if" scenarios using the developed QRA model. This model enables the exploration of other intervention strategies, including the use of higher-quality water for irrigation, assessing the impact of organic fertilizers, reducing contamination during the dicing of RTE cantaloupes, removing highly contaminated lots through targeted sampling plans, shortening shelf life, and exploring combinations of these scenarios, among others.

The recent publication of a dose–response relationship based on the virulence categories of *L. monocytogenes* [97] suggests an improvement in the accuracy of assessments, particularly as initial data are now available on the genomic diversity of *L. monocytogenes* in agricultural soils [102] and production environments [103,104]. An important limitation of the current model is that it does not account for biofilm formation and strain-specific persistence throughout the processing chain. Future modeling efforts incorporating biofilm dynamics and strain tracking could better distinguish the relative contribution of each contamination source and enable more targeted intervention strategies. In conclusion, we have developed a robust set of functions to evaluate the risk of listeriosis associated with the consumption of RTE diced cantaloupe. This QRA has highlighted the potential variations in *L. monocytogenes* risks under different agricultural and processing conditions. The findings from our model underscore the importance of maintaining good agricultural practices in the fields, adhering to rigorous hygiene practices during harvest and processing, and ensuring proper storage conditions by consumers to mitigate the risk of listeriosis. Additionally, the adaptability of our QRA model to different products and processing steps provides significant flexibility for food safety authorities and food business operators around the world, enabling them to tailor interventions specifically to their regional needs.

Supplementary Materials: The following supporting information can be downloaded at https://www.mdpi.com/article/10.3390/foods14132212/s1: Table of parameters allowing us to define temperatures and duration for the transport module.

Author Contributions: Conceptualization, U.G.-B., R.P., L.G., V.C. and M.S.; methodology, U.G.-B., R.P., L.G. and M.S.; software, R.P., U.G.-B., L.G., M.S. and V.C.; validation, U.G.-B. and R.P.; formal analysis, R.P., U.G.-B., L.G., V.C. and M.S.; investigation, U.G.-B., R.P., J.D.O.M., A.H., V.C., L.G. and M.S.; resources, M.S.; data curation, U.G.-B. and L.G.; writing—original draft preparation, L.G., U.G.-B. and R.P.; writing—review and editing, L.G., U.G.-B., R.P., J.D.O.M., A.H., A.A., C.G., A.F., H.A.-Q., Q.D., J.K., V.C. and M.S.; visualization, U.G.-B., R.P., L.G., V.C. and M.S.; supervision, M.S.; project administration, M.S.; funding acquisition, M.S. All authors have read and agreed to the published version of the manuscript.

Funding: This research was funded by the European Commission Directorate-General for Health and Food Safety (DG Sante).

Institutional Review Board Statement: Not applicable.

Informed Consent Statement: Not applicable.

Data Availability Statement: The data presented in this study are openly available in [https://github.com/WorldHealthOrganization/qraLm, accessed on 8 June 2025]. The data employed to build the quantitative risk assessment model are published elsewhere, and can be directly extracted from the articles cited.

Acknowledgments: The FAO/WHO Joint Expert Meeting on Microbiological Risk Assessment (JEMRA) on *Listeria monocytogenes* in foods (Part 1), held in Rome, Italy, from 24 to 28 October 2022, recommended the development of a quantitative risk assessment model for *Listeria monocytogenes* in non-ready-to-eat frozen vegetables that considers processing and consumer handling. The model was finalized during the JEMRA meeting on *Listeria monocytogenes* in foods Part 2: Risk Assessment Models, which was held in Geneva, Switzerland, from 29 May to 2 June 2023. The authors thank the JEMRA experts for Part 1 (https://www.who.int/publications/m/item/jemra-of-listeria-monocytogenes-in-foods, accessed on 8 June 2025) and Part 2 (https://www.who.int/publications/m/item/jemra-of-listeria-monocytogenes-in-foods-part-2-risk-assessment-models, accessed on 8 June 2025), as well as the FAO/WHO secretariat for their suggestions and feedback.

Conflicts of Interest: Author R.P. is self-employed. He participated in this work in his own personal capacity. The remaining authors declare that the research was conducted in the absence of any commercial or financial relationships that could be construed as a potential conflict of interest. The sponsors had no role in the design, execution, interpretation, or writing of the study. Disclaimer/Author's Note: J.D.O.M., A.H. and M.S. are a staff member(s) of the World Health Organization. The authors alone are responsible for the views expressed in this publication and they do not necessarily represent the official position, decisions, policy, or views of the World Health Organization.

References

1. Chan, Y.-W.; Hoban, A.; Moore, H.; Greig, D.R.; Painset, A.; Jorgensen, F.; Chattaway, M.A.; Jenkins, C.; Balasegaram, S.; McCormick, J. Two outbreaks of foodborne gastrointestinal infection linked to consumption of imported melons, United Kingdom, March to August 2021. *J. Food Prot.* **2023**, *86*, 100027. [CrossRef]
2. Jenkins, E.; Gardenhire, I.; Whitney, B.M.; Martin, K.B.; Schwensohn, C.; Gieraltowski, L.; Leeper, M.M.; McCurdy, V.; McClure, M.; Wellman, A. An investigation of an outbreak of *Salmonella* Newport infections linked to melons–United States, 2020. *Food Control* **2023**, *152*, 109833. [CrossRef]
3. McGeoch, L.J.; Hoban, A.; Sawyer, C.; Rabie, H.; Painset, A.; Browning, L.; Brown, D.; McCarthy, C.; Nelson, A.; Firme, A. *Salmonella* Saintpaul outbreak associated with cantaloupe consumption, the United Kingdom and Portugal, September to November 2023. *Epidemiol. Infect.* **2024**, *152*, e78. [CrossRef]
4. Federman, S.S.; Jenkins, E.; Wilson, C.; DeLaGarza, A.; Schwensohn, C.; Schneider, B.; Nsubuga, J.; Literman, R.; Wellman, A.; Whitney, B.M. An investigation of an outbreak of *Salmonella* Typhimurium infections linked to cantaloupe–United States, 2022. *Food Control* **2024**, *166*, 110733. [CrossRef]
5. Walsh, K.A.; Bennett, S.D.; Mahovic, M.; Gould, L.H. Outbreaks associated with cantaloupe, watermelon, and honeydew in the United States, 1973–2011. *Foodborne Pathog. Dis.* **2014**, *11*, 945–952. [CrossRef]

6. McCollum, J.T.; Cronquist, A.B.; Silk, B.J.; Jackson, K.A.; O'Connor, K.A.; Cosgrove, S.; Gossack, J.P.; Parachini, S.S.; Jain, N.S.; Ettestad, P. Multistate outbreak of listeriosis associated with cantaloupe. *N. Engl. J. Med.* **2013**, *369*, 944–953. [CrossRef]
7. NSW. Listeria Outbreak Investigation. In *Summary Report for the Melon Industry, October 2018*; NSW Department of Primary Industries, Biosecurity & Food Safety: Orange, Australia, 2018; p. 12; ISBN 978-1-76058-267-8.
8. EFSA Panel on Biological Hazards; Ricci, A.; Allende, A.; Bolton, D.; Chemaly, M.; Davies, R.; Fernández Escámez, P.S.; Girones, R.; Herman, L.; Koutsoumanis, K. *Listeria monocytogenes* contamination of ready-to-eat foods and the risk for human health in the EU. *EFSA J.* **2018**, *16*, e05134. [CrossRef] [PubMed]
9. Leclercq, A.; Kooh, P.; Augustin, J.-C.; Guillier, L.; Thébault, A.; Cadavez, V.; Gonzales-Barron, U.; Sanaa, M. Risk factors for sporadic listeriosis: A systematic review and meta-analysis. *Microb. Risk Anal.* **2021**, *17*, 100128. [CrossRef]
10. Locatelli, A.; Spor, A.; Jolivet, C.; Piveteau, P.; Hartmann, A. Biotic and abiotic soil properties influence survival of *Listeria monocytogenes* in soil. *PLoS ONE* **2013**, *8*, e75969. [CrossRef]
11. Vivant, A.-L.; Garmyn, D.; Piveteau, P. *Listeria monocytogenes*, a down-to-earth pathogen. *Front. Cell. Infect. Microbiol.* **2013**, *3*, 87. [CrossRef] [PubMed]
12. Kljujev, I.; Raicevic, V.; Jovicic-Petrovic, J.; Vujovic, B.; Mirkovic, M.; Rothballer, M. *Listeria monocytogenes*–Danger for health safety vegetable production. *Microb. Pathog.* **2018**, *120*, 23–31. [CrossRef] [PubMed]
13. Beuchat, L.R. *Listeria monocytogenes*: Incidence on vegetables. *Food Control* **1996**, *7*, 223–228. [CrossRef]
14. Domenichini, G.; Fogliazza, D.; Pagani, M. Studies on the transmission of *Listeria* by means of arthropods. *Ital. J. Food Sci.* **1992**, *4*, 269–278.
15. Lapanje, A.; Zrimec, A.; Drobne, D.; Rupnik, M. Long-term Hg pollution-induced structural shifts of bacterial community in the terrestrial isopod (Porcellio scaber) gut. *Environ. Pollut.* **2010**, *158*, 3186–3193. [CrossRef] [PubMed]
16. Townsend, A.; Strawn, L.K.; Chapman, B.J.; Dunn, L.L. A systematic review of *Listeria* species and *Listeria monocytogenes* prevalence, persistence, and diversity throughout the fresh produce supply chain. *Foods* **2021**, *10*, 1427. [CrossRef]
17. Lagarde, J.; Feurer, C.; Denis, M.; Douarre, P.-E.; Piveteau, P.; Roussel, S. *Listeria monocytogenes* prevalence and genomic diversity along the pig and pork production chain. *Food Microbiol.* **2023**, *119*, 104430. [CrossRef] [PubMed]
18. Weller, D.; Wiedmann, M.; Strawn, L.K. Spatial and temporal factors associated with an increased prevalence of *Listeria monocytogenes* in spinach fields in New York State. *Appl. Environ. Microbiol.* **2015**, *81*, 6059–6069. [CrossRef]
19. Mahmoud, B.; Vaidya, N.; Corvalan, C.; Linton, R. Inactivation kinetics of inoculated *Escherichia coli* O157: H7, *Listeria monocytogenes* and *Salmonella* Poona on whole cantaloupe by chlorine dioxide gas. *Food Microbiol.* **2008**, *25*, 857–865. [CrossRef]
20. Ukuku, D.O.; Fett, W.F. Relationship of cell surface charge and hydrophobicity to strength of attachment of bacteria to cantaloupe rind. *J. Food Prot.* **2002**, *65*, 1093–1099. [CrossRef]
21. Ferreira, V.; Wiedmann, M.; Teixeira, P.; Stasiewicz, M. *Listeria monocytogenes* persistence in food-associated environments: Epidemiology, strain characteristics, and implications for public health. *J. Food Prot.* **2014**, *77*, 150–170. [CrossRef] [PubMed]
22. Carpentier, B.; Cerf, O. Persistence of *Listeria monocytogenes* in food industry equipment and premises. *Int. J. Food Microbiol.* **2011**, *145*, 1–8. [CrossRef] [PubMed]
23. U.S. FDA. Environmental Assessment: Factors Potentially Contributing to the Contamination of Fresh Whole Cantaloupe Implicated in a Multi-State Outbreak of Listeriosis. 2011. Available online: https://www.fda.gov/food/outbreaks-foodborne-illness/outbreak-investigation-reports (accessed on 8 June 2025).
24. U.S. FDA. External Summary Report: FY 2013 Inspection, Environmental Sampling and Sample Collection (Pre and Post-Process) at Cantaloupe Packinghouses Assignment. 2015. Available online: https://www.fda.gov/food/foodborne-pathogens/external-summary-report-fy-2013-inspection-environmental-sampling-and-sample-collection-pre-and-post#_ftnref1 (accessed on 8 June 2025).
25. Dunn, L.L.; Friedrich, L.M.; Strawn, L.K.; Danyluk, M.D. Prevalence of *Listeria monocytogenes* and indicator microorganisms in Florida cantaloupe packinghouses, 2013–2014. *Food Microbiol.* **2022**, *104*, 103970. [CrossRef]
26. EFSA Panel on Biological Hazards; Koutsoumanis, K.; Ordóñez, A.A.; Bolton, D.; Bover-Cid, S.; Chemaly, M.; De Cesare, A.; Herman, L.; Hilbert, F.; Lindqvist, R.; et al. Microbiological hazards associated with the use of water in the post-harvest handling and processing operations of fresh and frozen fruits, vegetables and herbs (ffFVHs). Part 1 (outbreak data analysis, literature review and stakeholder questionnaire). *EFSA J.* **2023**, *21*, e08332. [CrossRef] [PubMed]
27. Gil, M.I.; Truchado, P.; Tudela, J.A.; Allende, A. Environmental monitoring of three fresh-cut processing facilities reveals harborage sites for *Listeria monocytogenes*. *Food Control* **2024**, *155*, 110093. [CrossRef]
28. Sullivan, G.; Wiedmann, M. Detection and prevalence of *Listeria* in US produce packinghouses and fresh-cut facilities. *J. Food Prot.* **2020**, *83*, 1656–1666. [CrossRef]

29. Ukuku, D.O.; Olanya, M.; Geveke, D.J.; Sommers, C.H. Effect of native microflora, waiting period, and storage temperature on *Listeria monocytogenes* serovars transferred from cantaloupe rind to fresh-cut pieces during preparation. *J. Food Prot.* **2012**, *75*, 1912–1919. [CrossRef]
30. Bassett, J.; McClure, P. A risk assessment approach for fresh fruits. *J. Appl. Microbiol.* **2008**, *104*, 925–943. [CrossRef] [PubMed]
31. Danyluk, M.D.; Friedrich, L.M.; Schaffner, D.W. Modeling the growth of *Listeria monocytogenes* on cut cantaloupe, honeydew and watermelon. *Food Microbiol.* **2014**, *38*, 52–55. [CrossRef]
32. Li, D.; Friedrich, L.M.; Danyluk, M.D.; Harris, L.J.; Schaffner, D.W. Development and validation of a mathematical model for growth of pathogens in cut melons. *J. Food Prot.* **2013**, *76*, 953–958. [CrossRef]
33. Gonzales-Barron, U.; Cadavez, V.; De Oliveira Mota, J.; Guillier, L.; Sanaa, M. A critical review of risk assessment models for *Listeria monocytogenes* in produce. *Foods* **2024**, *13*, 1111. [CrossRef]
34. Guzel, M. *Quantitative Risk Assessment for Listeria monocytogenes on Fresh-Cut Lettuce and Fresh-Cut Cantaloupe*; Texas A&M University: College Station, TX, USA, 2015.
35. FAO; WHO. Summary and Conclusions. In Proceedings of the Joint FAO/WHO Expert Meeting on Microbiological Risk Assessment of *Listeria monocytogenes* in Foods, Rome, Italy, 24–28 October 2022.
36. FAO; WHO. Part 2: Risk Assessment Models. Summary and Conclusions. In Proceedings of the Joint FAO/WHO Expert Meeting on Microbiological Risk Assessment of *Listeria monocytogenes* in Foods, Rome, Italy, 29 May–2 June 2023.
37. Nauta, M.J. Modelling bacterial growth in quantitative microbiological risk assessment: Is it possible? *Int. J. Food Microbiol.* **2002**, *73*, 297–304. [CrossRef]
38. Dowe, M.J.; Jackson, E.D.; Mori, J.G.; Bell, C.R. *Listeria monocytogenes* survival in soil and incidence in agricultural soils. *J. Food Prot.* **1997**, *60*, 1201–1207. [CrossRef] [PubMed]
39. Locatelli, A.; Depret, G.; Jolivet, C.; Henry, S.; Dequiedt, S.; Piveteau, P.; Hartmann, A. Nation-wide study of the occurrence of *Listeria monocytogenes* in French soils using culture-based and molecular detection methods. *J. Microbiol. Methods* **2013**, *93*, 242–250. [CrossRef] [PubMed]
40. Szymczak, B.; Szymczak, M.; Sawicki, W.; Dąbrowski, W. Anthropogenic impact on the presence of *L. monocytogenes* in soil, fruits, and vegetables. *Folia Microbiol.* **2014**, *59*, 23–29. [CrossRef]
41. Girardin, H.; Morris, C.E.; Albagnac, C.; Dreux, N.; Glaux, C.; Nguyen-The, C. Behaviour of the pathogen surrogates *Listeria innocua* and *Clostridium sporogenes* during production of parsley in fields fertilized with contaminated amendments. *FEMS Microbiol. Ecol.* **2005**, *54*, 287–295. [CrossRef] [PubMed]
42. Miceli, A.; Settanni, L. Influence of agronomic practices and pre-harvest conditions on the attachment and development of *Listeria monocytogenes* in vegetables. *Ann. Microbiol.* **2019**, *69*, 185–199. [CrossRef]
43. Oliveira, M.; Usall, J.; Viñas, I.; Solsona, C.; Abadias, M. Transfer of *Listeria innocua* from contaminated compost and irrigation water to lettuce leaves. *Food Microbiol.* **2011**, *28*, 590–596. [CrossRef]
44. McLaughlin, H.P.; Casey, P.G.; Cotter, J.; Gahan, C.G.; Hill, C. Factors affecting survival of *Listeria monocytogenes* and *Listeria innocua* in soil samples. *Arch. Microbiol.* **2011**, *193*, 775–785. [CrossRef]
45. Nicholson, F.A.; Groves, S.J.; Chambers, B.J. Pathogen survival during livestock manure storage and following land application. *Bioresour. Technol.* **2005**, *96*, 135–143. [CrossRef]
46. Allende, A.; Monaghan, J. Irrigation water quality for leafy crops: A perspective of risks and potential solutions. *Int. J. Environ. Res. Public Health* **2015**, *12*, 7457–7477. [CrossRef]
47. EFSA Panel on Biological Hazards (BIOHAZ). Scientific Opinion on the risk posed by pathogens in food of non-animal origin. Part 2 (*Salmonella* and Norovirus in leafy greens eaten raw as salads). *EFSA J.* **2014**, *12*, 3600.
48. Weller, D.; Shiwakoti, S.; Bergholz, P.; Grohn, Y.; Wiedmann, M.; Strawn, L.K. Validation of a previously developed geospatial model that predicts the prevalence of *Listeria monocytogenes* in New York State produce fields. *Appl. Environ. Microbiol.* **2016**, *82*, 797–807. [CrossRef]
49. Nyarko, E.; Kniel, K.E.; Millner, P.D.; Luo, Y.; Handy, E.T.; Reynnells, R.; East, C.; Sharma, M. Survival and growth of *Listeria monocytogenes* on whole cantaloupes is dependent on site of contamination and storage temperature. *Int. J. Food Microbiol.* **2016**, *234*, 65–70. [CrossRef] [PubMed]
50. Sharma, M.; Handy, E.T.; East, C.L.; Kim, S.; Jiang, C.; Callahan, M.T.; Allard, S.M.; Micallef, S.; Craighead, S.; Anderson-Coughlin, B. Prevalence of *Salmonella* and *Listeria monocytogenes* in non-traditional irrigation waters in the Mid-Atlantic United States is affected by water type, season, and recovery method. *PLoS ONE* **2020**, *15*, e0229365. [CrossRef] [PubMed]
51. Acheamfour, C.L.; Parveen, S.; Hashem, F.; Sharma, M.; Gerdes, M.E.; May, E.B.; Rogers, K.; Haymaker, J.; Duncan, R.; Foust, D. Levels of *Salmonella enterica* and *Listeria monocytogenes* in alternative irrigation water vary based on water source on the Eastern Shore of Maryland. *Microbiol. Spectr.* **2021**, *9*, e00669-21. [CrossRef]

52. Iwu, C.D.; Iwu-Jaja, C.J.; Elhadi, R.; Semerjian, L.; Okoh, A.I. Modelling the potential risk of infection associated with *Listeria monocytogenes* in irrigation water and agricultural soil in two district municipalities in South Africa. *Microorganisms* **2022**, *10*, 181. [CrossRef] [PubMed]
53. Raschle, S.; Stephan, R.; Stevens, M.J.; Cernela, N.; Zurfluh, K.; Muchaamba, F.; Nüesch-Inderbinen, M. Environmental dissemination of pathogenic *Listeria monocytogenes* in flowing surface waters in Switzerland. *Sci. Rep.* **2021**, *11*, 9066. [CrossRef]
54. Richards, G.M.; Beuchat, L.R. Attachment of *Salmonella* Poona to cantaloupe rind and stem scar tissues as affected by temperature of fruit and inoculum. *J. Food Prot.* **2004**, *67*, 1359–1364. [CrossRef]
55. Strawn, L.K.; Gröhn, Y.T.; Warchocki, S.; Worobo, R.W.; Bihn, E.A.; Wiedmann, M. Risk factors associated with *Salmonella* and *Listeria monocytogenes* contamination of produce fields. *Appl. Environ. Microbiol.* **2013**, *79*, 7618–7627. [CrossRef]
56. Strawn, L.K.; Fortes, E.D.; Bihn, E.A.; Nightingale, K.K.; Gröhn, Y.T.; Worobo, R.W.; Wiedmann, M.; Bergholz, P.W. Landscape and meteorological factors affecting prevalence of three food-borne pathogens in fruit and vegetable farms. *Appl. Environ. Microbiol.* **2013**, *79*, 588–600. [CrossRef]
57. Ukuku, D.O.; Fett, W. Behavior of *Listeria monocytogenes* inoculated on cantaloupe surfaces and efficacy of washing treatments to reduce transfer from rind to fresh-cut pieces. *J. Food Prot.* **2002**, *65*, 924–930. [CrossRef]
58. Hoelzer, K.; Pouillot, R.; Gallagher, D.; Silverman, M.B.; Kause, J.; Dennis, S. Estimation of *Listeria monocytogenes* transfer coefficients and efficacy of bacterial removal through cleaning and sanitation. *Int. J. Food Microbiol.* **2012**, *157*, 267–277. [CrossRef] [PubMed]
59. Nauta, M.J. Microbiological risk assessment models for partitioning and mixing during food handling. *Int. J. Food Microbiol.* **2005**, *100*, 311–322. [CrossRef]
60. Baranyi, J.; Roberts, T.A. A dynamic approach to predicting bacterial growth in food. *Int. J. Food Microbiol.* **1994**, *23*, 277–294. [CrossRef] [PubMed]
61. Fang, T.; Liu, Y.; Huang, L. Growth kinetics of *Listeria monocytogenes* and spoilage microorganisms in fresh-cut cantaloupe. *Food Microbiol.* **2013**, *34*, 174–181. [CrossRef] [PubMed]
62. Ding, T.; Iwahori, J.i.; Kasuga, F.; Wang, J.; Forghani, F.; Park, M.-S.; Oh, D.-H. Risk assessment for *Listeria monocytogenes* on lettuce from farm to table in Korea. *Food Control* **2013**, *30*, 190–199. [CrossRef]
63. Sant'Ana, A.S.; Franco, B.D.; Schaffner, D.W. Risk of infection with *Salmonella* and *Listeria monocytogenes* due to consumption of ready-to-eat leafy vegetables in Brazil. *Food Control* **2014**, *42*, 1–8. [CrossRef]
64. Derens-Bertheau, E.; Osswald, V.; Laguerre, O.; Alvarez, G. Cold chain of chilled food in France. *Int. J. Refrig.* **2015**, *52*, 161–167. [CrossRef]
65. Carrasco, E.; Pérez-Rodríguez, F.; Valero, A.; García-Gimeno, R.M.; Zurera, G. Risk assessment and management of *Listeria monocytogenes* in ready-to-eat lettuce salads. *Compr. Rev. Food Sci. Food Saf.* **2010**, *9*, 498–512. [CrossRef]
66. Nauta, M.J.; Litman, S.; Barker, G.C.; Carlin, F. A retail and consumer phase model for exposure assessment of *Bacillus cereus*. *Int. J. Food Microbiol.* **2003**, *83*, 205–218. [CrossRef]
67. Jeyaletchumi, P.; Tunung, R.; Selina, P.M.; Chai, L.; Radu, S.; Farinazleen, M.; Cheah, Y.; Mitsuaki, N.; Yoshitsugu, N.; Kumar, M.P. Evaluation of *Listeria* spp. and *Listeria monocytogenes* in selected vegetable farms. *J. Trop. Agric. Food Sci.* **2011**, *39*, 255–266.
68. Linke, K.; Rückerl, I.; Brugger, K.; Karpiskova, R.; Walland, J.; Muri-Klinger, S.; Tichy, A.; Wagner, M.; Stessl, B. Reservoirs of *Listeria* species in three environmental ecosystems. *Appl. Environ. Microbiol.* **2014**, *80*, 5583–5592. [CrossRef]
69. Soni, D.K.; Singh, R.K.; Singh, D.V.; Dubey, S.K. Characterization of *Listeria monocytogenes* isolated from Ganges water, human clinical and milk samples at Varanasi, India. *Infect. Genet. Evol.* **2013**, *14*, 83–91. [CrossRef] [PubMed]
70. Stea, E.C.; Purdue, L.M.; Jamieson, R.C.; Yost, C.K.; Truelstrup Hansen, L. Comparison of the prevalences and diversities of *Listeria* species and *Listeria monocytogenes* in an urban and a rural agricultural watershed. *Appl. Environ. Microbiol.* **2015**, *81*, 3812–3822. [CrossRef] [PubMed]
71. Jongman, M.; Korsten, L. Assessment of irrigation water quality and microbiological safety of leafy greens in different production systems. *J. Food Saf.* **2017**, *37*, e12324. [CrossRef]
72. Lyautey, E.; Lapen, D.R.; Wilkes, G.; McCleary, K.; Pagotto, F.; Tyler, K.; Hartmann, A.; Piveteau, P.; Rieu, A.; Robertson, W.J. Distribution and characteristics of *Listeria monocytogenes* isolates from surface waters of the South Nation River watershed, Ontario, Canada. *Appl. Environ. Microbiol.* **2007**, *73*, 5401–5410. [CrossRef]
73. Cooley, M.B.; Quiñones, B.; Oryang, D.; Mandrell, R.E.; Gorski, L. Prevalence of shiga toxin producing *Escherichia coli*, *Salmonella enterica*, and *Listeria monocytogenes* at public access watershed sites in a California Central Coast agricultural region. *Front. Cell. Infect. Microbiol.* **2014**, *4*, 30. [CrossRef]
74. Ijabadeniyi, O.A.; Debusho, L.K.; Vanderlinde, M.; Buys, E.M. Irrigation water as a potential preharvest source of bacterial contamination of vegetables. *J. Food Saf.* **2011**, *31*, 452–461. [CrossRef]

75. Fu, Y.; Bhunia, A.K.; Yao, Y. Abrasive brushing reduces pathogen biofilms at cantaloupe rind surface. *Int. J. Food Microbiol.* **2020**, *329*, 108685. [CrossRef]
76. Guzel, M.; Moreira, R.G.; Omac, B.; Castell-Perez, M.E. Quantifying the effectiveness of washing treatments on the microbial quality of fresh-cut romaine lettuce and cantaloupe. *LWT* **2017**, *86*, 270–276. [CrossRef]
77. Ukuku, D.O.; Bari, M.; Kawamoto, S.; Isshiki, K. Use of hydrogen peroxide in combination with nisin, sodium lactate and citric acid for reducing transfer of bacterial pathogens from whole melon surfaces to fresh-cut pieces. *Int. J. Food Microbiol.* **2005**, *104*, 225–233. [CrossRef] [PubMed]
78. FAO; WHO. *Prevention and Control of Microbiological Hazards in Fresh Fruits and Vegetables: Part 4: Specific Commodities: Meeting Report*; FAO: Rome, Italy; WHO: Geneva, Switzerland, 2023. [CrossRef]
79. FDA. *FDA-iRISK 4.2 Food Safety Modeling Tool: Technical Document*; U.S. Food and Drug Administration: Silver Spring, MD, USA, 2021. Available online: https://irisk.foodrisk.org/Documents/FDAiRISKTechnicalDocumentation.pdf (accessed on 8 June 2025).
80. Selma, M.V.; Ibáñez, A.M.; Allende, A.; Cantwell, M.; Suslow, T. Effect of gaseous ozone and hot water on microbial and sensory quality of cantaloupe and potential transference of *Escherichia coli* O157: H7 during cutting. *Food Microbiol.* **2008**, *25*, 162–168. [CrossRef] [PubMed]
81. Patil, R.D. Transfer of *Listeria monocytogenes* During Cutting, Slicing, Dicing, and Subsequent Storage of Cantaloupe and Honeydew Melons. Master's Thesis, Michigan State University, East Lansing, MI, USA, 2017.
82. Pouillot, R.; Chen, Y.; Van Doren, J.M. Elucidating the influence of the lower and upper microbiological limits: When is a 3-class sampling plan useful to test for pathogens in food? *Food Control* **2024**, *163*, 110544. [CrossRef]
83. Mercier, S.; Villeneuve, S.; Mondor, M.; Uysal, I. Time–temperature management along the food cold chain: A review of recent developments. *Compr. Rev. Food Sci. Food Saf.* **2017**, *16*, 647–667. [CrossRef] [PubMed]
84. McKellar, R.C.; LeBlanc, D.I.; Lu, J.; Delaquis, P. Simulation of Escherichia coli O157: H7 behavior in fresh-cut lettuce under dynamic temperature conditions during distribution from processing to retail. *Foodborne Pathog. Dis.* **2012**, *9*, 239–244. [CrossRef]
85. Scolforo, C.Z.; Bairros, J.V.; Rezende, A.C.B.; Silva, B.S.; Alves, R.B.; Costa, D.S.; Andrade, N.J.; Sant'Ana, A.S.; Pena, W.E. Modeling the fate of *Listeria monocytogenes* and *Salmonella enterica* in the pulp and on the outer rind of Canary melons (Cucumis melo (Indorus Group)). *LWT* **2017**, *77*, 290–297. [CrossRef]
86. Farber, J.; Wang, S.; Cai, Y.; Zhang, S. Changes in populations of *Listeria monocytogenes* inoculated on packaged fresh-cut vegetables. *J. Food Prot.* **1998**, *61*, 192–195. [CrossRef]
87. Hong, Y.K.; Yoon, W.B.; Huang, L.; Yuk, H.G. Predictive Modeling for Growth of Non-and Cold-adapted *Listeria monocytogenes* on Fresh-cut Cantaloupe at Different Storage Temperatures. *J. Food Sci.* **2014**, *79*, M1168–M1174. [CrossRef]
88. Moreira, J.; Mera, E.; Singh Chhetri, V.; King, J.M.; Gentimis, T.; Adhikari, A. Effect of storage temperature and produce type on the survival or growth of *Listeria monocytogenes* on peeled rinds and fresh-cut produce. *Front. Microbiol.* **2023**, *14*, 1151819. [CrossRef]
89. FSA. *National Microbiological Survey and Consumer Habits in relation to Frozen Vegetables, Fruits and Herbs (19NS6)*; Monitoring & Surveillance Series; Food Safety Authority of Ireland, safeFood: Dublin, Ireland, 2022; p. 62. ISBN 978-1-910348-54-3.
90. Pouillot, R.; Miconnet, N.; Afchain, A.-L.; Delignette-Muller, M.L.; Beaufort, A.; Rosso, L.; Denis, J.-B.; Cornu, M. Quantitative Risk Assessment of *Listeria monocytogenes* in French Cold-Smoked Salmon: I. Quantitative Exposure Assessment. *Risk Anal.* **2007**, *27*, 683–700. [CrossRef]
91. O'Connor-Shaw, R.; Roberts, R.; Ford, A.; Nottingham, S. Shelf life of minimally processed honeydew, kiwifruit, papaya, pineapple and cantaloupe. *J. Food Sci.* **1994**, *59*, 1202–1206. [CrossRef]
92. Bonanno, L.; Bergis, H.; Gnanou-Besse, N.; Asséré, A.; Danan, C. Which domestic refrigerator temperatures in Europe?-Focus on shelf-life studies regarding *Listeria monocytogenes* (Lm) in ready-to-eat (RTE) foods. *Food Microbiol.* **2024**, *123*, 104595. [CrossRef] [PubMed]
93. Roccato, A.; Uyttendaele, M.; Membré, J.-M. Analysis of domestic refrigerator temperatures and home storage time distributions for shelf-life studies and food safety risk assessment. *Food Res. Int.* **2017**, *96*, 171–181. [CrossRef] [PubMed]
94. Silva, D.; Celidonio, F.; Oliveira, K. Verification of the temperature of domestic refrigerators to minimize the deterioration and possible foodborne illnesses. *Hig. Aliment.* **2008**, *22*, 42–45.
95. Amaro, A.L.; Spadafora, N.D.; Pereira, M.J.; Dhorajiwala, R.; Herbert, R.J.; Müller, C.T.; Rogers, H.J.; Pintado, M. Multitrait analysis of fresh-cut cantaloupe melon enables discrimination between storage times and temperatures and identifies potential markers for quality assessments. *Food Chem.* **2018**, *241*, 222–231. [CrossRef]
96. Luciano, W.A.; Griffin, S.; Targino de Souza Pedrosa, G.; Alvarenga, V.; Valdramidis, V.; Magnani, M. Growth behavior of low populations of *Listeria monocytogenes* on fresh-cut mango, melon and papaya under different storage temperatures. *Food Microbiol.* **2022**, *102*, 103930. [CrossRef]

97. Pouillot, R.; Kiermeier, A.; Guillier, L.; Cadavez, V.; Sanaa, M. Updated Parameters for *Listeria monocytogenes* Dose–Response Model Considering Pathogen Virulence and Age and Sex of Consumer. *Foods* **2024**, *13*, 751. [CrossRef]
98. FAO; WHO. *Risk Assessment of Listeria monocytogenes in Ready-to-Eat Foods: Technical Report*; World Health Organization and Food and Agriculture Organization of the United Nations: Geneve, Switzerland, 2004; pp. 1–269.
99. R Core Team. *R: A Language and Environment for Statistical Computing*; R Foundation for Statistical Computing: Vienna, Austria, 2021. Available online: https://www.R-project.org/ (accessed on 8 June 2025).
100. Johnston, L.M.; Jaykus, L.-A.; Moll, D.; Martinez, M.C.; Anciso, J.; Mora, B.; Moe, C.L. A field study of the microbiological quality of fresh produce. *J. Food Prot.* **2005**, *68*, 1840–1847. [CrossRef]
101. Zhang, H.; Yamamoto, E.; Murphy, J.; Locas, A. Microbiological safety of ready-to-eat fresh-cut fruits and vegetables sold on the Canadian retail market. *Int. J. Food Microbiol.* **2020**, *335*, 108855. [CrossRef]
102. Félix, B.; Sevellec, Y.; Palma, F.; Douarre, P.E.; Felten, A.; Radomski, N.; Mallet, L.; Blanchard, Y.; Leroux, A.; Soumet, C.; et al. A European-wide dataset to uncover adaptive traits of *Listeria monocytogenes* to diverse ecological niches. *Sci. Data* **2022**, *9*, 190. [CrossRef]
103. Jorgensen, J.; Bland, R.; Waite-Cusic, J.; Kovacevic, J. Diversity and antimicrobial resistance of *Listeria* spp. and *L. monocytogenes* clones from produce handling and processing facilities in the Pacific Northwest. *Food Control* **2021**, *123*, 107665. [CrossRef]
104. Sullivan, G.; Orsi, R.H.; Estrada, E.; Strawn, L.; Wiedmann, M. Whole-genome sequencing-based characterization of *Listeria* isolates from produce packinghouses and fresh-cut facilities suggests both persistence and reintroduction of fully virulent *L. monocytogenes*. *Appl. Environ. Microbiol.* **2022**, *88*, e01177-22. [CrossRef] [PubMed]

foods

MDPI

Article

A Quantitative Risk Assessment Model for *Listeria monocytogenes* in Ready-to-Eat Smoked and Gravad Fish

Ursula Gonzales-Barron [1,2], Régis Pouillot [3,*], Taran Skjerdal [4], Elena Carrasco [5], Paula Teixeira [6], Matthew J. Stasiewicz [7], Akio Hasegawa [8], Juliana De Oliveira Mota [8], Laurent Guillier [9], Vasco Cadavez [1,2] and Moez Sanaa [8,*]

1 Centro de Investigação de Montanha (CIMO), Instituto Politécnico de Bragança, Campus de Santa Apolónia, 5300-253 Bragança, Portugal; ubarron@ipb.pt (U.G.-B.); vcadavez@ipb.pt (V.C.)
2 Laboratório para a Sustentabilidade e Tecnologia em Regiões de Montanha, Instituto Politécnico de Bragança, Campus de Santa Apolónia, 5300-253 Bragança, Portugal
3 Independent Researcher, 18 rue Mohamed Al Ghazi, Rabat 10170, Morocco
4 Norwegian Veterinary Institute, Section of Bacteriology—Food and GMO, Postbox Sentrum 750, N-0106 Oslo, Norway; taran.skjerdal@vetinst.no
5 Departamento de Ciencia y Tecnología de los Alimentos, UIC Zoonosis y Enfermedades Emergentes (ENZOEM), Campus de Excelencia Internacional en Agroalimentación (CeiA3), Universidad de Córdoba, Campus Rabanales, Edificio Darwin-Anexo, E-14071 Córdoba, Spain; bt2cajie@uco.es
6 CBQF—Centro de Biotecnologia e Química Fina—Laboratório Associado, Escola Superior de Biotecnologia, Universidade Católica Portuguesa, Rua Diogo Botelho 1327, 4169-005 Porto, Portugal; pcteixeira@ucp.pt
7 Department of Food Science and Human Nutrition, University of Illinois at Urbana-Champaign, 905 S Goodwin Ave., Urbana, IL 61801, USA; mstasie@illinois.edu
8 Nutrition and Food Safety Department, World Health Organization, 1202 Geneva, Switzerland; hasegawaa@who.int (A.H.)
9 Risk Assessment Department, French Agency for Food, Environmental and Occupational Health & Safety (Anses), 14 rue Pierre et Marie Curie, 94701 Maisons-Alfort, France; laurent.guillier@anses.fr
* Correspondence: rpouillot.work@gmail.com (R.P.); sanaam@who.int (M.S.)

Abstract: This study introduces a quantitative risk assessment (QRA) model aimed at evaluating the risk of invasive listeriosis linked to the consumption of ready-to-eat (RTE) smoked and gravad fish. The QRA model, based on published data, simulates the production process from fish harvest through to consumer intake, specifically focusing on smoked brine-injected, smoked dry-salted, and gravad fish. In a reference scenario, model predictions reveal substantial probabilities of lot and pack contamination at the end of processing (38.7% and 8.14% for smoked brined fish, 34.4% and 6.49% for smoked dry-salted fish, and 52.2% and 11.1% for gravad fish), although the concentrations of *L. monocytogenes* are very low, with virtually no packs exceeding 10 CFU/g at the point of sale. The risk of listeriosis for an elderly consumer per serving is also quantified. The lot-level mean risk of listeriosis per serving in the elderly population was 9.751×10^{-8} for smoked brined fish, 9.634×10^{-8} for smoked dry-salted fish, and 2.086×10^{-7} for gravad fish. Risk reduction strategies were then analyzed, indicating that the application of protective cultures and maintaining lower cold storage temperatures significantly mitigate listeriosis risk compared to reducing incoming fish lot contamination. The model also addresses the effectiveness of control measures during processing, such as minimizing cross-contamination. The comprehensive QRA model has been made available as a fully documented qraLm R package. This facilitates its adaptation for risk assessment of other RTE seafood, making it a valuable tool for public health officials to evaluate and manage food safety risks more effectively.

Keywords: brined fish; salted fish; gravlax; listeriosis; exposure assessment; simulation

1. Introduction

Cold-smoked and gravad fish are products with considerable public health implications in regards to listeriosis since they are not heat-treated, are generally eaten with no prior heating, and have a long shelf life. Over more than a decade, many reports and surveys have highlighted that ready-to-eat (RTE) seafood products are prone to contamination with *L. monocytogenes* [1–6]. Data have shown that fish used as raw material can be contaminated with *L. monocytogenes* cells before harvesting, which then multiply previous to smoking or marination when in the processing plant [7]. Implementation of whole genome sequencing (WGS) has revealed that different genotypes dominate in the environment, processing, and in different countries, both in sea and on land [8]. According to the latest European Union (EU) zoonoses report [9], in 2022 RTE fish and fishery products was the food category with the highest occurrence of *L. monocytogenes*, with an overall mean prevalence of 7.1% (N = 9727), varying from 0.0 to 20.0% between Member States (MS).

While qualitative methods provide valuable insights for the risk ranking of pathogens in food chains, a quantitative approach is essential for precisely quantifying the efficacy of control measures, enabling the assessment of risk levels and the mathematical modeling necessary to effectively predict and manage public health outcomes. Quantitative risk assessment (QRA) is a systematic approach used to estimate the risk of infection and illness from exposure to pathogens through various environmental mediums like food, water, or air [10].

Recently, using a "generic quantitative risk assessment" model, EFSA [11] contrasted the risk of listeriosis in the EU elderly population linked to foods products such as RTE fish, pâté, cooked meats, sausages, soft and semi-soft cheeses, and blanched frozen vegetables and determined that gravad fish in normal atmosphere packaging and hot-/cold-smoked fish in reduced-oxygen packaging were the top-most high-risk products. In terms of reported outbreaks, according to EU surveillance data [12], in the period between 2010 and 2020, fish and fish products (namely, crab meat, crustaceans, shellfish, mollusks, smoked fish, and non-specified seafood) caused 23% of the 53 strong-evidence total outbreaks in the EU, whereas, in 2022, EFSA [9] ranked *L. monocytogenes* in fish and fish products as one of the top ten pathogen/food vehicle pairs causing the highest number of deaths in strong-evidence outbreaks in the reporting EU MS. The high share of RTE seafood as a causative agent of listeriosis was also purported by a recent genomic-based epidemiological study [13], which estimated that 27% of 228 listeriosis cases in Germany between 2010 and 2020 were most likely caused by smoked or gravad salmon products.

More recently, ECDC and EFSA [14] investigated a prolonged cross-border outbreak of *L. monocytogenes* ST173 that, between 2017 and 2024, caused 73 cases of listeriosis, including 14 deaths, in Belgium, the Czech Republic, Germany, Finland, the Netherlands, Italy, and the UK. WGS cluster analysis and tracing evidence indicated that the strain spread in Europe originated from a past single source in the fish production chain. Contrary to Europe, where seafood like gravad and smoked fish in certain packaging types pose high risks, the attribution of outbreaks of listeriosis to seafood in the USA appears to be lower [15].

In view of the fact that most of the listeriosis cases occur sporadically [16], it is also pertinent to point out the outcomes of a meta-analysis on case-control studies of sporadic listeriosis [17]. These authors found that, among the RTE food categories, seafood, processed meats, cheese, and composite foods, RTE seafood presented the highest association with sporadic listeriosis, with pooled odd ratios of 10.75 ($p < 0.001$) for non-perinatal populations (immunocompromised and the elderly) and 6.273 ($p < 0.001$) for all susceptible populations (perinatal/non-perinatal, immunocompromised, and the elderly). In order to provide valuable insights on practices and strategies to reduce the current risk of listeriosis associated with RTE seafood products, various QRA models have been developed. A recent critical review on listeriosis QRA models [18] revealed that, although 12 out of the 13 seafood models retrieved (published between 1998 and 2022) investigated the food products specifically considered as high risk—cold-smoked or gravad fish—all of them represented short supply chains only, either from end processing/retail to table [16,19–24] or the sole consumption module [25–30]. None of the available QRA models included a processing module nor at least a qualitative assessment of the most relevant opportunities of cross-contamination during processing, probably due to the insufficiency of data at the time of their development. In 2022, the Joint FAO/WHO Expert meeting on microbiological risk assessment of *Listeria monocytogenes* in foods recommended the development of a full primary production (harvest and farming) to consumption risk assessment for the risk of *L. monocytogenes* for RTE seafood (hot- and cold-smoked fish and gravad fish) [31].

In light of new data and predictive microbiology models, the objective of this study was to build a QRA model of longer scope for RTE smoked and gravad fish capable of representing the growth, inactivation, and potential cross-contamination from processing to consumption as well as the retarding effect of background microbiota on the development of *L. monocytogenes*.

This study addresses the critical need for an updated quantitative risk assessment model that incorporates the latest data and technological advancements in predictive microbiology. By developing a comprehensive model that evaluates the entire chain of production from primary processing to consumption, this paper aims to provide actionable insights and strategies to significantly reduce the health risks associated with the consumption of RTE smoked and gravad fish.

The model was designed to be able to assess the contribution of the initial contamination of the incoming fish, the contribution of cross-contamination during the filleting of fish and slicing prior to packaging, the effect of the sampling schemes at the end of processing, the effect of time and temperature throughout the logistics of the end-product, the effect of improved practices throughout the supply chain, and the effect of lactic acid bacteria (LAB) cultures added for the biocontrol of *L. monocytogenes*. The model's structure for both RTE smoked and gravad fish was developed according to the recent Expert Panel recommendation of the Joint FAO/WHO Expert meeting on microbiological risk assessment [31]. The present article aims to describe in detail the QRA model and subsequently illustrate its functionality by reference and what-if scenario analysis.

2. Materials and Methods

2.1. Exposure Assessment

The exposure assessment model was developed for two RTE seafood products, smoked fish and gravad fish, and both are presented in this article since they share most of the processing operation units. The processing stages are schematized in Figure 1. The model starts from whole fish units and considers the storage before filleting, which can represent the transport to the primary processing facility or waiting time at the processing facility. After filleting, there is a holding-off time at the facility before preparation, yet it can alternatively correspond to the transport of fish fillets to a secondary processing facility.

At this point, there is a differentiation between smoked fish and gravad fish. For the processing of smoked fish, the stages modelled are salting/brining and smoking/maturation, whereas for the gravad fish the stages are the smearing of fillets with condiments, followed by maceration. The stages that follow are common to both products—the slicing of fillets, packaging, within-lot testing, cold chain distribution (i.e., transport to retail, display at retail, and transport from retail to home), and home storage and portioning—although the parameters feeding the model are product-specific.

Each of the stages shown in Figure 1 was coded as a function stochastically estimating the microbial prevalence and numbers after a process of microbial growth, inactivation, cross-contamination, or partitioning [32], except for within-lot testing, which did not conform to any of the aforementioned processes. Table 1 presents a synthesis of the modules, the sequence of stages and processes they consist of, the assumption and data sources employed, and the corresponding functions programmed in the R software version 4.4.2.

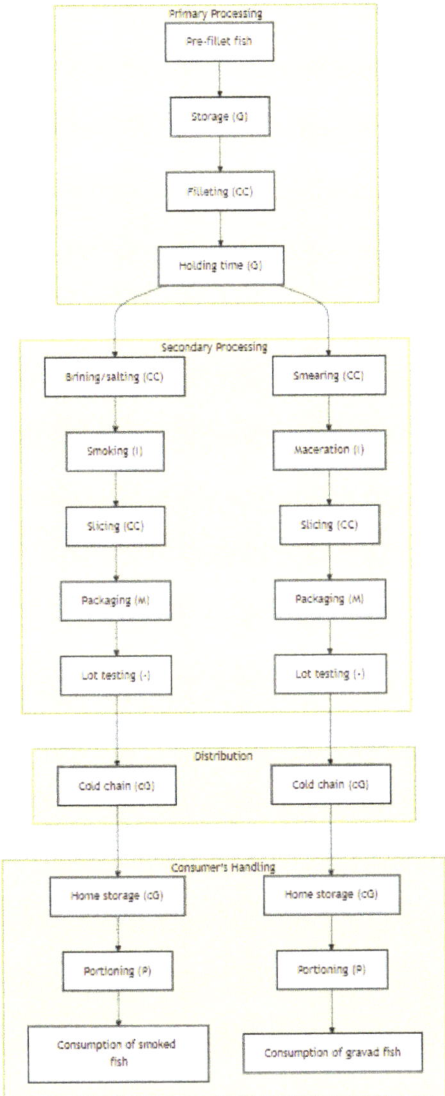

Figure 1. Schematic of the four-module exposure assessment of *L. monocytogenes* in smoked fish (left) and gravad fish (right), with indications of the modelled processes: CC, cross-contamination; G, growth; cG, growth in competition with lactic acid bacteria; M, mixing; I, inactivation; P, partitioning.

Table 1. Sequence of stages, microbial processes represented, data sources, assumptions, and corresponding functions coded in R for the construction of the exposure assessment model of *Listeria monocytogenes* (LM) in RTE smoked and gravad fish.

Module	Stage	Microbial Process	Assumptions	Sources	Function in R
Primary processing	Generation of contaminated lots of pre-fillet (pre-processed) fish	None	LM prevalence in fish units is modelled from data found in incoming fish sampled at primary processing. Since LM numbers in incoming fish were generally low (<10 CFU/g), LM concentration in fish was calculated from the expected proportion of non-zeroes under a Poisson distribution.	Autio et al. [33], Cruz et al. [34], Dass [29], Di Ciccio et al. [35], Markkula et al. [36], Medrala et al. [37], Miettinen et al. [38], Rorvik et al. [39], Vogel et al. [40], Jarvis [41]	Lot2LotGen()
	Storage	Growth	LM in pre-filleted fish is assumed to follow the kinetics of a cocktail of CICC 21632 (serotype 1/2a), CICC 21633 (serotype 1/2a), CICC 21635 (serotype 4b), and CICC 21639 (serotype 1/2a) LM inoculated in raw salmon flesh. Lag phase is considered.	Jia et al. [42]	sfRawFishStorage(), served by sfGrowthLDP()
	Filleting	Cross-contamination	LM is transferred to the fillets during the slicing process of the raw whole fish using a compartmental model defined by two variability distributions: "a", the transfer rate between slicer blade (or filleting knife) and product, and "e", the transfer rate from the original contamination to the slicing system. Distribution parameters were obtained from published data.	Hoelzer et al. [43], Aarnisalo et al. [44]	sfSlicer()
	Holding-off time	Growth	The same predictive microbiology model as in Storage, yet the LM growth and the exhaustion of the lag phase were followed up in this stage.	Jia et al. [42]	sfRawFishStorage(), served by sfGrowthLDP()
Secondary processing	Brining or salting (smoked fish)	Cross-contamination	Fish fillets can be salted either through brine injection or by dry salting. Internal contamination by brining may occur during the injection of saline solution due to brine containers or the brine itself serving as reservoirs for LM, at a given probability. For dry-salted fillets, it is assumed that external contamination can occur during the procedure of smearing salt on fish fillets if surfaces are contaminated with LM, at a given probability. A published distribution for LM transfer rate is assumed.	Gudbjornsdottir et al. [45], Gudmundsdottir et al. [46] Hoelzer et al. [43]	sfBrineORsaltCC(), served by sfBriningCC() and sfSmearingCC()
	Smoking and maturation (smoked fish)	Inactivation	Salting, drying, and smoking produce a slight reduction in LM, which is different if fish fillets were brine-injected or dry-salted. The LM $\log_{10}$ reduction in brine-injected fish fillets was assumed to be that of the combined results of inoculation experiments in smoked salmon, first salted through brine injection, and then submitted to cycles of cold-smoking between 6 and 8 h until a total maturation time of 18–24 h. The LM $\log_{10}$ reduction in dry-salted fish was assumed to be that of the combined results of inoculation experiments on the surface of dry-salted salmon, determined before and after a total maturation time between 18 and 24 h.	Eklund et al. [47], Porsby et al. [48] Eklund et al. [47], Porsby et al. [48], Neunlist et al. [49]	sfSmoking()
	Smearing (gravad fish)	Cross-contamination	During smearing of fish fillets with salt, sugar, and spices, external contamination can occur if surfaces are contaminated with LM, at a given probability. A published distribution for LM transfer rate is assumed.	Hoelzer et al. [43]	sfSmearingCC()

Table 1. *Cont.*

Module	Stage	Microbial Process	Assumptions	Sources	Function in R
	Maceration (gravad fish)	Inactivation	The maceration of fish smeared with gravlax curing agents is assumed to reduce the populations of LM, according to the results of a study where inoculated raw salmon was smeared with salt, brown sugar, black pepper, and dill and left to macerate at 4.5 °C for 72 h.	Lopes et al. [50]	sfMaceration()
	Slicing	Cross-contamination	The same compartmental model as in Filleting, but used to produce slices from smoked or gravad fish fillets.	Hoelzer et al. [43], Aarnisalo et al. [44]	sfSlicer()
	Packaging	Mixing	No cross-contamination is assumed during packaging. Consecutive slices of recently-sliced RTE smoked or gravad fish are gathered into packages of end-product.	-	sfPackaging()
	Within-lot testing	None	At a given probability, any lot of RTE smoked or gravad fish can be subjected to sampling and testing, according to a two-class or three-class microbiological sampling plan.	-	sfTesting()
Distribution	Cold chain	Growth	LM in smoked or gravad fish, as affected by populations of lactic acid bacteria (LAB), are assumed to grow during the various cold chain logistics stages, including transportation to retail, display at retail, and transportation to home. Specific growth rates for LM and LAB are estimated from secondary models, using validated kinetic parameters. LM numbers follow an extended Jameson-effect competition model, which uses an interaction Gamma parameter.	Mejlholm and Dalgaard [51–55], Mejlholm et al. [56] Gimenez et al. [57], Moller et al. [58]	sfColdChain(), served by sfMejlholmDalgaard(), sfMejlholmDalgaardLAB(), sfGrowthJameson()
Consumer's handling	Home storage	Growth	The same as in Cold chain, following up the growth.	Mejlholm and Dalgaard [51–55], Mejlholm et al. [56], Gimenez et al. [57], Moller et al. [58]	sfColdChain(), served by sfMejlholmDalgaard(), sfMejlholmDalgaardLAB(), sfGrowthJameson()
	Portioning	Partitioning	The consumer is assumed to take a number of RTE fish slices from the pack. LM cells present in a contaminated pack are assumed to be moderately clustered within the package.	Nauta [59]	sfPortioning()

2.1.1. Contaminated Lots of Pre-Fillet Fish

The prevalence of *L. monocytogenes* in pre-processed fish was modeled using published data, assuming that the variability between different lots of fish follows a *Beta* distribution. The model was expressed as

$$s_j \sim Binomial(n_j, p_j)\, p_j \sim Beta(\alpha, \beta)$$

This distribution is defined by parameters α and β, derived from the survey data listed in Table 2. Each sampling result, from a total of 12 fish lots, is considered a part of a binomial distribution, representing the unobserved prevalence in each lot.

The model employs Bayesian methods with non-informative priors for α and β, set to follow Gamma distributions. After 20,000 iterations in a Markov chain Monte Carlo (MCMC) simulation, the average values for α and β were calculated, with the mean prevalence of *L. monocytogenes* in any given fish lot modeled as Beta (0.8741, 5.880). This represents an average prevalence rate of approximately 14.86%, with a 95% confidence range from 0.037 to 55.41%.

Since, to the best of the authors' knowledge, the numbers of *L. monocytogenes* in gutted and cleaned fish are very low, mostly below the limit of quantification (10 CFU/g) [7,60], the Poisson assumption was employed to approximate the concentration of *L. monocytogenes* per gram of raw fish from the unobservable prevalence of the lot *j* (p_j). Such a procedure calls for two assumptions: (1) that *L. monocytogenes* cells are randomly distributed in fish (i.e., are not clustered), and (2) that the analytical weight was the same (25 g) in all detection assays carried out by the sources shown in Table 2. A further assumption if that a clean, degutted fish unit weighs 3800 g.

Table 2. *L. monocytogenes* prevalence in raw fish sampled at processing facilities.

Country	Product	Sample Size, *n*	Positive Enrichment, *s*	Prevalence (%)	Source *
Finland	Raw rainbow trout	35	0	0.00	Autio et al. [33]
Brazil	Raw salmon	255	105	41.2	Cruz et al. [34]
Ireland	Raw salmon	60	17	28.3	Dass [29]
Italy	Raw salmon	21	5	23.8	Di Ciccio et al. [35]
Finland	Raw fish	45	2	4.40	Markkula et al. [36]
	Fish before processing	212	9	4.20	
Poland	Incoming salmon	46	2	4.34	Medrala et al. [37]
	Incoming seatrout	26	4	15.4	
Finland	Raw fish	18	2	11.1	Miettinen et al. [38]
Norway	Salmon pre filleting	24	4	16.6	Rorvik et al. [39]
Denmark	Raw fish	12	0	0.00	Vogel et al. [40]
	Raw fish	18	0	0.00	

(*) Data extracted from the Pathogens-in-Foods Database [61], except for Cruz et al. [34].

A function, Lot2LotGen(), was built to generate a contamination matrix, *N*, of dimensions $r \times c$, whose number of rows, *r*, represents the number of lots and can therefore be understood as number of iterations that will correspond to between-lot variability; the number of columns, *c*, represents the number of pre-filleting fish units of weight $Unit_{size}$ (3800 g) to be produced in the lot, considered as fixed and equal for all lots (cf. Appendix A.1).

2.1.2. Storage of Pre-Fillet Fish

The primary processing module begins with storing gutted, clean fish prior to filleting. Even under very cold storage conditions, *L. monocytogenes* can still grow. Jia et al. [42] have provided models for this growth based on their studies, where they inoculated salmon with various strains of *L. monocytogenes* and stored them at temperatures ranging from 4 to 35 °C. They used a Lotka–Volterra-based equation to determine how temperature affects the growth rate and the initial delay before growth starts, considering the competitive effects of the fish's natural microbiota.

The QRA model applies these findings to predict the growth rate and the initial delay period of the pathogen during storage, using a straightforward log-linear model to estimate the pathogen's concentration at the end of the storage period. Jia et al. [42] also provided the maximum population density of *L. monocytogenes* in raw fish (9.20 $\log_{10}$ CFU/g). The model assumes a brief holding period for the fish before filleting, with storage temperature and duration modeled to vary according to Pert distributions. Specifically, temperature and storage time for each lot are sampled from Pert (−2, 0, 4) and Pert (0.5, 2.0, 6.0), respectively.

Additionally, an auxiliary function, sfGrowthLDP(), calculates the number of *L. monocytogenes* in a unit of raw fish after storage, based on constant temperature and time settings using validated models for maximum growth rate and delay duration [42,62]. This function aids in multiple storage assessments for raw fish, as detailed in Appendix A.2. The function sfRawFishStorage() then uses these calculations to stochastically simulate the growth of *L. monocytogenes* during cold storage, pulling data from a contamination matrix and the storage variability parameters listed in Appendix A.3.

2.1.3. Filleting of Raw Fish

As early as 1995, Eklund et al. [47] and Rorvik et al. [39] observed that *L. monocytogenes* could be transferred from the exterior of fish to cut surfaces of fillets or sides, and concluded that filleting is a critical stage, since filleting tables, knives, and gloves of personnel could further spread the contamination. The representation of cross-contamination during fish slicing was regarded as relevant, because *L. monocytogenes* present on tables and cutting surfaces can adhere strongly after a short period of time, which entails the possibility that filleted fish become contaminated during the first stages of processing.

Aarnisalo et al. [44] carried out a study to investigate the transfer of *L. monocytogenes* from an inoculated slicing blade (slicer) to slices of gravad salmon, and from inoculated salmon fillet to the slicing machine and subsequently to slices of uninoculated fillets. A marked reduction in the counts from the slicing blade (5.9–9.0 $\log_{10}$ CFU/blade) to the fillets (1.6 $\log_{10}$ CFU/g) was observed after 39 slices; nonetheless, the first slices contained higher counts. Hoelzer et al. [43] utilized this bacterial transfer dataset, along with others produced from slicing meat and RTE meat products, to populate a compartment model consisting of four elements (slicer, chub, slice, and a bacterial loss bin). From this deterministic model, they produced various estimates of coefficients for slicing transfer (a) and for transfer from original contamination to the system (e) and derived thereof variability distributions regarding the transfer coefficients a and e. Additionally, Hoelzer et al. [43] proposed a probabilistic model of microbial transfer during a generic slicing process. This model and its parameters are employed in the present QRA to represent the transfer of *L. monocytogenes* cells during the filleting operation. A further assumption is that two fillets of weight 1300 g each can be obtained from one clean fish unit.

A function, sfSlicer(), was written to stochastically simulate the transfer of bacteria during this step. The function is fed by the outputs of the function sfRawFishStorage(), the weight of the fillet (w_{Fillet}), the load of *L. monocytogenes* cells on the slicer (knife or blade) ($Init_{Slicer}$), the parameters of the logistic distribution regarding the transfer coefficient a ($location_a$, $scale_a$), and the parameters of the normal distribution of the $\log_{10}$ coefficient e (μ_{loge}, σ_{loge}). The outputs of the function sfSlicer() are as follows: the contamination matrix of *L. monocytogenes* numbers in fish fillets (N_{Fillet}), the previous expenditure of *L. monocytogenes* lag phase corresponding to those fillets ($WorkDone_s$), the (unchanged) probability vector of contaminated lots after filleting $Prob_{UnitPos}$, and the mean prevalence of contaminated lots after filleting (P_{Fillet}) (cf. Appendix A.4).

2.1.4. Holding-Off Time of Fish Fillets

The growth of *L. monocytogenes* in fish fillets during the holding-off stage is simulated. The objective of introducing a storage time after filleting is to allow for the transportation of fish fillets from the primary to the secondary processing facility, or, alternatively, for a short period of time before commencing processing (salting in the case of smoked fish or smearing with curing agents in the case of gravad fish). The latter case is represented in the present QRA model, assuming that the holding-off temperature ($Temp_{hold}$, °C) and time ($time_{hold}$, h) follow Pert distributions, *Pert (−2, 0, 4)* and *Pert (1, 2.0, 6.0)*, respectively. The maximum population density (MPD) of *L. monocytogenes* in raw fish fillets is assumed to be the same as in raw clean fish (9.20 $\log_{10}$ CFU/g).

The function sfRawFishStorage() and its auxiliary function sfGrowthLPD(), explained in detail in Appendices A.2 and A.3, are reused in the holding-off stage to stochastically estimate the growth of *L. monocytogenes* in raw fish fillets. In a lot, the fish fillets are considered to be exposed to the same $Temp_{hold}$ and $time_{hold}$, which are sampled from the aforementioned Pert distributions. In addition to these parameters, the inputs of the function sfRawFishStorage() are the contamination matrix, N_{Fillet}, the microbial work done after the first storage ($WorkDone_s$, a matrix), the weight of the fillet (w_{Fillet}, a scalar), and the maximum population density of *L. monocytogenes* in the fish fillet (*MPD*, a scalar), whereas the outputs are the contamination matrix after *L. monocytogenes* growth (N_{Hold}) and the total microbial work done ($WorkDone_h$, a matrix).

$$\left\{N_{Hold\ ij},\ WorkDone_{h\ ij}\right\} = sfGrowthLPD\left(N_{Fillet\ ij},\ time_{hold\ i},\ Temp_{hold\ i},\ w_{Fillet},\ MPD,\ WorkDone_s\right)\ j = 1, 2\ldots, c_f$$

sfRawFishStorage() returns unaffected the mean prevalence of contaminated lots of fish fillets (P_{Hold}, a scalar) and the probability that the sampled lot is contaminated ($Prob_{UnitPos}$, a vector).

2.1.5. Brining or Salting of Fish Fillets (Relevant to Smoked Fish)

In the processing of smoked fish, fish fillets can be salted in two ways: by dry-salting or by brining (i.e., injection of a saturated NaCl solution). Given that *L. monocytogenes* is able to survive in the salted fish due to its halotolerance, no microbial reduction or inactivation process is represented at this stage. Instead, opportunities for external or internal contamination during salting are contemplated. In the case of brining, recirculating brine is a source for contamination since *L. monocytogenes* can survive in NaCl solutions of up to 10% [63]. Gudmundsdottir et al. [46] and Gudbjornsdottir et al. [45] found *L. monocytogenes* at frequencies of 21.4% (3/14) and 8.7% (2/23), respectively, in injection brines, which supported the fact that brine containers and the brine itself may serve as reservoirs for *L. monocytogenes*. Based on these data, the QRA model assumed that, on a lot basis, the probability that the brine solution is contaminated (Pcc_{Brine}) is 13.5% (5/37). When this event occurs, an internal contamination of the fish fillet is produced through the brine injected. Nonetheless, no data were found on the likely *L. monocytogenes* levels in contaminated brine solution, and therefore the numbers were assumed.

With regards to dry-salting, it is considered that cross-contamination can take place through tables or other surfaces during the smearing of fish fillets with salt/sugar/spices, at a certain probability, Pcc_{Smear}. The probability Pcc_{Smear} is assumed to be 3.9% [46]. Every fillet is subjected to the same Pcc_{Smear} probability, and if the contamination event takes place, cells are partially transferred according to a transfer coefficient. The normal distribution parameters of the $\log_{10}$ transfer coefficient of *L. monocytogenes* (TR_{Smear}) were taken from Hoelzer et al. [43] to represent the transfer from board to meat. Nonetheless, no data were found for the likely levels of *L. monocytogenes* on environmental elements in contact with fish fillets ($N_{Surface}$) while dry-salting or smearing with other ingredients (gravlax curing).

Two auxiliary functions were built, sfBriningCC() and sfSmearingCC(), to model the cross-contamination during brining and dry-salting, respectively. The function sfBriningCC() stochastically simulates the potential internal contamination of fish fillets during brining by the injection of salt solution [64] (cf. Appendix A.5). The function sfSmearingCC() stochastically simulates the potential external contamination of fish fillets during dry salting when producing smoked fish, or during smearing with gravlax curing agents when producing gravad or any macerated fish (cf. Appendix A.6). Those two auxiliary functions could be used on their own, if we were to assume that all the lots of fish fillets were processed either through brine injection or through dry salting. The main function sfBrineORsaltCC() was conceived to allow for the inclusion of both types of salting in the QRA model. Thus, some of the lots of fish fillets will be subjected to brining (at the probability p_{Brine}) and others to dry-salting (at the probability 1-p_{Brine}) (cf. Appendix A.7).

2.1.6. Smoking and Maturation of Salted Fish Fillets (Relevant to Smoked Fish)

The combination of hurdles that occur in the processing of smoked fish (salting, drying, and smoking) affect the homeostasis of *L. monocytogenes*; as they struggle to maintain their energy balance, they become metabolically exhausted during the subsequent step of maturation, causing the populations to drop [65]. Although the temperature during smoking is too low to eliminate *L. monocytogenes*, the phenolic compounds from smoke at a concentration of 20 ppm can inhibit *L. monocytogenes* [66].

In the literature, few studies investigating the overall effects of the individual processing steps for the production of smoked salmon have been undertaken for both types of salting (brine injection or dry-salting). Table 3 compiles the available information on the combined effect of cold-smoking and maturation on *L. monocytogenes*.

Table 3. Reduction in *L. monocytogenes* concentration in salted salmon fillets due to smoking and maturation, taken from challenge tests.

Type of Salting	Source	Conditions of Smoking and Maturation	Before Treatment (log_{10} CFU/g)	After Treatment (log_{10} CFU/g)	Reduction (log_{10} CFU/g)
Brine injection	Porsby et al. [48]	Cold smoking, 24 °C in cycles of 6 h	3.00 ± 0.3	2.50 ± 1.1	0.50
			3.30 ± 0.4	1.80 ± 0.9	1.50
			2.90 ± 0.2	1.00 ± 0.0	1.90
			3.30 ± 0.1	1.80 ± 0.6	1.50
			3.00 ± 0.1	1.10 ± 0.3	1.90
	Elkund et al. [47]	Cold smoking, 17–21 °C × 18 h	2.36	1.41	0.95
			3.34	2.63	0.71
		Cold smoking, 22–30 °C × 18 h	2.11	2.34	−0.23
			3.28	3.38	−0.10
			4.57	4.49	0.08
Dry-salting	Porsby et al. [48]	Cold smoking, 24 °C in cycles of 6 h	3.30 ± 0.2	1.30 ± 0.7	1.80
			3.10 ± 0.2	1.40 ± 0.7	1.70
	Neunlist et al. [49])	Liquid smoking and maturing, 4 °C × 24 h	5.70	4.70	1.00
	Elkund et al. [47]	Cold smoking, 17–21 °C × 18 h	2.04	0.83	1.21
			2.56	1.11	1.45
		Cold smoking, 22–30 °C ×18 h	0.11	−0.39	0.50
			1.04	0.54	0.50
			2.42	1.84	0.58

In each of those challenge studies, the conditions of smoking and the treatment prior to smoking were different. In the work of Neunlist et al. [49], smoking dry-salted salmon reduced *L. monocytogenes* counts by 0.5 log_{10} CFU/g; however, during subsequent maturation, the effect of salt and smoke continued, and the mean number of *L. monocytogenes* was further reduced by another 0.5 log_{10} CFU/g. Thus, the combination of the two steps significantly reduced the concentration of *L. monocytogenes* by 1.0 log_{10} CFU/g.

In the paper by Porsby et al. [48], the effect of smoking and maturation was carried out in salmon fillets that were salted either by brine injection or by dry-salting and in fresh unsalted salmon fillets as the control. From different challenge studies, they determined that smoking and maturation could reduce *L. monocytogenes* in up to 1.9 log_{10} after 24 h in brined fish, although they could not find statistical differences with salmon fillets that were dry-salted. Eklund et al. [47] inoculated *L. monocytogenes* in two different manners:

on the surface, to simulate external contamination, and internally, to simulate contamination through brine injection; cold-smoking was the carried out at two distinct ranges of temperature: a lower temperature (17–21 °C) and a higher temperature (22–30 °C). The results of Elkund et al. [47] suggested that smoking and maturation may be more effective in reducing *L. monocytogenes* populations in dry-salted salmon fillets than in brine injected ones. In the experiment where cold-smoking was conducted at the higher temperatures of 22–30 °C on salmon samples that were inoculated (internally) via brine injection, smoking and maturation did not appear to have any effect on *L. monocytogenes*. These results agreed with Niedziela et al. [67], who earlier pointed out that dry-salting is more effective in controlling *L. monocytogenes* than brine injection. Dry-salting pulls water out of the flesh due to osmotic pressure. On the contrary, a brine-injected fillet will keep the water distributed in the flesh, and the surface water on the fish may even be pulled towards the higher salt concentration within the fish. As the surface is the part of the fish that tends to have higher concentrations of *L. monocytogenes*, the difference of losing surface water and keeping surface water will cause a different microbial concentration change in the fish.

Thus, the present QRA model estimates the $\log_{10}$ reduction factors due to smoking and maturation for brine-injected fish fillets and dry-salted fish fillets by calculating the mean and sample standard deviations of the values compiled in Table 3. The mean (μ_{rBrine}) and standard deviation (σ_{RBrine}) of the normally distributed $\log_{10}$ reduction factor for brined fish fillets (R_{Brine}) are 0.871 and 0.807 $\log_{10}$, respectively, while the mean ($\mu_{rDrysalt}$) and standard deviation ($\sigma_{RDrysalt}$) of the normally distributed $\log_{10}$ reduction factor for dry-salted fish fillets ($R_{Drysalt}$) are 1.093 and 0.532 $\log_{10}$, respectively.

A function, sfSmoking(), was written to simulate the microbial process of inactivation of *L. monocytogenes* in the salted fish fillets subject to smoking and 18–24 h maturation (cf. Appendix A.8). Its inputs are the outputs of the function sfBrineORsaltCC() and the normal distribution parameters μ_{RBrine}, σ_{RBrine}, $\mu_{RDrysalt}$, and $\sigma_{RDrysalt}$ that describe the microbial $\log_{10}$ reduction in *L. monocytogenes* for brined or dry-salted fish fillets. These parameters are pre-defined, as described above.

2.1.7. Smearing of Fish Fillets with Ingredients (Relevant to Gravad Fish)

Different procedures have been reported for the elaboration of gravad fish. According to Wiernasz [68], for gravlax production in France, salmon fillets are cured with a mix of salt, sugar, pepper, and dill during a period of 14 h at 6 °C, whereas Peiris et al. [63] explain that, in Sweden, salmon or rainbow trout are rubbed with a mixture of sugar, salt, and pepper and are covered with dill and stored in plastic bags at low temperature for 48 h. A longer procedure has been described by Cruz et al. [34]; salmon fillets are hand-rubbed with a commercial mixture of NaCl, sodium nitrate, and sodium nitrite and stored at 4 °C for 24 h in high-density polypropylene boxes. After the excess of salt is washed out with chlorinated water and the fillets are drained, a mixture of sugar, NaCl, ground white pepper, and dried dill is hand-rubbed into the flesh side of the fillets, and they are then stored for 24 h at 4 °C in the boxes. Fillets are then layered on stainless steel supports, sprayed with sweet wine, and ripened for 48 h at 4 °C.

In a processing line for gravlax, Cruz et al. [34] found that, amongst the food contact surfaces, the pathogen was present in 40% of the samples from salting boxes, ripening trolleys after 48 h in the cold room, and weighting trays, whereas a lower frequency of positive samples was found in the salting table (30%). Therefore, as for dry-salting in the production of smoked fish, during smearing with salt/sugar/spices in the production of gravad fish it is assumed that bacterial transfer can occur from tables or other surfaces, at a probability Pcc_{Smear} (3.9% obtained from Gudmundsdottir et al. [46]). As with the smearing of fish with salt (Section 2.1.5), for the smearing of fish fillets with gravlax curing agents the same parameter values were employed: $\mu_{TRsmear} = -0.29$ and $\sigma_{TRSmear} = 0.31$ for the normal distribution truncated on $]\infty, 0]$ of the $\log_{10}$ transfer coefficient of *L. monocytogenes* (TR_{Smear}) [43], and the load of *L. monocytogenes* on environmental elements in contact with a fish fillet ($N_{Surface}$ = 10 CFU).

The function sfSmearingCC() is used to stochastically simulate the potential external contamination of fish fillets during smearing with salt, sugar, and condiments. The function is described in detail in Appendix A.6. In the context of the processing of gravad fish, sfSmearingCC() is fed by the outputs of the holding-off storage function (N_{Hold}, $Prob_{UnitPos}$, and P_{Hold}) and Pcc_{Smear} (0.039), $N_{surface}$ (100 CFU), $\mu_{TRSmear}$ (−0.29), and $\sigma_{TRSmear}$, (0.31). The function provides as outputs the contamination matrix N_{Smear}, the vector $Prob_{UnitPos\ Smear}$, and the scalar P_{Smear}.

2.1.8. Maceration or Curing of Fish Fillets (Relevant to Gravad Fish)

Lopes et al. [50] performed a challenge test on traditional gravlax salmon to determine the effect of processing on *L. monocytogenes* counts. Salmon fillets, contaminated at a level of 5.5 $\log_{10}$ CFU/g, were premoistened with lemon juice and covered with a mixture of salt, brown sugar, black pepper, and dill and then cured at 4.5 °C. After 72 h of curing, the population of *L. monocytogenes* was reduced by 0.9 $\log_{10}$ CFU/g. In the work of Neunlist et al. [49], the salting step had a weaker effect on *L. monocytogenes* culturability of 0.5 $\log_{10}$ CFU/g. From this information, a $\log_{10}$ reduction factor for gravlax maceration of fish fillets (R_{Gravad}) was derived, assuming that it followed a normal distribution with mean $\mu_{RGravad}$ = 0.70 and standard deviation $\sigma_{RGravad}$ = 0.283 $\log_{10}$ CFU/g.

A function, sfMaceration(), was built to stochastically represent the reduction of *L. monocytogenes* concentration in fish fillets smeared with gravlax curing agents during cold maceration or curing. The inputs of this function are the contamination matrix N_{Smear} and the normal distribution parameters $\mu_{RGravad}$ and $\sigma_{RGravad}$, defining the microbial $\log_{10}$ reduction in *L. monocytogenes* in macerating fish fillets smeared with curing ingredients.

2.1.9. Slicing of Processed Fish Fillets

In the literature, a few environmental microbiological surveys have evidenced that slicing machines in RTE seafood processing facilities can be a source of *L. monocytogenes* spread. For instance, Di Ciccio et al. [35] repeatedly isolated *L. monocytogenes* serotypes 1/2a and 1/2b from slicer belts, distribution trays, slicing machines, and slicing covers for three years in a smoked-salmon production facility: out of 95 environmental samples tested, slicing machines (37%) and working tables (43%) had the highest frequencies of detection. In the USA, in a processing plant of catfish fillets, Chen et al. [69] determined that in 15% (7/45) of the sampling times, skinning, slicing, or blending equipment were contaminated with *L. monocytogenes*. In Ireland, Dass [29] isolated MLVA types *L. monocytogenes* serotypes c and b in the slicing machines over a one-year survey (2 positives out of 36). Thus, it was deemed necessary to represent cross-contamination during the slicing of processed fish fillets in the present QRA model. The function sfSlicer(), developed from the compartmental model of Hoelzer et al. [43] and used to simulate cross-contamination during fish filleting, is also employed at the slicing stage. In this case, the unit to be sliced is the processed fish fillet of weight w_{Fillet} and the sliced unit is the smoked fish slice or the gravad fish slice, of weight w_{Slice}.

The function sfSlicer(), already used during the filleting process and described in Appendix A.4, takes the outputs of the function sfSmoking() in the case of the QRA for smoked fish, or the outputs of the function sfMaturation() in the case of the QRA for gravad fish. In any case, the weight of a fish fillet (w_{Fillet}) is 1300 g and the weight of a slice (w_{Slice}) is 32.5 g. The parameters of the logistic distribution about the transfer coefficient *a* ($location_a$ = 0.07; $scale_a$ = 0.03), and the parameters of the normal distribution of the $\log_{10}$ coefficient *e* (μ_{loge} = −2.12; σ_{loge} = 0.85) remain the same as those used in the stage of filleting [43], whereas the load of *L. monocytogenes* cells on the slicer ($Init_{Slicer}$) is set to 0.

2.1.10. Packaging of Processed Fish Slices

There is no assumption of cross-contamination during the packaging of processes fish slices. A function, sfPackaging(), was written. It consists of a simple grouping of consecutive slices to produce a pack of $Slices_{pack}$ slices. It adds up the microbial load

N_{Slice}, corresponding to the group of slices, and the weight of the slices (pack). $Slices_{pack}$ is assumed to be 8 slices per pack; thus, the grouping produces a pack of weight $Unit_{SizePack}$, which equals $Slices_{pack} \times w_{Slice}$ ($8 \times 32.5 = 260$ g). The inputs of the function sfPackaging() are N_{Slice} and $Slices_{pack}$, and the output is the contamination matrix N_{Pack}. The number of packs of end-product (c_p) produced in a lot will then be $c_s/Slices_{pack}$. Thus, the output N_{Pack} is a matrix of dimensions r lots by c_p packs. The mean prevalence of contaminated packs, P_{Pack}, returns the value of P_{Smoked} for smoked fish packs, or the value of P_{Smear} for gravad fish packs.

2.1.11. Within-Lot Testing

The QRA model enables the testing of *L. monocytogenes* in food unit samples taken from a lot, according to a two-class or a three-class mixed sampling plan. In the two-class plan, n samples are randomly extracted and analyzed per lot. For each sample, a sub-sample of g grams is used in the enrichment essay, and the lot is rejected if more than c samples are positive in detection. In a three-class mixed sampling plan, samples are also enumerated, and the lot is rejected if more than c units are positive in detection or if at least one unit is found to have a concentration greater than a predefined concentration M [70]. It is assumed that the enumeration assay is conducted only on positive samples in detection, by direct plating of $g_{TestedEnum}$ g taken from the same sample.

A function, sfTesting(), was built to consider this selection step. Contaminated lots detected after testing are not removed from the matrix, and so the input matrix output N_{Pack} is returned unchanged. The function fvTesting() only updates the probability for each lot to be put on the market, using the Bayes' theorem (cf. Appendix A.10).

2.1.12. Cold Chain

The characteristics of smoked fish and gravad fish—namely, intrinsic factors (pH, aw, preserving compounds, lactic acid protective cultures) and extrinsic factors (temperature, atmosphere in the package) do not preclude *L. monocytogenes* from growing [26]. If the pathogen is present on processed fish fillets, salting will not be sufficient to inhibit its growth. The slow development of *L. monocytogenes* relies upon cold temperatures and vacuum packaging, which reduces the pathogen's growth due to anaerobic conditions and the inhibitory effect of competitive LAB. The cold chain step in the present QRA models encompasses the transportation from end processing to retail, the time at retail, and the transportation from retail to home.

In the cold chain segment of this study, the growth of *L. monocytogenes* and lactic acid bacteria (LAB), which can inhibit the pathogen, is modeled under various conditions. This module encompasses several key steps: Predictive models are utilized to determine the growth rates of both the pathogen and LAB in seafood, taking into account internal factors like pH and external factors such as temperature (refer to Appendices A.11 and A.12 for model details). A comprehensive growth model that includes initial growth delays is employed to estimate the concentrations of the pathogen and LAB at the conclusion of the cold chain (see Appendix A.13). The characteristics of the seafood that influence microbial growth rates are evaluated through a specific function (see Appendix A.14). The principal cold chain function integrates this data to simulate the growth of both microbes as the seafood progresses from production to retail (refer to Appendix A.15). Table 4 provides a list of the parameters used in these models, detailing their significance and data sources.

Table 4. Parameters used for modelling the simultaneous growth of *L. monocytogenes* and lactic acid bacteria (LAB) in RTE seafood during cold chain distribution and home storage.

Type of Parameter	Parameter	Definition (Unit)	Value or Distribution	Source
Relative to kinetic parameters of LM in RTE seafood	$\mu_{LM\ ref}$	Optimum growth rate of LM (h^{-1})	0.419	Mejlholm and Dalgaard [51]
	MPD_{LM}	Maximum population density of LM (log_{10} CFU/g)	$MPD_{LM\ min} = 6.60$ $MPD_{LM\ mode} = 7.36$ $MPD_{LM\ max} = 8.20$ $MPD_{LM} \sim$ Pert ($MPD_{LM\ min}$, $MPD_{LM\ mode}$, $MPD_{LM\ max}$)	Pérez-Rodríguez et al. [24]
	$h_{0\ LM}$	Parameter regarding the initial physiological state of LM (-)	$\mu_{h0} = 2.8$ $\sigma_{h0} = 4.6$ $h_{0\ LM} \sim$ Normal (μ_{h0}, σ_{h0}), $h_0 > 0$	Couvert et al. [71]: q_0 then obtained by $(1/(\exp(h_0)-1))$
Relative to kinetic parameters of LAB in RTE seafood	$\mu_{LAB\ ref}$	Optimum growth rate of LAB (h^{-1})	0.583	Mejlholm and Dalgaard [54]
	MPD_{LAB}	Maximum population density of LAB (log_{10} CFU/g)	$MPD_{LAB\ min} = 8.0$ $MPD_{LAB\ mode} = 8.5$ $MPD_{LAB\ max} = 9.0$ $MPD_{LAB} \sim$ Pert ($MPD_{LAB\ min}$, $MPD_{LAB\ mode}$, $MPD_{LABM\ max}$)	$MPD_{LAB\ min}$ from Mejlholm and Dalgaard [54]
	$q_{0\ LAB}$	Parameter regarding the initial physiological state of LAB (-)	$\ln q_{0\ LABmin} = -12$ $\ln q_{0\ LABmode} = 2.73$ $\ln q_{0\ LABmax} = 1.26$ $q_{0\ LAB} \sim \exp\{$ Pert ($\ln q_{0\ LABmin}$, $\ln q_{0\ LABmode}$, $\ln q_{0\ LABmax}$)$\}$	Couvert et al. [71]
Relative to smoked fish characteristics	$\overline{C}_{0\ LAB\ SF}$	Between-lot mean concentration of LAB in smoked fish after packaging (log_{10} CFU/g)	$\overline{\overline{C}}_{0\ LABminSF} = -1.00$ $\overline{C}_{0\ LABmodeSF} = 0.28$ $\overline{C}_{0\ LABmaxSF} = 1.60$ $\overline{C}_{0\ LAB\ SF}$ $\sim$ Pert ($\overline{\overline{C}}_{0\ LABminSF}$, $\overline{\overline{C}}_{0\ LABmodeSF}$, $\overline{C}_{0\ LABmaxSF}$)	Wiernasz et al. [72]
	pH_{SF}	pH of smoked fish (-)	$pH_{minSF} = 5.8$ $pH_{modeSF} = 6.1$ $pH_{maxSF} = 6.5$ $pH_{SF} \sim$ Pert (pH_{minSF}, pH_{modeSF}, pH_{maxSF})	pH_{minSF} from Mejlholm and Dalgaard [51] pH_{modeSF} from Porsby et al. [48] pH_{maxSF} from Hwang and Sheen [73]
	$NaCl_{SF}$	NaCl content in smoked fish (% wb)	$NaCl_{minSF} = 1.5$ $NaCl_{modeSF} = 3.4$ $NaCl_{maxSF} = 5.3$ $NaCl_{SF} \sim$ Pert ($NaCl_{minSF}$, $NaCl_{modeSF}$, $NaCl_{maxSF}$)	$NaCl_{minSF}$ from FAO-WHO [26] $NaCl_{modeSF}$ from Porsby et al. [48], Mejlholm and Dalgaard [51], FAO-WHO [26], and Orozco [74] $NaCl_{maxSF}$ from Mejlholm and Dalgaard [51]
	Phe_{SF}	Phenol compound in smoked fish (ppm)	$Phe_{minSF} = 5$ $Phe_{modeSF} = 10$ $Phe_{maxSF} = 22$ $Phe_{SF} \sim$ Pert (Phe_{minSF}, Phe_{modeSF}, Phe_{maxSF})	Phe_{minSF} from Leblanc et al. [72] Phe_{modeSF} from Hwang and Sheen [73], Porsby et al. [48], Mejlholm and Dalgaard [51], FAO/WHO [26], Leblanc et al. [75], and Eklund et al. [47] Phe_{maxSF} from Porsby et al. [48]
	$CO2_{equi\ SF}$	CO_2 concentration at equilibrium in the packaging of smoked fish (fraction)	$CO2_{equi\ SF\ min} = 0.25$ $CO2_{equi\ SF\ mode} = 0.25$ $CO2_{equi\ SF\ max} = 0.30$ $CO2_{equi\ SF} \sim$ Pert ($CO2_{equi\ SF\ min}$, $CO2_{equi\ SF\ mode}$, $CO2_{equi\ SF\ max}$)	Mejlholm and Dalgaard [51]
	Others: Nit_{SF}, $LA_{tot\ GF}$, $AA_{tot\ SF}$, $BA_{tot\ SF}$, $CA_{tot\ SF}$, $DA_{tot\ SF}$, $SA_{tot\ SF}$	Nitrites concentration (ppm) and lactic acid, acetic acid, benzoic acid, citric acid, diacetate, lactic acid, and sorbic acid concentrations in water phase (ppm)	Allow for minimum, mode, and maximum for each compound to be sampled from Pert distribution. Values of zero set to all.	
Relative to gravad fish characteristics	$\overline{C}_{0\ LAB\ GF}$	Between-lot mean concentration of LAB in gravad fish after packaging (log_{10} CFU/g)	$\overline{\overline{C}}_{0\ LABminGF} = -1.00$ $\overline{C}_{0\ LABmodeGF} = 0.28$ $\overline{C}_{0\ LABmaxGF} = 1.60$ $\overline{C}_{0\ LAB\ GF}$ $\sim$ Pert ($\overline{\overline{C}}_{0\ LABminGF}$, $\overline{\overline{C}}_{0\ LABmodeGF}$, $\overline{C}_{0\ LABmaxGF}$)	Wiernacz et al. [72]
	pH_{GF}	pH of gravad fish (-)	$pH_{minGF} = 6.1$ $pH_{modeGF} = 6.2$ $pH_{maxGF} = 6.3$ $pH_{SF} \sim$ Pert (pH_{minSF}, pH_{modeSF}, pH_{maxSF})	Mejlholm and Dalgaard [51] and Orozco [74]

Table 4. *Cont.*

Type of Parameter	Parameter	Definition (Unit)	Value or Distribution	Source
	$NaCl_{GF}$	NaCl content in gravad fish (% wb)	$NaCl_{minGF} = 3.0$ $NaCl_{modeGF} = 3.2$ $NaCl_{maxGF} = 3.4$ $NaCl_{GF} \sim Pert\ (NaCl_{minGF}, NaCl_{modeGF}, NaCl_{maxGF})$	Mejlholm and Dalgaard [51] and Aarnisalo et al. [44]
	Phe_{GF}	Phenol compound in gravad fish (ppm)	$Phe_{minSF} = 0$ $Phe_{modeSF} = 0$ $Phe_{maxSF} = 5$ $Phe_{GF} \sim Pert\ (Phe_{minGF}, Phe_{modeGF}, Phe_{maxGF})$	Mejlholm and Dalgaard [51]
	$CO2_{equi\ GF}$	CO_2 concentration at equilibrium in the packaging of gravad fish (fraction)	$CO2_{equi\ GF\ min} = 0.25$ $CO2_{equi\ GF\ mode} = 0.25$ $CO2_{equi\ GF\ max} = 0.30$ $CO2_{equi\ GF} \sim Pert\ (CO2_{equi\ SF\ min}, CO2_{equi\ SF\ mode}, CO2_{equi\ SF\ max})$	Mejlholm and Dalgaard [51]
	Others: Nit_{GF}, $LA_{tot\ GF}$, $AA_{tot\ GF}$, $BA_{tot\ GF}$, $CA_{tot\ GF}$, $DA_{tot\ GF}$, $SA_{tot\ GF}$	Nitrite concentration and lactic acid, acetic acid, benzoic acid, citric acid, diacetate, lactic acid, and sorbic acid concentrations in water phase (ppm)	Allow for minimum, mode and maximum for each compound to be sampled from Pert distribution. Values of zero set to all.	-
Relative to cold chain distribution	$time_{CC}$	Time elapsed between end of production and arrival of the product at home (h)	$time_{CC\ min} = 12$ $time_{CC\ mode} = 144$ $time_{CC\ max} = 720$ $time_{CC} \sim Pert\ (time_{CC\ min}, time_{CC\ mode}, time_{CC\ max})$	FDA-FSIS [23]
	$Temp_{CC}$	Average temperature between end of production and arrival of the product at home (°C)	$Temp_{CC\ min} = 0.28$ $Temp_{CC\ mode} = 4.60$ $Temp_{CC\ max} = 7.00$ $Temp_{CC} \sim Pert\ (Temp_{CC\ min}, Temp_{CC\ mode}, Temp_{CC\ max})$	From Normal (4.6, 2.2) °C in Pouillot et al. [20]
	$CorTimeTemp_{CC}$	Correlation between time and temperature during cold chain	−0.16	Pouillot et al. [19]
Relative to home storage	$Time_{Home}$	Storage time at home (h)	$Time_{Home\ min} = 0.73$ $Time_{Home\ mode} = 70$ $Time_{Home\ max} = 840$ for smoked fish and 528 for gravad fish $Time_{Home} \sim Pert\ (time_{Home\ min}, time_{Home\ mode}, time_{Home\ max})$	Minimum and mode values from Weibull (shape = 1.14, scale = 18.39) days in Endrikat et al. [76] Maximum is best guess: 35 days for smoked fish and 22 days for gravad fish
	$Temp_{Home}$	Storage temperature at home (°C)	$Temp_{Home\ min} = 1.12$ $Temp_{Home\ mode} = 7.0$ $Temp_{Home\ max} = 12.9$ $Temp_{CC} \sim Pert\ (Temp_{Home\ min}, Temp_{Home\ mode}, Temp_{Home\ max})$	From Normal (7, 3) °C in Pouillot et al. [20]
	$CorTimeTemp_{Home}$	Correlation between time and temperature at home storage	−0.12	Pouillot et al. [19]

Briefly (see Appendix A for details), the function sfMejlholmDalgaard() deterministically computes the growth rate of *L. monocytogenes* in RTE seafood at given intrinsic and extrinsic characteristics, according to the secondary model based on the Gamma concept for lightly preserved seafood developed and validated by Mejlholm and Dalgaard [51–56] using the. Similarly, the function sfMejlholmDalgaardLAB() deterministically estimates the growth rate of LAB in RTE seafood at given environmental characteristics, according to the secondary model developed in [54]. The function sfGrowthJameson() estimates the numbers of *L. monocytogenes* and LAB (CFU) in RTE seafood at given environmental characteristics after a certain time period. The function is based on the Baranyi–Roberts-based Jameson-effect competition model put forward by Giménez and Dalgaard [57], supplemented with the Gamma (γ) interaction parameter later proposed by Moller et al. [58]. By default, γ is set to 1. Eventually, assuming Jameson-effect microbial competition and allowing for the presence of a lag phase, the function sfColdChain() stochastically estimates the concentration of *L. monocytogenes* and LAB populations in the RTE seafood product at the end of the cold chain distribution (i.e., arrival at consumer's home).

2.1.13. Home Storage

During home storage, *L. monocytogenes* and LAB continue to grow. Endrikat et al. [76] represented the home storage time in days by a Weibull distribution with shape parameter 1.14 and scale parameter 18.39. This information was translated into a Pert distribution for the minimum value $Time_{Home\ min}$ (0.73 h as the 2.5% percentile of the Weibull distribution) and the mode $Time_{Home\ mode}$ (70 h as the mode of the Weibull distribution). The maximum home storage time, $Time_{Home\ max}$, consists of a best guess of 35 days (840 h). Likewise, a Pert distribution regarding the home storage temperature ($Temp_{Home}$) was approximated from the normal distribution used in Pouillot et al. [20], truncating at the 2.5% and 97.5% percentiles to obtain the minimum ($Temp_{Home\ min}$ = 1.12) and the maximum ($Temp_{Home\ max}$ = 12.9) parameters, respectively. The mean of the normal distribution was set as the mode ($Temp_{Home\ mode}$ = 7.0). In addition, a low correlation between home storage time and temperature was quantified as rank correlation by Pouillot et al. [19], which will be used in the present QRA model ($CorTimeTemp_{Home}$ = −0.12).

The simultaneous growth of *L. monocytogenes* and LAB are simulated using the function ColdChain() and its auxiliary functions (cf. previous section), as performed for the distribution module of cold chain, including interactions [77]. Values of home storage time ($time_{Home\ ij}$) and temperature ($Temp_{Home\ ij}$) are sampled for every RTE seafood pack (i.e., different consumers), targeting the rank correlation value of $CorTimeTemp_{Home}$ [78].

$$time_{Home\ ij} \sim Pert\ (time_{Home\ min}, time_{Home\ mode}, time_{Home\ max})\ i = 1, 2, \ldots, r; j = 1, 2, \ldots, c_p$$
$$Temp_{Home\ ij} \sim Pert\ (Temp_{Home\ min}, Temp_{Home\ mode}, Temp_{Home\ max})\ i = 1, 2, \ldots, r; j = 1, 2, \ldots, c_p$$

The growth rates of *L. monocytogenes* ($\mu_{LM\ ij}$) and LAB ($\mu_{LAB\ ij}$) in RTE seafood for every pack ij stored at the temperature $Temp_{Home\ ij}$ are then computed by the functions sfMejlholmDalgaard() and sfMejlholmDalgaardLAB(), respectively, using the product's environmental characteristics already generated in the cold chain stage by the function sfCharacteristics().

Next, the function sfGrowthJameson() is executed to produce the numbers of *L. monocytogenes* and LAB in the RTE seafood units at the point of consumption ($N_{Home\ LM}$, $N_{Home\ LAB}$ in CFU).

$$\left\{N_{Home\ LM\ ij}, N_{Home\ LAB\ ij}, \ln q_{Home\ LM\ ij}, \ln q_{Home\ LAB\ ij}\right\} = sfGrowthJameson$$
$$\left(N_{cc\ LM\ ij}, N_{cc\ LAB\ ij}, time_{Home\ ij}, \exp(q_{cc\ LM\ ij}), \exp(q_{cc\ LAB\ ij}), \mu_{LM\ ij}, \mu_{LAB\ ij}, MPD_{LM\ i}, MPD_{LAB\ i}, \gamma, Unit_{SizePack}\right)$$
$$i = 1, 2, \ldots, r;\ j = 1, 2 \ldots, c_p$$

2.1.14. Portioning Before Consumption

During serving at home, it is assumed that the consumer removes a serving size $Serv_{size}$ (g) from the pack of RTE seafood of net weight $Unit_{SizePack}$ g. Then, the number of *L. monocytogenes* cells in this small unit can be considered to be a sample from a Beta-binomial distribution. Furthermore, a moderate clustering of *L. monocytogenes* cells in the RTE seafood contained in the pack is assumed (dispersion, b = 1; [59]). The dispersion parameter and the number of portions that can be obtained from a pack are assumed to be independent of the microbial numbers. A function, sfPortioning(), was developed to simulate the microbial process of partitioning (cf. Appendix A.16).

2.2. Hazard and Risk Characterisation

Several dose-response relationships based on the exponential model are available for *L. monocytogenes* [16,21,26,79,80]. The dose-response model chosen to estimate the risk of listeriosis per serving of RTE seafood is that of Pouillot et al. [79] for the elderly population (>65 years old) with unknown underlying conditions. According to this model, the probability r that an ingested *L. monocytogenes* cell causes an invasive listeriosis follows a $\log_{10}$ normal distribution, with mean −12.83 and standard deviation 1.62. The function DRQuick() from the `doseresponsemodels` R package [79] was employed to estimate the marginal probabilities of invasive listeriosis in the elderly population, $RiskServing_{ij}$, from the input matrix of doses, $N_{Portion\ ij}$. The mean risk for every lot i ($RiskLot_i$) was calculated as a risk averaged across servings, j, and weighed by the lot-specific probability, $Prob_{UnitPos\ Tested\ i}$:

$$RiskLot_i = \frac{\sum_{j=1}^{c_p} RiskServing_{ij} \times Prob_{UnitPos\ Tested\ i}}{c_p}$$

2.3. QRA Model's Ouputs

The model's outputs were summarized at three stages: end of processing, point of consumption, and risk characterization. The descriptors at the *end of processing* were computed from the prevalence and the contamination matrix outputs of the function sfTesting() and included the following: (1) descriptive statistics (mean, median, and 95% confidence interval) of the mean concentration of *L. monocytogenes* (CFU/g) in the fraction of contaminated lots; (2) the prevalence of contaminated packs; and (3) the probability that a contaminated pack contains more than 10 CFU *L. monocytogenes* per g RTE seafood.

The model's descriptors at the *point of consumption* were estimated from the outputs of the function sfColdChain(), applied to depict home storage, and encompassed the following: (1) descriptive statistics (mean, median, and 95% confidence interval) of the concentration of *L. monocytogenes* in any serving; (2) the prevalence of contaminated servings; (3) the probability that a contaminated serving contains more than 10 CFU *L. monocytogenes* per g RTE seafood; and (4) the probability that a contaminated serving contains more than 100 CFU *L. monocytogenes* per g RTE seafood.

The descriptors for risk characterization include the summary statistics, mean, median, and 2.5, 97.5, and 99.5 percentiles of the lot-level mean risk per serving *RiskLot*.

2.4. QRA Model's Functionality: Reference and What-If Scenarios

To illustrate the utility of the QRA model, reference scenarios were separately run for the three types of RTE seafood studied: smoked brine-injected fish, smoked dry-salted fish, and gravad fish. In addition, five what-if scenarios were run for each type of RTE seafood, which assessed the following: (1) reduction of the between-lot mean prevalence of *L. monocytogenes*; (2) absence of initial contamination load on filleting knives; (3) no transfer of contamination during salting (smoked fish) or smearing with ingredients (gravad fish); (4) application of protective LAB cultures; and (5) reduction of storage temperature at home.

(a) Reference, constituted by the baseline scenarios of the three RTE seafood products, with parameter values supported as much as possible by current data and, in their absence, by reasonable assumptions.
(b) Reduction of *L. monocytogenes* prevalence in a lot of incoming fish, assessed by setting the parameter $\hat{\alpha}$ of the Beta distribution regarding prevalence to half its original value (0.437).
(c) Reduction in cross-contamination during filleting, assessed by assuming that filleting knives are cleaned/disinfected after filleting every fish, and therefore setting the parameter $Init_{Slicer}$ of the function sfSlicer() to zero.
(d) Absence of contamination during salting or smearing of fish fillets, represented by a zero probability of contamination during brine injection (Pcc_{Brine} = 0%) for smoked brined fish and zero probability of contamination during smearing with salt or curing agents (Pcc_{Smear} = 0%) for both smoked dry-salted fish and gravad fish.
(e) Application of protective cultures when processing fish fillets, represented by an increase in the mean lot concentration of LAB in RTE seafood after packaging ($\overline{C}_{0\ LAB}$) by 5 $\log_{10}$ CFU/g. Therefore, the minimum ($\overline{C}_{0\ LAB\ min}$), mode ($\overline{C}_{0\ LAB\ mode}$), and maximum ($\overline{C}_{0\ LAB\ max}$) parameters defining the Pert distribution regarding $\overline{C}_{0\ LAB}$ were set to 4.00, 5.28, and 6.60 $\log_{10}$ CFU/g, respectively, for the RTE products.
(f) Lower home storage temperature, represented by a decrease of 1.5 °C in the mode and maximum storage temperature at home ($Temp_{Home\ mode}$ = 5.5 and $Temp_{Home\ max}$ = 11.4).

The reference and model scenarios for the three RTE seafood products were run by setting an initial contamination matrix size of r = 5000 lots and c =100 fish units in the function Lot2LotGen().

2.5. QRA Model's Implementation

All the functions described in Section 2.1 were programmed in R (R Core Team [81]) and compiled in the package qraLm, which can be installed from the Github repository: https://github.com/WorldHealthOrganization/qraLm, accessed on 10 September 2024. The reference manual can be found at: https://WorldHealthOrganization.github.io/qraLm/reference/, accessed on 10 September 2024.

3. Results and Discussion

3.1. Reference Scenario and Comparison with Other QRA Models

The outcomes of the references scenarios suggested that, despite the high prevalence of lots contaminated with *L. monocytogenes* at the end of processing—38.7% for smoked brined fish, 34.4% for smoked dry-salted fish, and 52.2% for gravad fish—the mean concentrations in the fraction of contaminated lots would be very low (0.0028 CFU/g for smoked brined fish, 0.0021 CFU/g for smoked dry-salted fish, and 0.0029 CFU/g for gravad fish; Table 5). The prevalences of contaminated packs leaving the processing facilities were estimated at 8.14%, 6.49%, and 11.14%, respectively, although virtually no pack would contain numbers higher than 10 CFU/g (Table 5). Higher levels of contamination in cold-smoked and salt-cured salmon produced in Finland were predicted by a Markov-chain-based model developed by Pasonen et al. [30]. They estimated that ~10% of the contaminated packs exceeded the 100 CFU/g limit, whereas the prevalence of contaminated packs was 22% (95% CI: 20–25%). Although the assumptions of such a QRA differed in many instances from our model, it is plausible that their higher exposure estimates primarily obey the greater *L. monocytogenes* initial prevalence levels from the Finnish survey data used as inputs in their QRA model (in the range of 15.7–31.8% in the period between 2004 and 2010). By contrast, in our QRA model the mean of the between-lot *L. monocytogenes* prevalence was 14.85%, as modelled by a Beta (0.8741, 5.880) distribution, built upon survey data from multiple countries. Nonetheless, despite these differences, our model did coincide with that of Pasonen et al. [30] in the prediction that *L. monocytogenes* contaminated packs tend to have low bacterial concentrations, but on a few rare occasions the concentrations at the point of consumption would be in the order of hundreds of CFU/g (Table 6).

According to the present QRA model, at the end of processing a production lot of gravad fish would be more likely to be contaminated with *L. monocytogenes* than a production lot of smoked fish, and, in turn, if the smoked fish was salted by brine-injection the probability of lot contamination would be higher than if dry-salted. The prevalence of *L. monocytogenes* predicted by our QRA model for smoked (6.40–8.14%) and gravad fish packs (11.15%) are fairly in concordance with the joint results of EFSA's EU-wide baseline survey and monitoring data, which encountered overall prevalence values of 8.42% (219/2602) in cold-smoked fish and 11.11% (30/270) in gravad fish, as compiled by Pérez-Rodríguez et al. [24].

Outcomes from other epidemiological surveys are also comparable to our simulation findings for gravad fish; namely, the incidence of 15% for cold stored gravlax from a Brazilian salmon processing plant (Cruz et al. [34]) and the incidence of 8% in gravlax fish from Icelandic markets [82]. In terms of enumeration estimates at the point of consumption (Table 6), the simulations predicted mean concentrations of *L. monocytogenes* in any serving (contaminated or not) of 103.6 CFU/g for smoked brined fish and 125.9 CFU/g for smoked dry-salted fish, and a higher mean concentration, at 162.6 CFU/g, for gravad fish, although it is worth mentioning that these high mean estimates are a result of a few highly contaminated servings, as implied by their 95% confidence intervals ([0–0.2261 CFU/g], [0–0.1197 CFU/g], and [0–3.3787 CFU/g], respectively). In the same order of magnitude of these intervals fell the median values of *L. monocytogenes* concentration found for cold-smoked fish (2.48 CFU/g) and for gravad fish (3.34 CFU/g) predicted by Pérez-Rodríguez et al. [24].

Table 5. Simulation outcomes of the exposure assessment model of *L. monocytogenes* in RTE seafood products at the end of processing for the reference and selected what-if scenarios. Simulations ran for 5000 lots consisting of 100 fish units each, where packs of final product weigh 260 g. Prevalence estimates are shown as proportions (scale 0–1).

Scenario	Mean Counts (CFU/g) in Contaminated Lots (Mean, Median, [95% CI])	Prevalence of Contaminated Lots	Prevalence of Contaminated Packs	P (N > 10 CFU/g in a Contaminated Pack) *
Smoked brined fish				
Reference	0.0028; 0.0017 [1.18×10^{-4}–0.0130]	0.3870	0.0814	0
Lower prevalence of contaminated lots	0.0023; 0.0014 [9.49×10^{-5}–0.0097]	0.2343	0.0497	0
No initial load on filleting knives	0.0023; 0.0012 [7.69×10^{-5}–0.0117]	0.3298	0.0652	0
No contamination during brining	0.0028; 0.0017 [1.20×10^{-4}–0.0129]	0.3767	0.0792	0
Addition of LAB culture	0.0028; 0.0017 [1.18×10^{-4}–0.0130]	0.3870	0.0814	0
Lower home storage temperature	0.0028; 0.0017 [1.18×10^{-4}–0.0130]	0.3870	0.0814	0
Smoked dry-salted fish				
Reference	0.0021; 0.0013 [1.28×10^{-4}–0.0095]	0.3443	0.0649	0
Lower prevalence of contaminated lots	0.0016; 0.0009 [7.18×10^{-5}–0.0072]	0.1611	0.0311	0
No initial load on filleting knives	0.0018; 0.0010 [8.71×10^{-5}–0.0085]	0.2927	0.0512	0
No contamination during smearing	0.0012; 0.0005 [3.58×10^{-5}–0.0063]	0.3554	0.0406	0
Addition of LAB culture	0.0021; 0.0013 [1.28×10^{-4}–0.0095]	0.3443	0.0649	0
Lower home storage temperature	0.0021; 0.0013 [1.28×10^{-4}–0.0095]	0.3443	0.0649	0
Gravad fish				
Reference	0.0029; 0.0020 [4.25×10^{-4}–0.0102]	0.5215	0.1114	0
Lower prevalence of contaminated lots	0.0022; 0.0016 [2.66×10^{-4}–0.0080]	0.3124	0.0542	0
No initial load on filleting knives	0.0023; 0.0016 [2.41×10^{-4}–0.0091]	0.4521	0.0886	0
No contamination during smearing	0.0026/0.0018 [3.97×10^{-4}–0.0100]	0.5031	0.1137	0
Addition of LAB culture	0.0029; 0.0020 [4.25×10^{-4}–0.0102]	0.5215	0.1114	0
Lower home storage temperature	0.0029; 0.0020 [4.25×10^{-4}–0.0102]	0.5215	0.1114	0

(*) In all scenarios, P (N > 100 CFU/g in a contaminated pack) = 0.

Table 6. Simulation outcomes of the exposure assessment model of *L. monocytogenes* in RTE seafood products at the point of consumption for the reference and selected what-if scenarios. Simulations ran for 5000 lots consisting of 100 fish units each, where packs of final product weigh 260 g and a serving is a slice of product (32.5 g). Prevalence estimates are shown as proportions (scale 0–1).

Scenario	Counts (CFU/g) in Any Serving (Mean, Median, [95% CI])	Prevalence of Contaminated Servings	P (N > 10 CFU/g in a Contaminated Serving)	P (*N* > 100 CFU/g in a Contaminated Serving)
Smoked brined fish				
Reference	103.6; 0.00 [0–0.2261]	0.0459	0.0122	0.0040
Lower prevalence of contaminated lots	103.4; 0.00 [0–0.0605]	0.0267	0.0110	0.0037
No initial load on filleting knives	76.39; 0.00 [0–0.1195]	0.0453	0.0111	0.0035
No contamination during brining	100.9; 0.00 [0–0.2139]	0.0444	0.0120	0.0039
Addition of LAB culture	0.0548; 0.00 [0–0.0886]	0.0416	0.0021	0.0002
Lower home storage temperature	6.4923; 0.00 [0–0.0994]	0.0413	0.0064	0.0016
Smoked dry-salted fish				
Reference	125.9; 0.00 [0–0.1197]	0.0352	0.0100	0.0032
Lower prevalence of contaminated lots	58.58; 0.00 [0–0.0206]	0.0165	0.0082	0.0027
No initial load on filleting knives	104.0; 0.00 [0–0.0642]	0.0274	0.0092	0.0029
No contamination during smearing	74.26; 0.00 [0–0.0763]	0.0210	0.0060	0.0019
Addition of LAB culture	0.0340; 0.00 [0–0.0508]	0.0320	0.0015	8.1×10^{-5}
Lower home storage temperature	3.8733; 0.00 [0–0.0567]	0.0318	0.0052	0.0012
Gravad fish				
Reference	162.6; 0.00 [0–3.3787]	0.0735	0.0260	0.0104
Lower prevalence of contaminated lots	85.84; 0.00 [0–0.4191]	0.0353	0.0213	0.0085
No initial load on filleting knives	124.9; 0.00 [0–1.8050]	0.0579	0.0237	0.0094
No contamination during smearing	156.5; 0.00 [0–3.3617]	0.0744	0.0255	0.0101
Addition of LAB culture	0.2100; 0.00 [0–0.4503]	0.0660	0.0057	0.0005
Lower home storage temperature	12.486; 0.00 [0–1.0267]	0.0675	0.0157	0.0052

In their short-scope QRA model, the consumption of gravad fish was linked to greater exposure than smoked fish, in agreement with our model. Nonetheless, our model's estimate for the level of exposure from a contaminated serving of smoked fish ($\log_{10}$ (103.6/0.0459) = 3.35 $\log_{10}$ CFU/g, worked out from Table 5) was considerably higher than the estimates from the QRA models of *L. monocytogenes* in contaminated servings of French cold-smoked salmon (mean: 1.38 $\log_{10}$ CFU/g; Pouillot et al. [19]) and smoked salmon consumed in Navarra (mean: 2.25 $\log_{10}$ CFU/g; Garrido et al. [27]).

On the other hand, our mean concentrations in any serving of smoked fish (103.6 CFU/g and 125.9 CFU/g) are close to the output of a QRA model of *L. monocytogenes* in Irish cold-smoked salmon, which predicted a mean concentration in any serving of 71.6 CFU/g (as determined by the product of the mean prevalence of contaminated servings of 18% and the mean concentration in contaminated servings of $10^{2.6}$ CFU/g) (Dass [29]).

The probability that a serving is contaminated with *L. monocytogenes* followed a decreasing order: 7.35% for gravad fish, 4.59% for smoked brined fish, and 3.52% for smoked dry-salted fish (Table 6). The same rank order was observed for the probabilities that a contaminated serving contains more than 10 CFU/g and 100 CFU/g. These were in the ranges 1.00–2.60% and 0.32–1.04%, respectively. At least three-fold higher probabilities of *L. monocytogenes* in RTE fish servings were found by the EFSA's generic short-scope QRA model [16], which estimated that proportions of 8.01% cold-smoked fish servings and 4.66% gravad fish servings would exceed a concentration of 100 CFU/g. These considerable deviations may stem from the following assumptions: (1) initial prevalences of *L. monocytogenes* (at end of processing) that were higher (17.4% for cold-smoked fish and 12.2% for gravad fish) than the throughputs of our model (6.49–8.14% for smoked fish and 11.14% for gravad fish); (2) initial concentrations of *L. monocytogenes* in the contaminated fraction (at end of processing) that were higher (0.867 $\log_{10}$ CFU/g for cold-smoked fish and 1.011 $\log_{10}$ CFU/g for gravad fish) than the throughputs of our model (0.0021–0.0028 CFU/g for smoked fish and 0.0029 CFU/g for gravad fish); and (3) serving sizes (49–66 g for smoked fish and 129–154 g for gravad fish in the elderly population) that were higher than the serving size of one slice (32.5 g) assumed in our model.

The present QRA model revealed a high between-lot variability in the risk of listeriosis associated with the consumption of a slice of RTE fish product (Table 7). The heterogeneity in the mean risk from lot to lot is represented in Figures 2–4 for smoked brine-injected fish, smoked dry-salted fish, and gravad fish, respectively. The mean lot risks of listeriosis per serving were estimated at 9.751×10^{-8} (median 6.572×10^{-11}) for smoked brined fish, 9.634×10^{-8} (median 5.352×10^{-11}) for smoked dry-salted fish, and 2.086×10^{-7} (1.376×10^{-9}) for gravad fish (Table 7). On a serving basis of one slice, the mean risk linked to gravad fish was 0.33 or 0.34 $\log_{10}$ higher than those of smoked brined or dry-salted fish. Similarly, the EFSA's generic QRA model [16] calculated a 1.12 $\log_{10}$ higher risk of listeriosis for sliced reduced-oxygen packaged gravad fish in comparison to sliced reduced-oxygen packaged cold-smoked fish in the elderly population, using the same dose-response model (Pouillot et al. [80]). Nevertheless, the median risk values were considerably higher in the EFSA model (4.37×10^{-9} for cold-smoked fish and 5.86×10^{-8} for gravad fish), which is likely to have arisen from the higher values of the model inputs discussed above.

Table 7. Statistics of the mean risk of invasive listeriosis per lot of RTE seafood products in the elderly population for the reference and selected what-if scenarios. The logarithm base 10 of the mean risk reduction attained by each scenario in comparison to the reference one is shown ($\log_{10}$ RR).

Scenario	Mean	Median	2.5 pct	97.5 pct	99.5 pct	$\log_{10}$ RR
Smoked brined fish						
Reference	9.751×10^{-8}	6.572×10^{-11}	3.693×10^{-15}	3.064×10^{-7}	3.836×10^{-6}	-
Lower prevalence of contaminated lots	8.778×10^{-8}	1.478×10^{-11}	3.911×10^{-16}	1.422×10^{-7}	2.881×10^{-6}	0.05
No initial load on filleting knives	6.920×10^{-8}	2.431×10^{-11}	7.494×10^{-16}	1.233×10^{-7}	2.272×10^{-6}	0.15
No contamination during brining	9.272×10^{-8}	6.164×10^{-11}	6.477×10^{-15}	2.566×10^{-7}	3.287×10^{-6}	0.02
Addition of LAB culture	2.718×10^{-10}	5.399×10^{-12}	2.409×10^{-15}	1.918×10^{-9}	8.972×10^{-9}	2.55
Lower home temperature	1.112×10^{-8}	1.769×10^{-11}	2.842×10^{-15}	3.082×10^{-8}	2.113×10^{-7}	0.94
Smoked dry-salted fish						
Reference	9.634×10^{-8}	5.352×10^{-11}	1.183×10^{-14}	1.703×10^{-7}	4.087×10^{-6}	-
Lower prevalence of contaminated lots	5.113×10^{-8}	9.071×10^{-12}	1.868×10^{-15}	5.483×10^{-8}	1.513×10^{-6}	0.28
No initial load on filleting knives	7.428×10^{-8}	2.415×10^{-11}	4.059×10^{-15}	8.223×10^{-8}	2.386×10^{-6}	0.11
No contamination during smearing	5.984×10^{-8}	6.953×10^{-11}	1.835×10^{-13}	9.772×10^{-8}	2.608×10^{-6}	0.20
Addition of LAB culture	1.693×10^{-10}	4.462×10^{-12}	6.363×10^{-15}	1.502×10^{-9}	5.888×10^{-9}	2.76
Lower home temperature	6.899×10^{-9}	1.607×10^{-11}	9.780×10^{-15}	1.991×10^{-8}	1.817×10^{-7}	1.14

Table 7. *Cont.*

Scenario	Mean	Median	2.5 pct	97.5 pct	99.5 pct	log_{10} RR
Gravad fish						
Reference	2.086×10^{-7}	1.376×10^{-9}	4.863×10^{-13}	1.402×10^{-6}	9.436×10^{-6}	-
Lower prevalence of contaminated lots	1.133×10^{-7}	1.900×10^{-10}	3.531×10^{-14}	5.372×10^{-7}	3.815×10^{-6}	0.27
No initial load on filleting knives	1.623×10^{-7}	6.810×10^{-10}	1.344×10^{-13}	9.013×10^{-7}	7.841×10^{-6}	0.10
No contamination during smearing	2.037×10^{-7}	1.347×10^{-9}	6.725×10^{-13}	1.368×10^{-6}	1.036×10^{-5}	0.01
Addition of LAB culture	1.019×10^{-9}	2.940×10^{-11}	7.731×10^{-14}	7.457×10^{-9}	2.977×10^{-8}	2.31
Lower home temperature	2.761×10^{-8}	2.343×10^{-10}	1.328×10^{-13}	1.500×10^{-7}	9.657×10^{-7}	0.88

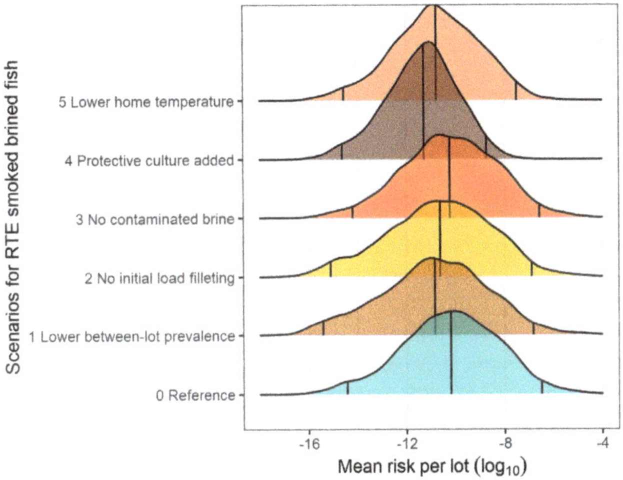

Figure 2. Lot-level mean risk (log_{10}) associated with the consumption of a 32.5-g serving (slice) of RTE smoked brine-injected fish, as evaluated for the reference and selected scenarios. Vertical lines on density plots indicate the median and interquartile range limits.

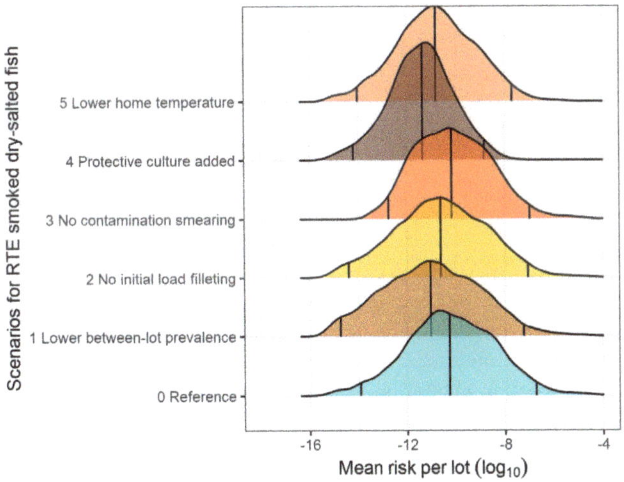

Figure 3. Lot-level mean risk (log_{10}) associated with the consumption of a 32.5-g serving (slice) of RTE smoked dry-salted fish, as evaluated for the reference and selected scenarios. Vertical lines on density plots indicate the median and interquartile range limits.

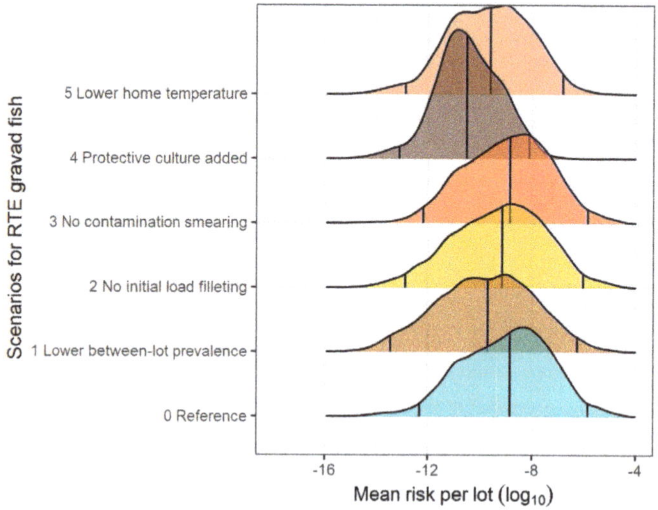

Figure 4. Lot-level mean risk ($\log_{10}$) associated with the consumption of a 32.5-g serving (slice) of RTE gravad fish, as evaluated for the reference and selected scenarios. Vertical lines on density plots indicate the median and interquartile range limits.

In comparison to our QRA simulation, other models attained very different mean risk estimates for cold-smoked salmon, namely, the model of Pouillot et al. [20] (1.3×10^{-6} in the French elderly population) and the model of Garrido et al. [27] (4.17×10^{-10} in the normal and immunocompromised population of Navarra, Spain). Dass [29], in his listeriosis QRA model from Irish cold-smoked salmon, estimated a median risk in the high-risk population (6.165×10^{-11}), which was in closer agreement with our median risk estimates for smoked fish.

For comparison with the EFSA's model outputs of number of annual illnesses [16], the mean risk estimates of our model were used to compute the mean number of invasive listeriosis in the 28 EU MS per year, based on the same estimated number of annual servings in the elderly population (1.99×10^{10} servings of smoked fish and 8.35×10^{9} servings of gravad fish). According to these values, our QRA model would predict a mean of 1940 annual cases of listeriosis linked to the consumption of smoked fish and 1741 annual cases linked to gravad fish. Using the same consumption data, EFSA [16] estimated a considerably lower number of listeriosis cases, at 201 and 230 cases in the elderly EU population linked to cold-smoked fish and gravad fish, respectively.

It is worth highlighting that the listeriosis dose-response model chosen has a high impact on the final risk estimate, and hence on the number of cases predicted. For instance, if the dose-response model of FAO-WHO [26] for increased susceptible population had been used, the present QRA model would have produced lower mean risk per serving estimates of 3.566×10^{-9} (median 4.511×10^{-13}) for the smoked brined fish and 5.599×10^{-9} (median 9.776×10^{-12}) for the gravad fish. These would have in turn predicted 72 and 48 annual listeriosis cases in the EU, 28 linked to these two products, respectively, assuming a consumption of 2.03×10^{10} servings of smoked fish and 8.53×10^{9} servings of gravad fish in the increased susceptible population (elderly plus pregnant population, taken from Pérez-Rodríguez et al. [24]).

3.2. What-If Scenarios

Among the what-if scenarios evaluated, reducing the prevalence of *L. monocytogenes* in fish entering the processing facilities would produce a decrease in the proportion of contaminated lots at the end of processing in about 40–53% for the three RTE seafood products, followed by the use of disinfected filleting knives (13–15%; Table 5). The absence of bacterial transfer during brine injection or smearing with ingredients would generate a lower effect on the prevalence of contaminated lots (decreasing it in 2.5–3.5% in smoked brined fish and gravad fish) because the frequency of contamination during salting assumed in the reference scenario is already low. Likewise, the what-if scenario that would reduce the prevalence of contaminated packs at the end of processing the most would be the reception of incoming fish with a lower prevalence of *L. monocytogenes*; this would reduce the proportion of outgoing contaminated packs from 8.14% to 4.97% in smoked brined fish, from 6.49% to 3.11% in smoked dry-salted fish, and from 11.14% to 5.42% in gravad fish (Table 5). The effectiveness of the what-if scenarios in decreasing the *L. monocytogenes* concentration in contaminated lots is, however, less evident in the mean and median estimates, given the high variability in *L. monocytogenes* contamination from lot to lot; however, the lowering effect can be appreciated in the 97.5th percentile. A scenario of absence of contamination during brine injection or smearing with ingredients would not practically lower the mean concentration of *L. monocytogenes* in contaminated lots at the end of processing, as it would lower the prevalence of contaminated lots and contaminated packs. As deduced by the 97.5th percentile in the concentration of contaminated lots (0.2261 CFU/g in smoked brined fish, 0.1197 CFU/g in smoked dry-salted fish, and 3.3787 CFU/g in gravad fish in the reference scenario), halving the prevalence of incoming fish into processing lines would be far more effective (to 0.0605 CFU/g, 0.0206 CFU/g, and 0.4191 CFU/g, respectively) than keeping filleting knives free of *L. monocytogenes* at all times (to 0.1195 CFU/g, 0.0642 CFU/g, and 1.8050 CFU/g, respectively; Table 5).

The effectiveness of the use of protective LAB cultures and the lower storage temperature at home can be compared with the other scenarios through the exposure assessment simulation outputs at the point of consumption (Table 6). Whereas reducing the contamination prevalence of incoming fish was the scenario that produced the highest reduction (52–54%) in the prevalence of contaminated servings (from 4.59% to 2.67% in smoked brined fish, from 3.52% to 1.65% in smoked dry-salted fish, and from 7.35% to 3.53% in gravad fish), in terms of *L. monocytogenes* numbers, the greatest control was achieved by the use of protective cultures, which led to the lowest *L. monocytogenes* concentration in any serving, and, as a consequence, the lowest probabilities of finding counts greater than 10 CFU/g or 100 CFU/g in a contaminated serving. However, the efficacy of the use of LAB cultures was found to depend on the RTE seafood product: for the smoked brined/dry-salted fish, the application of protective cultures produced a 2000/3000-fold decrease in the mean *L. monocytogenes* concentration in any serving and an 83/85% decrease in the probability that a contaminated serving contains more than 10 CFU/g (Table 6). In gravad fish, the efficacy of using protective culture was lower, causing a 780-fold decrease in the mean *L. monocytogenes* concentration in any serving and a 78% decrease in the probability that a contaminated serving contains more than 10 CFU/g. Nonetheless, considering the 100 CFU/g limit, there was no real difference in LAB culture effectiveness between the types of RTE product, since in all products the application of protective cultures led to a ~96% reduction in the probability of finding *L. monocytogenes* counts greater than 100 CFU/g in a contaminated serving.

After the use of protective cultures, the next "most effective scenario" to control *L. monocytogenes* up to the point of consumption was the maintenance of the RTE seafood product to lower storage temperatures at home. Reductions in the mean concentration in any serving and its 97.5th percentile were in the order of 13-to-32-fold and 2.1-to-3.3-fold, respectively. From *L. monocytogenes* mean concentrations at the point of consumption of 103.6 CFU/g (97.5th pct 0.2261 CFU/g), 125.9 CFU/g (97.5th pct 0.1197 CFU/g), and 162.6 CFU/g (97.5th pct 3.378 CFU/g) in the smoked brined fish, smoked dry-salted fish, and gravad fish, respectively, the sole proper storage practice taken by the consumers would reduce the concentrations to 6.4923 CFU/g (97.5th pct 0.0994 CFU/g), 3.8733 CFU/g (97.5th pct 0.0567 CFU/g), and 12.486 CFU/g (97.5th pct 1.0267 CFU/g) in the smoked brined fish, smoked dry-salted fish, and gravad fish, respectively. The frequency of contaminated servings above the 100 CFU *L. monocytogenes* per gram of serving would also be reduced by 50–60% in the three RTE products. In terms of prevalence of contaminated servings, however, the level of reduction attained by keeping colder home temperatures would be nearly the same as that of applying protective cultures (4.13% versus 4.16% for smoked brined fish, 3.18% versus 3.20% for smoked dry-salted fish, and 6.75% versus 6.60% for gravad fish; Table 6).

An inverse result on the relative importance of these two scenarios (colder home temperatures and added protective LAB cultures) in the concentration of *L. monocytogenes* in the servings of smoked salmon was purported by the listeriosis QRA model of Pouillot et al. [19], where they found that the mean storage temperature at the consumer phase ($p = 10^{-20}$) was a stronger determinant of exposure than the initial background microbiota (LAB) concentration ($p = 0.002$). Other QRA models whose outcomes helped to reinforce the relevance of improving consumer's awareness of the correct storage conditions of RTE seafood and avoid temperature abuse were the ones developed by Dass [29], Garrido et al. [27], Pérez-Rodríguez et al. [24], and Pasonen et al. [30].

In our QRA model, the two scenarios relative to decreasing cross-contamination during secondary processing—namely, no initial contamination load on filleting knives and no bacterial transfer during brine injection or smearing with ingredients—were effective but, among the five what-if scenarios evaluated, yielded the least extent of reduction in *L. monocytogenes* prevalence and concentrations in RTE seafood servings (Table 6). In both smoked brined fish and gravad fish, maintaining filleting knives free of *L. monocytogenes* was more effective than the absence of contamination during brining/smearing in decreasing the mean concentration in any serving (23–25% reduction against 3.4–3.7%), the 97.5th percentile in the mean concentration in any serving (47% reduction against 0.5–1.2%), the proportion of contaminated servings with more than 10 CFU/g *L. monocytogenes* (9.0% reduction versus 1.6–1.9%), and the proportion of contaminated servings with more than 100 CFU/g *L. monocytogenes* (7.1–9.6% reduction versus 2.5–2.8%). The opposite outcome was encountered for the smoked dry-salted fish, where avoiding bacterial transfer during dry-salting would be more effective than keeping filleting knives disinfected, producing a reduction by 41% in the mean concentration in any serving versus 17%, a reduction by 44% in the prevalence of contaminated servings versus 22%, and a reduction by 40% in the proportion of servings with *L. monocytogenes* concentrations greater than 10 or 100 CFU/g against 8.0% or 9.3%, respectively.

In the three RTE seafood products, it was evident—within the constraints of the model assumptions—that the cross-contamination that could occur during processing (filleting and salting or smearing with curing agents) would be a less important determinant of the consumer's exposure to *L. monocytogenes* when compared to the maintenance of proper cold storage at home or to the pathogen's occurrence in fish entering processing.

The impact of the what-if scenarios on the between-lot mean risk per serving of RTE seafood product can be visualized in the density curves of Figures 2–4, which also show the shifts in the 2.5th and the 97.5th percentiles. Among the scenarios evaluated, the application of protective LAB cultures (which increases the initial LAB counts by 5 $\log_{10}$ CFU/g) would be the most effective strategy to hinder the growth of *L. monocytogenes*, thereby reaching a mean risk reduction of 2.55 $\log_{10}$ for smoked brined fish, 2.76 $\log_{10}$ for smoked dry-salted fish, and 2.31 $\log_{10}$ for gravad fish. This strategy also reduced, to the greatest extent, the median (by >1.0 $\log_{10}$) and the highest (97.5 and 99.5th by >2.2 $\log_{10}$) percentiles of the mean lot risk posed by the three RTE products (Table 7). Decreasing the home storage temperature by 1.5 °C (mode and maximum) turned out to be the second most important risk control strategy, causing a risk reduction level in the range of 0.52–0.76 $\log_{10}$ for the 50th percentile, 0.93–1.00 $\log_{10}$ for the 97.5th percentile, and 0.98–1.25 $\log_{10}$ for the 99.5th percentile among the three RTE products. In terms of mean risk reduction, keeping the RTE products at lower temperatures would bring about a mean risk reduction of 0.94 $\log_{10}$ in smoked brined fish, 1.14 $\log_{10}$ in smoked dry-salted fish, and 0.88 $\log_{10}$ in gravad fish. A very similar prediction was computed in a QRA model in Irish cold-smoked salmon

(Dass [29]), where, by fixing the storage temperature from 3–10 °C to 4 °C, the mean risk was decreased by 1.10 $\log_{10}$ in the increased susceptibility population. In comparison to our findings, lower mean risk reductions have been reported by QRA models in analogous scenarios, such as decreasing the distribution temperature profiles by 1–2 °C caused a risk reduction of 0.60 $\log_{10}$ (Pérez-Rodríguez et al. [24]), decreasing home storage temperature from 7 °C to 4 °C caused a median risk reduction of 0.52 $\log_{10}$ (Pasonen et al. [30]), and all domestic temperatures with a mean temperature of 5.4 °C led to a risk reduction of 0.46 $\log_{10}$ (Garrido et al. [27]).

In smoked dry-salted fish and gravad fish, decreasing the mean lot prevalence by half was the what-if scenario that followed in effectiveness, as attested by the mean risk reductions of 0.28 and 0.27 $\log_{10}$, respectively. The risk lot medians linked to these two RTE products were also reduced by 0.77 and 0.86 $\log_{10}$, respectively, whereas the high percentiles were reduced by 0.40- 0.50 $\log_{10}$. This would imply that, if contamination was assumed to be carried over between lots (within the same processing facility), the mean lot risk would in turn be increased.

Nonetheless, in smoked brine-injected fish, the risk reduction attained by decreasing the prevalence of contaminated lots was low (0.05 $\log_{10}$) and comparable to the effect of avoiding contamination during brine injection (0.02 $\log_{10}$ risk reduction), although the former provided a slightly higher risk reduction, as also attested by the median (1.478×10^{-11} versus 6.164×10^{-11}), the 97.5th percentile (1.422×10^{-7} versus 2.566×10^{-7}), and the 99.5th percentile (2.881×10^{-6} versus 3.287×10^{-6}). The relatively lower importance of the scenario of reduced initial prevalence, found in the present QRA, is in agreement with the findings of other authors, who determined that the strategies that reduced growth (such as the use of protective cultures or the reduced storage temperature tested in our simulations) present a greater impact on the final risk than strategies that reduce the occurrence [19,22,83]. Furthermore, measures aimed at reducing the initial prevalence of *L. monocytogenes* in fish are not well documented.

The strategy of keeping filleting knives free of *L. monocytogenes* at all times (i.e., the very ideal scenario of having no initial load on knives before filleting every fish) rendered a mean risk reduction that was comparable across the three RTE products evaluated—0.15 $\log_{10}$ in smoked brined fish, 0.11 $\log_{10}$ in smoked dry-salted fish, and 0.10 $\log_{10}$ in gravad fish—with fairly important reductions in the median risk (0.30–0.43 $\log_{10}$) and high percentiles (0.10–0.40 $\log_{10}$). In relation to the absence of bacterial transfer during fillet smearing with salt or with curing agents, the effectiveness of this scenario to diminish the final risk of listeriosis depended on the type of product, which in gravad fish provided a very negligible mean risk reduction of 0.01 $\log_{10}$, while in dry-salted fish the mean risk reduction was higher at 0.20 $\log_{10}$ (very close to the effect of lowering the prevalence of contaminated lots). This difference is explained by the fact that, although the parameters assumed for smearing were the same in both products, gravad fish has intrinsic characteristics that are more favorable for *L. monocytogenes* growth. Thus, in gravad fish, the growth of *L. monocytogenes* after the product leaves the factory has a contribution to final risk that nearly overcomes the risk reduction effect of the absence of contamination during fillet smearing.

Taking together the results of the five scenarios across the three RTE seafood products, it became clear that the consumer's conventional measure of maintaining the product at the recommended cold temperatures is very effective in decreasing the risk of listeriosis (overall mean risk reduction of ~1.00 $\log_{10}$). Nevertheless, since this measure is at the same time the hardest to implement (and impossible to control), strategies that are placed at the processing level must be considered, namely reducing the prevalence of *L. monocytogenes* in incoming un-filleted fish (overall mean risk reduction of ~0.20 $\log_{10}$) and/or slowing down the pathogen's growth in the product. The latter can be achieved by the use of protective cultures (overall mean risk reduction of ~2.56 $\log_{10}$) or even the modification of the physicochemical properties of the product (lower water activity, addition of organic acids or other growth inhibitors [84]). Avoiding cross-contamination is an effective control measure; however, the model has evidenced that, once *L. monocytogenes* has entered the processing lot, minimizing the chances for cross-contamination during filleting, salting, or smearing with curing agents would result in a comparatively small risk reduction, which is lower for gravad fish (0.01–0.10 $\log_{10}$) compared to smoked fish (0.02–0.15 $\log_{10}$).

4. Conclusions and Perspectives

We have developed a robust four-module quantitative risk assessment (QRA) model of *L. monocytogenes* in ready-to-eat (RTE) fish products. This model, composed of 18 functions and a total of 128 parameters, is freely available and designed with flexibility in mind. It provides, for the first time, the ability to check for the efficacy of batch testing on risk. Our findings highlight the critical importance of strict hygiene practices during processing, the benefits of using protective LAB cultures, and the necessity of proper storage conditions by consumers to effectively mitigate the risk of listeriosis. The model's inherent adaptability ensures that food safety authorities can tailor interventions to meet specific needs, enhancing food safety measures for various RTE seafood products.

As new data become available, the parameters can be updated, and the functions, programmed in R software, can be reassembled to represent different seafood products. For example, to adapt the model for exposure assessment of sashimi, the smoking function (sfSmoking()) can be omitted, while for ceviche, it would be appropriate to remove the smoking function and add the maceration function (sfMaceration()), adjusting the parameters to suit these specific raw fish dishes.

This QRA model is intended to assist food safety authorities worldwide in obtaining risk estimates and evaluating risk management options tailored to their specific RTE seafood chains and product characteristics. Moreover, the model facilitates the evaluation of a variety of intervention strategies (such as the use of bacteriocinogenic cultures, organic acids, extended smoking periods) by allowing adjustments to interaction parameters in the Jameson model or modifying growth rates for *L. monocytogenes* and lactic acid bacteria (LAB). This QRA model not only provides a framework for assessing the risk of listeriosis from the consumption of smoked and gravad fish but also offers the potential to apply its methodology to a broader range of seafood products, thereby supporting global efforts in improving food safety.

Author Contributions: Conceptualization, U.G.-B., R.P. and M.S.; methodology, U.G.-B., R.P. and M.S.; software, R.P., U.G.-B., M.S., L.G. and V.C.; validation, U.G.-B. and R.P.; formal analysis, R.P., U.G.-B., L.G., V.C. and M.S.; investigation, U.G.-B., R.P., V.C., L.G. and M.S.; resources, M.S.; data curation, U.G.-B., R.P., J.D.O.M. and A.H.; writing—original draft preparation, U.G.-B.; writing—review and editing, U.G.-B., R.P., J.D.O.M., A.H., L.G., V.C. and M.S.; visualization, U.G.-B., R.P., T.S., E.C., P.T., M.J.S., J.D.O.M., A.H., L.G. and M.S.; supervision, M.S.; project administration, M.S.; funding acquisition, M.S. All authors have read and agreed to the published version of the manuscript.

Funding: This research received no external funding.

Institutional Review Board Statement: Not applicable.

Informed Consent Statement: Not applicable.

Data Availability Statement: The original contributions presented in the study are included in the article; further inquiries can be directed to the corresponding author.

Acknowledgments: The FAO/WHO Joint Expert Meeting on Microbiological Risk Assessment (JEMRA) on *Listeria monocytogenes* in foods (Part 1), held in Rome, Italy from 24–28 October 2022, recommended the development of a quantitative risk assessment model for *Listeria monocytogenes* in non-ready-to-eat frozen vegetables that considers processing and consumer handling. The model was finalized during the JEMRA meeting on *Listeria monocytogenes* in foods Part 2: Risk Assessment Models, which was held in Geneva, Switzerland from 29 May to 2 June 2023. The authors thank the JEMRA experts for Part 1 (https://www.who.int/publications/m/item/jemra-of-listeria-monocytogenes-in-foods, accessed on 1 September 2024) and Part 2, (https://www.who.int/

publications/m/item/jemra-of-listeria-monocytogenes-in-foods-part-2-risk-assessment-models, accessed on 1 September 2024) as well as the FAO/WHO secretariat for their suggestions and feedback.

Conflicts of Interest: Author R.P. is self-employed. He participated in this work in his own personal capacity. The remaining authors declare that the research was conducted in the absence of any commercial or financial relationships that could be construed as a potential conflict of interest.

Appendix A.

Appendix A.1. The Function Lot2LotGen

For every lot $i = \{1, 2, \ldots, r\}$, a value of prevalence P_i is sampled from the between-lot prevalence distribution $Beta\ (\hat{\alpha},\ \hat{\beta})\ (\hat{\alpha} = 0.8741,\ \hat{\beta} = 5.880)$. The probability that lot *i* is contaminated ($Prob_{UnitPos}$) is estimated as

$$Prob_{UnitPos\ i} = 1 - P(s_i = 0)$$

assuming that the number of contaminated fish units, s_i, in a given lot, *i*, follows a Binomial distribution:

$$s_i \sim Binomial(c, P_i)$$

The mean prevalence of contaminated lots (*P*) is calculated as the complement of the Beta-binomial probability evaluated at $s = 0$,

$$P = 1 - \frac{\Gamma(c+\beta)}{\Gamma(c+\alpha+\beta)}$$

where $\Gamma(x)$ is the Gamma function. To produce the contaminated fish units, the mean numbers of *L. monocytogenes* per gram of fish (λ_i, CFU/g) is calculated for every lot *i* from the Poisson assumption,

$$\lambda_i = -\frac{\ln(1-P_i)}{w}$$

where w is the analytical sample weight used for detection analysis, assumed to be 25 g. The total number of *L. monocytogenes* in a contaminated lot ($N_{tot\ i}$) is then sampled from a Poisson distribution,

$$N_{tot\ i} \sim Poisson(\lambda_i \times Unit_{size} \times c),\ \ N_{tot\ i} > 0$$

Such total bacterial cells are then randomly distributed among the fish units of a lot, according to a multinomial distribution, assuming that all fish units have the same probability $p_1 = p_2 = \ \ldots\ = p_c = 1/c$ of being contaminated,

$$\left[N_1\ N_2\ \ \ldots\ \ N_c\right]_i \sim Multinomial\left(N_{tot\ i},\ \left[p_1\ p_2\ \ \ldots\ \ p_c\right]_i\right)$$
$$i = 1,\ 2,\ \ldots,\ r$$

Thus, the resulting matrix *N* contains the numbers of *L. monocytogenes* cells in *c* units of fish (horizontal dimension) in a sample of *r* contaminated lots (i.e., where $\sum_i N_i > 0$) (vertical dimension). The outputs of the function Lot2LotGen() are as follows: the contamination matrix (*N*), the mean prevalence of contaminated lots (*P*, a scalar), and the probability that the sampled lot is contaminated ($Prob_{UnitPos}$, a vector).

Appendix A.2. The Function sfGrowthLDP

First, for a given temperature *Temp* and *time*, the maximum growth rate (μ, h^{-1}) and lag phase duration (λ, h) of *L. monocytogenes* in raw fish are computed from the equations

$$\mu = \left\{a_{Lm} \times (Temp - T_{min})^{0.75}\right\}^2$$

$$\lambda = \exp(A_{Lm}) \times \mu_{max}{}^{-m_{Lm}}$$

where T_{min} is the minimum growth temperature of *L. monocytogenes* in raw fish (1.3 °C) and a_{Lm}, A_{Lm}, and m_{Lm} are empirical coefficients, set at 0.0581, 0.84, and 1.11, respectively [42]. In order to deal with the lag phase over multiple storage steps, the work to be "potentially" done at the current storage stage (wk) is calculated as

$$wk = \mu \times \lambda$$

and the work "actually" done is computed as

$$wk_{actual} = wk - WorkDone_{s-1}$$

where $WorkDone_{s\text{-}1}$ is an input of the function and refers to the work done in a previous storage stage, *s-1*. At the first stage, $WorkDone_0$ takes the value of zero. The function updates the work done with the current expenditure of the lag phase, to produce $WorkDone_s$

$$WorkDone_s = WorkDone_{s-1} + \mu \times time$$

where $WorkDone_s$ is an output that enables the reusability of the function over more than one storage stage, keeping track of the remaining lag phase. Next, the effective lag phase duration (λ_{eff}) for the present storage stage is computed as

$$\lambda_{eff} = \begin{cases} 0, \ wk_{actual} < 0 \\ wk_{actual}/\mu, \ wk_{actual} \geq 0, \ \mu > 0 \end{cases}$$

and deducted from *time* to produce the effective time ($time_{eff}$) for loglinear growth,

$$time_{eff} = \max\left(time - \lambda_{eff}, 0\right)$$

Considering N_0 (CFU) as the initial number of *L. monocytogenes* in a raw fish unit of weight $Unit_{size}$ grams, the number after storage (N, CFU) is estimated as

$$N = \begin{cases} 10^{\left(\log_{10} N_0 + \frac{\mu}{\ln(10)} time_{eff}\right)}, \text{if } 10^{\left(\log_{10} N_0 + \frac{\mu}{\ln(10)} time_{eff}\right)} < Unit_{size} \times 10^{MPD} \\ Unit_{size} \times 10^{MPD}, \text{if } 10^{\left(\log_{10} N_0 + \frac{\mu}{\ln(10)} time_{eff}\right)} \geq Unit_{size} \times 10^{MPD} \end{cases}$$

The inputs and outputs of the function sfGrowthLDP() are represented by

$$\{N, \ WorkDone_t\} = sfGrowthLPD(N_0, \ time, \ Temp, \ Unit_{size}, \ MPD, \ WorkDone_{s-1})$$

Appendix A.3. The Function sfRawFishStorage

The function sfGrowthLPD is applied to the contaminated fish units of the contamination matrix N ($N_{ij} > 0$), setting the value of $WorkDone_{t\text{-}1}$ to zero,

$$\left\{N_{stored\ ij}, \ WorkDone_{t\ ij}\right\} = sfGrowthLPD\left(N_{ij}\ , \ time_{Storage\ i}, \ Temp_{Storage\ i}, \ Unit_{size}, \ MPD\right)$$
$$j = 1, \ 2 \ldots, \ c$$

sfRawFishStorage() outputs are the contamination matrix after *L. monocytogenes* growth (N_{stored}) and the work done after the first storage ($WorkDone_t$, a matrix), and returns unaffected the mean prevalence of contaminated lots (P, a scalar) and the probability that the sampled lot is contaminated ($Prob_{UnitPos}$, a vector).

Appendix A.4. The Function sfSlicer

The function starts by calculating the total number of slices or fillets (nb_{Slices}) as

$$nb_{Slices} = \left\lfloor \frac{Unit_{size}}{w_{Fillet}} \right\rfloor$$

Then, for each lot i, the transfer coefficients a_i and e_i are sampled from their variability distributions, whose values apply to all fish units produced in that lot $N_{i1}, N_{i2}, \ldots, N_{ic}$,

$$a_i \sim Logistic\,(location_a, scale_a)\; 0 \le a_i \le 0.5$$

$$log_{10}e_i \sim Normal\left(\mu_{loge}, \sigma_{loge}\right)\; log_{10}e_i \le 0$$

The parameters $location_a$, $scale_a$, μ_{loge}, and σ_{loge} are set to 0.07, 0.03, −2.12, and 0.85, respectively, according to Hoelzer et al. [43]. Following the assumption that the fish units present half the total contamination on one side, the transferable numbers of *L. monocytogenes* on the fish unit in the initial state 0 ($CFU_{fish\ 0}$), before slicing, is obtained,

$$CFU_{fish\ 0\ ij} \sim Binomial\left(N_{stored\ ij},\ p = 0.5\right)$$

and the bacterial numbers on the slicer blade in the initial state is set to $CFU_{slicer\ 0}$ (= $Init_{Slicer}$). $Init_{Slicer}$ is assumed to be 1000 CFU. Then, computations to represent the *transition during slicing* are made, as follows [43]: for the *k*-th slice (*k* = 1, 2, . . ., $nb_{Slices} - 1$), the *L. monocytogenes* cells from the fillet that are not lost are estimated:

$$M0_{k\ ij} \sim Binomial\left(CFU_{fish\ 0\ ij},\ 1 - e_i\right)$$

The cells that remain on the slicer during the slicer-to-fish transfer ($M1_k$) are estimated as

$$M1_{k\ ij} \sim Binomial\left(CFU_{slicer\ 0\ ij},\ 1 - 2a_i\right)$$

The cells that are transferred from the slicer to the fillet ($M2_k$) are estimated as

$$M2_{k\ ij} \sim Binomial\left(CFU_{slicer\ 0\ ij} - M1_{k\ ij},\ 0.5\right)$$

The *L. monocytogenes* cells that are transferred from the slicer to the slice ($M3_k$) are calculated as

$$M3_{k\ ij} = CFU_{slicer\ 0\ ij} - M1_{k\ ij} - M2_{k\ ij}$$

The *L. monocytogenes* cells that are transferred from the fillet to the slicer are sampled from Binomial distributions (S_k), as follows:

$$S_{k\ ij} \sim Binomial\left(M0_{k\ ij},\ a_j\right)$$

Next, the computations linked to the *time step k after slicing* (*k*=1, 2, . . ., $nb_{Slices} - 1$) are carried out,

$$CFU_{slice\ (k+1)ij} = M3_{k\ ij} + \left(M0_{k\ ij} - S_{k\ ij}\right)$$
$$CFU_{slicer\ (k+1)ij} = M1_{k\ ij} + S_{k\ ij}$$
$$CFU_{fish\ (k+1)ij} = M2_{k\ ij}$$

where $CFU_{slice\ (k+1)}$, $CFU_{slicer\ (k+1)}$, and $CFU_{fish\ (k+1)}$ are the numbers of *L. monocytogenes* on the slice, on the slicer blade, and on the fish unit, respectively, at time *k*. This procedure respects the mass balance, if one considers the $\left(CFU_{fish\ 0\ ij} - M0_{\ k\ ij}\right)$ bacteria lost in the process [43]. This iterative algorithm continues slice by slice until reaching the last slice,

$k = nb_{Slices}$. At that point, it is assumed that the last slice takes the remaining contamination of the fish plus the remaining half of the original,

$$CFU_{slice\ \mathrm{nbSlices}ij} = CFU_{fish\ (\mathrm{nbSlices}-1)ij} + \left(N_{ij} - CFU_{fish\ 0\ ij}\right)$$

CFU_{slice} then becomes N_{Fillet}, the numbers of *L. monocytogenes* on the fish fillets. It should be pointed out that, since one raw fish produces nb_{Slices} fillets, the horizontal dimension of the contamination matrix N_{Fillet} is enlarged, from c fish units to $c \times nb_{Slices}$ fillet units. Thus, N_{Fillet} is arranged to be a matrix of dimensions $r \times c_f$, where $c_f = c \times nb_{Slices}$. To keep track of the remaining microbial lag phase after the first storage of fish (output of the function sfRawFishStorage), the $r \times c$ matrix $WorkDone_t$ is also rearranged into a new matrix of dimensions $r \times c_f$, by attributing the value of $WorkDone_t$ of a fish unit to all of its nb_{Slices} fillets.

Before returning the function outputs, there is a verification as to whether or not filleting renders any lot free of contamination (i.e., all the fillets produced in a lot having 0 CFU). This event, although infrequent, could occur in the case of low contamination of fish with the subsequent transfer of all cells to the slicer blade, or lost, and none to the fillets. In that case, the proportion of non-contaminated lots (π_0) is estimated from the N_{Fillet} matrix, and the mean prevalence value is corrected as

$$P_{Fillet} = P \times (1 - \pi_0)$$

A value of one (1 CFU) is then ascribed in only one of the fillet units produced in the non-contaminated lots in order to avoid deletion of the "clean" lots and resampling of contaminated lots to round off the matrix size.

During the slicing of processed fish fillet, the function starts by calculating the total number of slices (nb_{Slices}) as

$$nb_{Slices} = \left\lfloor \frac{w_{Fillet}}{w_{Slice}} \right\rfloor$$

The algorithm proceeds exactly as during the filleting of raw fish. CFU_{slice} becomes N_{Slice}, the load of *L. monocytogenes* in processed fish slices. However, the horizontal dimension of the output contamination matrix N_{Slice} is enlarged, from c_f fillet units to $c_f \times nb_{Slices}$ slices. N_{Fillet} is now a matrix of dimensions $r \times c_s$, where $c_s = c_f \times nb_{Slices}$. In addition to N_{Slice}, the function returns the probability vector of contaminated lots after filleting ($Prob_{UnitPos\ Smoked}$ for smoked fish slices or $Prob_{UnitPos\ Gravad}$ for gravad fish slices, in any case of unaltered values) and the mean prevalence of contaminated lots after slicing (P_{Slice}).

Appendix A.5. The Function sfBriningCC

The function sfBriningCC accounts for three possible scenarios [64]: (1) no cross-contamination occurring in lots already contaminated ($Lot_{Pos}\_CC_{Neg}$); (2) cross-contamination occurring in lots already contaminated ($Lot_{Pos}\_CC_{Pos}$); and (3) cross-contamination occurring in lots that were not contaminated ($Lot_{Neg}\_CC_{Pos}$). The probability of each scenario taking place for every lot i is calculated as

$$P\left(Lot_{Pos}\_CC_{Neg\ i}\right) = Prob_{UnitPosi} \times (1 - Pcc_{brine})$$

$$P\left(Lot_{Pos}\_CC_{Pos\ i}\right) = Prob_{UnitPosi} \times Pcc_{brine}$$

$$P\left(Lot_{Neg}\_CC_{Pos\ i}\right) = (1 - Prob_{UnitPosi}) \times Pcc_{brine}$$

The status of every lot ($I = 1, 2, \ldots, r$) is then randomly sampled from the probabilities $\{P\left(Lot_{Pos}\_CC_{Neg\ i}\right),\ P\left(Lot_{Pos}\_CC_{Pos\ i}\right),\ P\left(Lot_{Neg}\_CC_{Pos\ i}\right)\}$. This strategy is adopted in order to maintain the dimension $[r, c_f]$ of the input contamination matrix. At the same time, the volume (mL) of brine solution injected in a fish fillet ($volInj_i$) and the concentration of

L. monocytogenes (CFU/mL) in contaminated brine solution ($conc_{Brine\ i}$) are sampled for every lot i from

$$volInj_i \sim Pert(volInj_{min}, volInj_{mode}, volInj_{max})$$

$$Conc_{Brine\ i} \sim Pert(Conc_{Brine\ min}, Conc_{Brine\ mode}, Conc_{Brine\ max})$$

where $volInj_{min}$, $volInj_{mode}$, and $volInj_{max}$ are, respectively, the minimum, most likely, and maximum parameters of the Pert distribution for brine volume injected, and $conc_{Brine\ min}$, $conc_{Brine\ mode}$, and $conc_{Brine\ max}$ are the minimum, most likely, and maximum parameters, respectively, of the Pert distribution for *L. monocytogenes* concentration in the salt solution. Values of 25, 35, and 100 mL for $volInj_{min}$, $volInj_{mode}$, and $volInj_{max}$ and 0, 0.0145, and 0.060 CFU/mL for $conc_{Brine\ min}$, $conc_{Brine\ mode}$, and $conc_{Brine\ max}$ were assumed as products of best guess.

Next, for each fillet, the potential number of cells transferred from the volume of brine (N_{transf}) is obtained from

$$N_{transf\ ij} \sim Poisson(volInj_i \times Conc_{Brine\ i})\ j = 1, 2 \ldots, c_f$$

Next, depending on the status the sampled lot i takes, $N_{transf\ i\cdot}$ is added to $N_{Hold\ i\cdot}$ (or not), in order to determine the numbers of *L. monocytogenes* in all fillets produced in lot i after brining (N_{Brine}), according to

$$N_{Brine\ i\cdot} = \begin{cases} N_{Hold\ i\cdot} & i \in Lot_{Pos}\_CC_{Neg} \\ N_{Hold\ i\cdot} + N_{transf\ i\cdot} & i \in Lot_{Pos}\_CC_{Pos} \\ N_{transf\ i\cdot} & i \in Lot_{Neg}\_CC_{Pos} \end{cases}$$

Finally, the probability vector of contaminated lots after brining, $Prob_{UnitPos\ Brine}$, and the mean prevalence of contaminated lots after brining, P_{Brine}, are updated, taking into account the probability that the brine solution is contaminated (Pcc_{Brine}).

$$Prob_{UnitPosBrinei} = 1 - (1 - Prob_{UnitPosi}) \times (1 - Pcc_{Brine})$$

$$P_{Brine} = 1 - (1 - P_{Hold}) \times (1 - Pcc_{Brine})$$

In summary, the inputs of the function sfBriningCC() are the outputs of sfRawFish-Storage(), Pcc_{Brine}, $volInj_{min}$, $volInj_{mode}$, $volInj_{max}$, $conc_{Brine\ min}$, $conc_{Brine\ mode}$, and $conc_{Brine\ max}$, whereas the outputs are the contamination matrix N_{Brine}, the vector $Prob_{UnitPos\ Brine}$, and the scalar P_{Brine}.

Appendix A.6. The Function sfSmearingCC

Since the contamination event is considered at the fillet level (probability of occurrence Pcc_{Smear}), two scenarios must be taken into account: (1) the increase in contamination of the already contaminated lots, which is a scenario that produces lots of status Lot_{Pos}_CC; and (2) the contamination of "clean" lots, which produces lots of status Lot_{Neg}_CC. Each scenario has its own probability of occurrence.

To represent scenario 1, a contamination status matrix, E_1, consisting of zeros (no contamination of the fish fillet during smearing) and ones (contamination of the fish fillet during smearing) is produced by evaluating the Bernoulli distribution at Pcc_{Smear},

$$E_{1ij} \sim Bernoulli(Pcc_{Smear})\ i = 1, 2 \ldots, r;\ j = 1, 2 \ldots, c_f$$

For the fish fillets $E_{1\ ij} = 1$, the load of *L. monocytogenes* cells (N_{TR1}) transferred to a fish fillet is obtained by

$$N_{TR1\ ij} \sim Binomial\left(N_{surface}, 10^{TR_{Smear\ i}}\right)\ j = 1, 2, \ldots c_f$$

where the $\log_{10}$ of the transfer coefficient $TR_{Smear\,i}$ is sampled for every lot i from a normal distribution with mean $\mu_{TRsmear} = -0.29$ and $\sigma_{TRSmear} = 0.31$ truncated on $]-\infty, 0]$ [43]. A value of $N_{surface}$ of 100 CFU *L. monocytogenes* is best guessed.

$$TR_{Smear\ i} \sim Normal\ (\mu_{TRsmear}, \sigma_{TRSmear}),\ TR_i > 0$$

Scenario 1 is finally represented by the sum of the initial contamination matrix, N_{Hold}, and the matrix of the newly contaminated fish fillets, N_{TR1}.

$$N_{1\ ij} = N_{Hold\ ij} + N_{TR1\ ij}$$

To represent scenario 2, a matrix of zeros N_0 (i.e., non-contaminated fish fillets) of dimensions $r_2 = \left\lceil r \times \left(\frac{1}{P} - 1\right)\right\rceil$ and $c_2 = c_f$ is first produced, and the same microbial transfer simulation on fish fillets, as explained before, is performed,

$$E_{2ij} \sim Bernoulli(Pcc_{Smear})\ i = 1,\ 2\ldots,\ r_2;\ j = 1,\ 2\ldots, c_2$$

$$TR_{Smear\ i} \sim Normal\ (-0.29,\ 0.31),\ TR_i > 0$$

$$N_{TR2\ ij} \sim Binomial\left(N_{surface}, 10^{TR_{Smear\ i}}\right)\ j = 1,\ 2,\ \ldots c_2$$

Scenario 2 is finally characterized by the sum of the "clean lots" matrix, N_0, and the matrix of the newly contaminated fish fillets, N_{TR2}.

$$N_{2\ ij} = N_{0\ ij} + N_{TR2\ ij}$$

However, it is possible that lots of fillets remain non-contaminated in the matrix N_2 (rows with zero counts). Those lots are identified and removed from N_2, and the proportion of clean (removed) lots estimated (π_{0N2}).

So far, two contamination matrices of different numbers of lots have been produced, N_1 and N_2, which occur at the probabilities P_1 and P_2, respectively.

$$P_1 = P_{Hold}$$

$$P_2 = (1 - P_{Hold}) \times (1 - \pi_{0N2})$$

For the function to return a joint contamination matrix output (N_{Smear}) of the same original size (r, c_f), the status of every lot, $i\ \{Lot_{Pos}\_CC,\ Lot_{Neg}\_CC\}$ $(i = 1,\ 2,\ \ldots,\ r)$, is then randomly sampled from the probabilities $\{P_1,\ P_2\}$.

$$N_{Smear\ i\cdot} = \begin{cases} N_{1\ i\cdot}\ i \in Lot_{Pos}\_CC \\ N_{2\ i\cdot}\ \ i \in Lot_{Neg}\_CC \end{cases} i = 1,\ 2,\ \ldots,\ c_f$$

The resulting matrix N_{Smear} contains the numbers of *L. monocytogenes* cells in c_f fish fillets (horizontal dimension) after a contamination event during smearing in a sample of r contaminated lots (vertical dimension). Finally, the probability vector of contaminated lots after smearing, $Prob_{UnitPos\ Smear}$, and the mean prevalence of contaminated lots after smearing, P_{Smear}, are computed, as follows:

$$Prob_{UnitPosSmeari} = Prob_{UnitPosi} + (1 - Prob_{UnitPosi}) \times (1 - \pi_{0N2})$$

$$P_{Smear} = P_1 + P_2$$

Summarizing, the function sfSmearingCC() is fed by the outputs of sfRawFishStorage(), Pcc_{Smear}, $N_{surface}$, $\mu_{TRSmear}$, and $\sigma_{TRSmear}$, whereas the outputs are the contamination matrix N_{Smear}, the vector $Prob_{UnitPos\ Smear}$, and the scalar P_{Smear}.

Appendix A.7. The Function sfBrineORsaltCC

The functions commence by constructing the lot status vector *Type* to assign the type of salting that will be applied to every lot i of the contamination matrix N_{Hold}. *Type* is coded 1 for brining and 0 for dry-salting,

$$Type_i \sim Bernoulli\ (p_{Brine})\ i = 1,\ 2,\ \ldots,\ r$$

The input contamination matrix N_{Hold} is partitioned into two contamination matrices (N_{Hold1} and N_{Hold0}), according to the type of salting. The same is carried out for the probability vector $Prob_{UnitPos}$ to produce two vectors, $Prob_{UnitPos1}$ and $Prob_{UnitPos0}$.

$$N_{Hold1} = N_{Hold\ i\cdot}\ \forall\, i \mid Type_i = 1$$

$$N_{Hold0} = N_{Hold\ i\cdot}\ \forall\, i \mid Type_i = 0$$

$$Prob_{UnitPos\ 1} = Prob_{UnitPos\ i}\ \forall\, i \mid Type_i = 1 Prob_{UnitPos\ 0} = Prob_{UnitPos\ i}\ \forall\, i \mid Type_i = 0$$

The auxiliary functions are then applied, according to the type of salting,

$$[N_{Brine},\ Prob_{UnitPosBrine}] = \text{sfBriningCC}\begin{pmatrix} N_{Hold\ 1},\ Prob_{UnitPos\ 1},\ P_{Hold},\ Pcc_{Smear}, N_{surface}, \\ volInj_{min}, volInj_{mode},\ volInj_{max}, Conc_{Brine\ min}, \\ Conc_{Brine\ mode},\ Conc_{Brine\ max} \end{pmatrix}$$

$$[N_{Smear},\ Prob_{UnitPosSmear}] = \text{sfSmearingCC}\begin{pmatrix} N_{Hold\ 0},\ Prob_{UnitPos\ 0},\ P_{Hold},\ Pcc_{brine}, N_{surface}, \\ \mu_{TRsmear},\ \sigma_{TRSmear} \end{pmatrix}$$

The contamination matrix after brining or salting, $N_{BrineORSalt}$, is arranged from the previous outputs, maintaining the index locations of *Type*,

$$N_{BrineORSalt\ i} = \begin{cases} N_{Brine\ i\cdot}\ \ \forall\, i \mid Type_i = 1 \\ N_{Smear\ i\cdot}\ \ \forall\, i \mid Type_i = 0 \end{cases} i = 1,\ 2,\ \ldots,\ r$$

In the same way, the probability vector of contaminated lots after brining or salting ($Prob_{UnitPos\ BrineORSalt}$) is then arranged from the $Prob_{UnitPos}$ outputs of the auxiliary functions, correcting them by p_{Brine}.

$$Prob_{UnitPos\ BrineORSalt\ i} = \begin{cases} Prob_{UnitPos\ Brine\ i} \times p_{Brine}\ \forall\, i \mid Type_i = 1 \\ Prob_{UnitPos\ Smear\ i} \times (1 - p_{Brine})\ \forall\, i \mid Type_i = 0 \end{cases} i = 1,\ 2,\ \ldots,\ r$$

The mean prevalence of contaminated lots after brining or salting ($P_{BrineORSalt}$) is computed as

$$P_{BrineORSalt} = 1 - \{p_{Brine} \times (1 - P_{Hold}) \times (1 - Pcc_{Brine}) + (1 - p_{Brine}) \times (1 - P_{Hold}) \times (1 - Pcc_{Smear})\}$$

The information contained in the vector *Type* is to be passed to the function sfSmoking(), since the extent of microbial reduction during smoking depends on the type of salting.

The inputs of the function sfBrineORsaltCC() are the probability of brining (p_{Brine}) and the inputs of the auxiliary functions sfBriningCC and/or sfSmearingCC, whereas the outputs are the contamination matrix $N_{BrineORSalt}$, the probability of contaminated lots ($Prob_{UnitPos\ BrineORSalt}$, a vector), the mean prevalence of contamination ($P_{BrineORSalt}$, a scalar), and the type of salting of the lots (*Type*, a vector).

Appendix A.8. The Function sfSmoking

The algorithm starts by sampling the extent of microbial $\log_{10}$ reduction $R_{Brine\ i}$ or $R_{Drysalt\ i}$ for every lot i, conditional to the type of salting the fish fillets of that lot were subjected to. This information is stored in r-length vector *Type*.

$$R_{Brine\ i} \sim Normal(\mu_{RBrine}, \sigma_{RBrine})\ R_{Brine\ i} > 0,\ \forall\, i \mid Type_i = \text{"1"}$$

$$R_{Drysalt\ i} \sim Normal\left(\mu_{RDrysalt}, \sigma_{RDrysalt}\right)\ R_{Brine\ i} > 0,\ \forall\, i \,\Big|\, Type_i = \text{"0"}$$

For every lot i, the probability of survival of a microbial cell, $p_{Survive\ i}$, is computed as

$$p_{Survive\ i} = \begin{cases} 10^{R_{Brine\ i}}\ \forall\, i \mid Type_i = \text{"1"} \\ 10^{R_{Dry\ salt\ i}}\ \forall\, i \,\Big|\, Type_i = \text{"0"} \end{cases}$$

and the number of surviving cells after smoking, N_{Smoked}, is obtained by

$$N_{Smoked\ ij} \sim Binomial\left(N_{BrineORSalt\ ij}, p_{Survive\ i}\right)\ j = 1, 2, \ldots, c_f$$

To update the prevalence estimates after smoking and maturation, the probability of at least one cell surviving in the lot i ($atLeastOne_i$) is calculated:

$$atLeastOne_i = 1 - (1 - p_{Survive\ i})^{\sum_{j=1}^{cf} N_{BrineORSalt\ ij}}$$

The probability vector of contaminated lots after smoking and maturation, $Prob_{UnitPos\ Smoked}$, and the mean prevalence of contaminated lots, P_{Smoked}, are then updated,

$$Prob_{UnitPosSmokedi} = Prob_{UnitPos\ BrineORSalti} \times atLeastOne_i$$

$$P_{Smoked} = P_{BrineORSalt} \times \frac{\sum_{i=1}^{r} atLeastOne_i}{r}$$

The function sfSmoking() returns the prevalences $Prob_{UnitPos\ Smoked}$, P_{Smoked}, and the contamination matrix N_{Smoked} containing the load of *L. monocytogenes* in the fish fillets originated from the contaminated lots.

Appendix A.9. The Function sfMaceration

The algorithm starts by estimating the probability of survival of a microbial cell, $p_{Survive\ i}$, in every lot i,

$$R_{Gravad\ i} \sim Normal(\mu_{RGravad}, \sigma_{RGravad})\ R_{Gravad\ i} > 0$$

$$p_{Survive\ i} = 10^{-R_{Gravad\ i}}$$

and the number of surviving cells after maceration or curing, $N_{Macerated}$, is obtained by

$$N_{Gravad\ ij} \sim Binomial(N_{Smear\ ij}, p_{Survive\ i})\ j = 1, 2, \ldots, c_f$$

The probability of at least one cell surviving in the lot i ($atLeastOne_i$) is calculated as

$$atLeastOne_i = 1 - (1 - p_{Survive\ i})^{\sum_{j=1}^{cf} N_{Smear\ ij}}$$

The probability vector of contaminated lots after maceration, $Prob_{UnitPos\ Gravad}$, and the mean prevalence of contaminated lots, P_{Gravad}, are then updated,

$$Prob_{UnitPosGravad\ i} = Prob_{UnitPos\ Smeari} \times atLeastOne_i$$

$$P_{Gravad} = P_{Smear} \times \frac{\sum_{i=1}^{r} atLeastOne_i}{r}$$

The function sfMaceration() returns the contamination matrix N_{Gravad} containing the load of *L. monocytogenes* in gravad fish fillets, the lot-specific probability of contaminated gravad fish fillets ($Prob_{UnitPos\ Gravad}$, a vector), and the mean prevalence of contaminated gravad fish fillets (P_{Gravad}, a scalar).

Appendix A.10. The Function sfTesting

The function sfTesting() receives the outputs of the function sfPackaging(): n, the number of tested units; g, the sub-sample weight in grams used for detection; c, the number of positive samples accepted (two-class or three-class mixed plan); M, the maximum limit concentration; $p_{lot\ tested}$, the proportion of tested lots; Se, the probability of the test (enumeration or detection) to detect, independently, each bacteria present in a sample; and $g_{TestedEnum}$, the sub-sample weight in grams used for the enumeration assay.

Assuming perfect homogenization of the sample, each of the bacterium present in each of the $r \times c_p$ units of $Unit_{SizePack}$ weight has a probability of being present in the g grams of the sub-sample *and* detected equal to $Se \times g/Unit_{SizePack}$. The number of bacteria detected in the detection assay is then

$$N_{Detected\ ij} \sim Binomial\left(N_{Pack\ ij}, Se \times g/Unit_{SizePack}\right)$$

and the detection test is positive if $N_{Detected\ ij} > 0$.

If the sampling plan is a three-class plan, an enumeration test (direct plating) is performed. The number of bacteria enumerated in the sample is

$$N_{Enumerated\ ij} \sim Binomial\left(N_{Pack\ ij}, Se \times g_{TestedEnum}/Unit_{SizePack}\right)$$

and the estimated concentration is $N_{Enumerated\ ij}/g_{TestedEnum}$ CFU/g. The algorithm assumes that the enumeration is performed only on samples positive in detection.

In order to evaluate the probability of each lot to be rejected, 1000 (by default) Monte Carlo samples of n samples are constituted for each lot and, for each of these Monte Carlo samples, the microbiological criteria is applied (i.e., in a two-class plan, the lot is rejected if $>c$ samples among n are detected, whereas in a three-class plan the lot is rejected if $>c$ samples among n are detected or if at least one sample has an estimated concentration $> M$ CFU/g). The mean number of times the lot is rejected among the Monte Carlo samples multiplied by the probability for the lot to be tested is an estimate of $P_{pos\ i}$, the probability of lot i being rejected.

Given the probability of lot i being contaminated pre-testing ($Prob_{UnitPos}$, which corresponds to $Prob_{UnitPos\ Smoked}$ for smoked fish slices, or to $Prob_{UnitPos\ Gravad}$ for gravad fish slices) and the probability for lot i to be rejected ($P_{pos\ i}$), the prevalence of contaminated lots given that they were not detected after testing ($Prob_{UnitPos\ tested\ i}$) is

$$Prob_{UnitPos\ tested\ i} = \frac{(1 - P_{pos\ i}) \times Prob_{UnitPos\ i}}{(1 - P_{pos\ i}) \times Prob_{UnitPos\ i} + (1 - Prob_{UnitPos\ i})}$$

The mean prevalence of contaminated lots after within-lot testing (P_{Tested}, a scalar) is therefore

$$P_{Tested} = \frac{\left(1 - \frac{\sum_{i=1}^{r} P_{pos\ i}}{r}\right) \times P_{Pack}}{\left(1 - \frac{\sum_{i=1}^{r} P_{pos\ i}}{r}\right) \times P_{Pack} + (1 - P_{Pack})}$$

Appendix A.11. The Function sfMejlholmDalgaard

The growth rate of *L. monocytogenes* (μ_{LM} [h^{-1}]) in RTE seafood is calculated as

$$\mu_{LM} = \mu_{LM\,ref} \times \gamma_T(T) \times \gamma_{a_w}(a_w) \times \gamma_{pH}(pH) \times \gamma_{LAC}([LAC]) \times \gamma_{Phe}(phe) \times \gamma_{nit}(nit)$$
$$\times \gamma_{CO2}([CO2]) \times \gamma_{DAC}([DAC]) \times \gamma_{AAC}([AAC]) \times \gamma_{BAC}([BAC]) \times \gamma_{CAC}([CAC]) \times \gamma_{SAC}([SAC])$$
$$\times \xi(T, a_w, pH, [LAC], phe, nit, [CO2], [DAC], [AAC], [BAC], [CAC], [SAC])_{int}$$

where $\mu_{LM\,ref}$ (=0.419 h^{-1}) is the optimum growth rate of *L. monocytogenes* in the RTE seafood at the reference temperature (T_{ref}) of 25 °C, $\gamma_x(x)$ are the cardinal terms for the independent effect of the environmental parameter x, and $\xi()$ is the term describing the effect of interactions between environmental parameters. Each of the terms are calculated as follows:

$$\gamma_T(T) = \begin{cases} \left(\frac{T-(-2.83)}{25-(-2.83)}\right)^2 & if\ T > -2.83°C \\ 0 & if\ T \leq -2.83°C \end{cases}$$

where T is the temperature in °C. The value of −2.83 belongs to *L. monocytogenes* minimum temperature for growth [53].

$$\gamma_{a_w}(a_w) = \begin{cases} \frac{(a_w-0.923)}{(1-0.923)} & if\ a_w > 0.923 \\ 0 & if\ a_w \leq 0.923 \end{cases}$$

where a_w is the water activity of the RTE seafood product. The minimum a_w for *L. monocytogenes* growth is 0.923 [53]. The value of a_w can be estimated from the percentage of salt content [NaCl] [%].

$$a_w = 0.999489 - 0.005179 \times [NaCl] - 0.0001272 \times [NaCl]^2$$

$$\gamma_{pH}(pH) = \begin{cases} 1 - 10^{(4.97-pH)} & if\ pH > 4.97 \\ 0 & if\ pH \leq 4.97 \end{cases}$$

where pH is the pH of the RTE seafood product. The value of 4.97 corresponds to *L. monocytogenes* minimum pH for growth [53].

$$\gamma_{LAC}([LAC_U]) = \begin{cases} \left(1 - \frac{[LAC_U]}{3.79}\right) & if\ [LAC_U] < 3.79\ mM \\ 0 & f\ [LAC_U] \geq 3.79\ mM \end{cases}$$

where $[LAC_U]$ is the concentration (mM) of undissociated lactic acid and the value of 3.79 belongs to the minimum inhibitory concentration (MIC) of LAC_U. $[LAC_U]$ can be estimated from the concentration of total lactic acid (LA_{tot}) in ppm,

$$[LAC_U] = \frac{LA_{tot}}{90.08 \times (1 + 10^{pH-3.86})}$$

$$\gamma_{Phe}(Phe) = \begin{cases} \left(\frac{32-Phe}{32}\right) & if\ PHE < 32\ ppm \\ 0 & if\ PHE \geq 32\ ppm \end{cases}$$

where *Phe* is the concentration of phenol compound (ppm). The value of 32 ppm belongs to the maximum *Phe* for *L. monocytogenes* growth [53].

$$\gamma_{nit}(nit) = \begin{cases} \left(\frac{350-nit}{350}\right)^2 & if\ nit < 350\ ppm \\ 0 & if\ nit \geq 350\ ppm \end{cases}$$

where *nit* is the concentration of nitrites (ppm). The maximum *nit* for *L. monocytogenes* growth is 350 ppm [53].

$$\gamma_{CO2}([CO2]) = \begin{cases} \left(\frac{3140-[CO2]}{3140}\right) & if\ [CO2] < 3140\ mM \\ 0 & if\ [CO] \geq 3140\ mM \end{cases}$$

where $[CO2]$ is the concentration of CO_2 in mM, and the value of 3140 mM belongs to the maximum $[CO2]$ for *L. monocytogenes* growth [53]. The value of $[CO2]$ can be estimated from the CO_2 concentration at equilibrium in % ($CO2_{equi}$),

$$[CO2] = \frac{CO2_{equi} \times 101,323 \times 2.4429}{\exp\left(-6.8346 + \frac{12,817}{T+273.15} - \frac{3,766,800}{(T+273.15)^2} + \frac{2.997E8}{(T+273.15)^3}\right)}$$

$$\gamma_{DAC}([DAC_U]) = \begin{cases} \left(1 - \sqrt{\frac{[DAC_U]}{4.8}}\right) & if\ [DAC_U] < 4.8\text{ mM} \\ 0 & f\ [DAC_U] \geq 4.8\text{ mM} \end{cases}$$

where $[DAC_U]$ is the concentration of undissociated diacetate (mM). The minimum inhibitory concentration of DAC_U is 4.8 mM [53]. $[DAC_U]$ can be estimated from the total diacetate DA_{tot} (in ppm),

$$[DAC_U] = \frac{DA_{tot}}{119.1 \times (1 + 10^{pH-4.76})}$$

$$\gamma_{AAC}([AAC_U]) = \begin{cases} \left(1 - \sqrt{\frac{[AAC_U]}{10.3}}\right) & if\ [AAC_U] < 10.3\text{ mM} \\ 0 & if\ [AAC_U] \geq 10.3\text{ mM} \end{cases}$$

where $[AAC_U]$ is the concentration of undissociated acetic acid (mM). The value of 10.3 mM corresponds to the minimum inhibitory concentration of AAC_U [53]. $[AAC_U]$ can be estimated from the total acetic acid (AA_{tot}) in ppm,

$$[AAC_U] = \frac{AA_{tot}}{60.05 \times (1 + 10^{pH-4.76})}$$

$$\gamma_{BAC}([BAC_U]) = \begin{cases} \left(1 - \frac{[BAC_U]}{0.349}\right) & if\ [BAC_U] < 0.349\text{ mM} \\ 0 & if\ [BAC_U] \geq 0.349\text{ mM} \end{cases}$$

where $[BAC_U]$ is the concentration of undissociated benzoic acid (mM). The minimum inhibitory concentration of BAC_U is 0.349 mM [53]. $[BAC_U]$ can be estimated from the total BA (BA_{tot}) in ppm,

$$[BAC_U] = \frac{BA_{tot}}{122.12 \times (1 + 10^{pH-4.19})}$$

$$\gamma_{CAC}([CAC_U]) = \begin{cases} \left(1 - \frac{[CAC_U]}{2.119}\right) & if\ [CAC_U] < 2.119\text{ mM} \\ 0 & if\ [CAC_U] \geq 2.119\text{ mM} \end{cases}$$

where $[CAC_U]$ is the concentration of undissociated citric acid (mM). The minimum inhibitory concentration of CAC_U is 2.119 mM [53]. $[CAC_U]$ can be computed from the total CA (CA_{tot}) in ppm,

$$[CAC_U] = \frac{CA_{tot}}{192.13 \times (1 + 10^{pH-3.13})}$$

$$\gamma_{SAC}([SAC_U]) = \begin{cases} \left(1 - \frac{[SAC_U]}{1.896}\right) & if\ [SAC_U] < 1.896\text{ mM} \\ 0 & if\ [SAC_U] \geq 1.896\text{ mM} \end{cases}$$

where $[SAC_U]$ is the concentration of undissociated sorbic acid (mM). The minimum inhibitory concentration of SAC_U is 1.896 mM [53]. $[SAC_U]$ can be estimated from the total SA (SA_{tot}) in ppm,

$$[SAC_U] = \frac{SA_{tot}}{112.1 \times (1 + 10^{pH-4.76})}$$

As for the interaction term, Mejlholm and Dalgaard [52] use the Le Marc [77] approach,

$$\xi\{\phi(T,a_w,pH,nit,phe,\ CO2,acids)\} = \begin{cases} 1 & if\ \psi \le 0.5 \\ 2(1-\psi) & if\ 0.5 < \psi < 1 \\ 0 & if\ \psi \ge 1 \end{cases}$$

where

$$\psi = \sum_i \left(\frac{\phi(i)}{2\prod_{j\neq i}(1-\phi(j))} \right)$$

and

$$\phi(T) = \left(1 - \tfrac{T+2.83}{25+2.83}\right)^2$$
$$\phi(a_w) = \left(1 - \sqrt{\tfrac{a_w-0.923}{1-0.923}}\right)^2$$
$$\phi(pH) = \left(1 - \sqrt{1-10^{(4.97-pH)}}\right)^2$$
$$\phi(phe) = \left(1 - \sqrt{\tfrac{32-Phe}{32}}\right)^2$$
$$\phi(nit) = \left(1 - \tfrac{350-nit}{350}\right)^2$$
$$\phi(CO2) = \left(1 - \sqrt{\tfrac{3140-CO2}{3140}}\right)^2$$

$$\phi(acids) = (1 - \left(1-\sqrt{\tfrac{[LAC_U]}{3.79}}\right) \times \left(1-\sqrt{\tfrac{[DAC_U]}{4.80}}\right) \times \left(1-\sqrt{\tfrac{[AAC_U]}{10.3}}\right) \times \left(1-\tfrac{[BAC_U]}{0.349}\right) \times \left(1-\tfrac{[CAC_U]}{2.119}\right) \times \left(1-\tfrac{[SAC_U]}{1.896}\right))^2$$

Appendix A.12. The Function sfMejlholmDalgaardLAB

The growth rate of lactic acid bacteria (μ_{LAB} [h^{-1}]) in RTE seafood is computed as,

$$\mu_{LAB} = \mu_{LAB\ ref} \times \gamma_T(T) \times \gamma_{a_w}(a_w) \times \gamma_{pH}(pH) \times \gamma_{LAC}([LAC]) \times \gamma_{Phe}(phe) \times \gamma_{nit}(nit)$$
$$\gamma_{CO2}([CO2]) \times \gamma_{DAC}([DAC]) \times \gamma_{AAC}([AAC]) \times \gamma_{BAC}([BAC]) \times \gamma_{CAC}([CAC]) \times \gamma_{SAC}([SAC])$$
$$\times \xi(T,a_w,pH,[LAC],phe,nit,[CO2],[DAC],[AAC],[BAC],[CAC],[SAC])_{int}$$

where $\mu_{LAB\ ref}$ (=0.583 h^{-1}) is the optimum growth rate of LAB in the RTE seafood at the reference temperature (T_{ref}) of 25 °C. Each of the cardinal terms, $\gamma_x(x)$, and the interaction, $\xi()$, are calculated in a similar way as for the *L. monocytogenes* growth rate equation, yet with cardinal parameters specific for LAB [54].

$$\gamma_T(T) = \begin{cases} \left(\frac{T-(-5.25)}{25-(-5.25)}\right)^2 & if\ T > -5.25\ °C \\ 0 & if\ T \le -5.25\ °C \end{cases}$$

where T is the temperature in °C.

$$\gamma_{a_w}(a_w) = \begin{cases} \frac{(a_w-0.928)}{(1-0.928)} & if\ a_w > 0.928 \\ 0 & if\ a_w \le 0.928 \end{cases}$$

where a_w is the water activity of the RTE seafood product.

$$\gamma_{pH}(pH) = \begin{cases} 1-10^{(4.24-pH)} & if\ pH > 4.24 \\ 0 & if\ pH \le 4.24 \end{cases}$$

where pH is the pH of the RTE seafood product.

$$\gamma_{LAC}([LAC_U]) = \begin{cases} \left(1 - \frac{[LAC_U]}{12}\right) & if\ [LAC_U] < 12 \text{ mM} \\ 0 & f\ [LAC_U] \geq 12 \text{ mM} \end{cases}$$

where $[LAC_U]$ is the concentration (mM) of undissociated lactic acid.

$$\gamma_{Phe}(Phe) = \begin{cases} \left(\frac{40.3 - Phe}{40.3}\right) & if\ PHE < 40.3 \text{ ppm} \\ 0 & if\ PHE \geq 40.3 \text{ ppm} \end{cases}$$

where *Phe* is the concentration of phenol compound (ppm).

$$\gamma_{nit}(nit) = \begin{cases} \left(\frac{2780 - nit}{2780}\right)^2 & if\ nit < 2780 \text{ ppm} \\ 0 & if\ nit \geq 2780 \text{ ppm} \end{cases}$$

where *nit* is the concentration of nitrites (ppm).

$$\gamma_{CO2}([CO2]) = \begin{cases} \left(\frac{6691 - [CO2]}{6691}\right) & if\ [CO2] < 6691 \text{ mM} \\ 0 & if\ [CO] \geq 6691 \text{ mM} \end{cases}$$

where *[CO2]* is the concentration of CO_2 in mM.

$$\gamma_{DAC}([DAC_U]) = \begin{cases} \left(1 - \sqrt{\frac{[DAC_U]}{33.3}}\right) & if\ [DAC_U] < 33.3 \text{ mM} \\ 0 & f\ [DAC_U] \geq 33.3 \text{ mM} \end{cases}$$

where $[DAC_U]$ is the concentration of undissociated diacetate (mM).

$$\gamma_{AAC}([AAC_U]) = \begin{cases} \left(1 - \sqrt{\frac{[AAC_U]}{10.3}}\right) & if\ [AAC_U] < 10.3 \text{ mM} \\ 0 & if\ [AAC_U] \geq 10.3 \text{ mM} \end{cases}$$

where $[AAC_U]$ is the concentration of undissociated acetic acid (mM).

$$\gamma_{BAC}([BAC_U]) = \begin{cases} \left(1 - \frac{[BAC_U]}{1.51}\right)^2 & if\ [BAC_U] < 1.51 \text{ mM} \\ 0 & if\ [BAC_U] \geq 1.51 \text{ mM} \end{cases}$$

where $[BAC_U]$ is the concentration of undissociated benzoic acid (mM).

$$\gamma_{CAC}([CAC_U]) = \begin{cases} \left(1 - \frac{[CAC_U]}{10.3}\right)^2 & if\ [CAC_U] < 10.3 \text{ mM} \\ 0 & if\ [CAC_U] \geq 10.3 \text{ mM} \end{cases}$$

where $[CAC_U]$ is the concentration of undissociated citric acid (mM).

$$\gamma_{SAC}([SAC_U]) = \begin{cases} \left(1 - \frac{[SAC_U]}{12.6}\right)^2 & if\ [SAC_U] < 12.6 \text{ mM} \\ 0 & if\ [SAC_U] \geq 12.6 \text{ mM} \end{cases}$$

where $[SAC_U]$ is the concentration of undissociated sorbic acid (mM).

The interaction term is determined as

$$\xi\{\phi(T, a_w, pH, nit, phe,\ CO2, acids)\} = \begin{cases} 1 & if\ \psi \leq 0.5 \\ 2(1-\psi) & if\ 0.5 < \psi < 1 \\ 0 & if\ \psi \geq 1 \end{cases}$$

where

$$\psi = \sum_i \left(\frac{\phi(i)}{2\prod_{j\neq i}(1-\phi(j))} \right)$$

and

$$\phi(T) = \left(1 - \frac{T+5.25}{25+5.25}\right)^2$$
$$\phi(a_w) = \left(1 - \sqrt{\frac{a_w-0.928}{1-0.928}}\right)^2$$
$$\phi(pH) = \left(1 - \sqrt{1-10^{(4.24-pH)}}\right)^2$$
$$\phi(phe) = \left(1 - \sqrt{\frac{40.3-Phe}{40.3}}\right)^2$$
$$\phi(nit) = \left(1 - \frac{2780-nit}{2780}\right)^2$$
$$\phi(CO2) = \left(1 - \sqrt{\frac{6691-CO2}{6691}}\right)^2$$
$$\phi(acids) = (1 - \left(1-\sqrt{\frac{[LAC_U]}{12}}\right) \times \left(1-\sqrt{\frac{[DAC_U]}{33.3}}\right) \times \left(1-\sqrt{\frac{[AAC_U]}{10.3}}\right) \times \left(1-\frac{[BAC_U]}{1.51}\right) \times \left(1-\frac{[CAC_U]}{10.3}\right) \times \left(1-\frac{[SAC_U]}{12.6}\right))^2$$

Appendix A.13. The Function sfGrowthJameson

The inputs of sfGrowthJameson() are the time (t in h), the initial numbers of *L. monocytogenes* and LAB in the food unit ($N_{0\,LM}$, $N_{0\,LAB}$ in CFU), the initial values of the ideal substance for *L. monocytogenes* and LAB ($q_{0\,LM}$, $q_{0\,LAB}$), the maximum population densities of *L. monocytogenes* and LAB (MPD_{LM}, MPD_{LAB} in $\log_{10}$ CFU/g), the food unit's weight ($Unit_{Size}$ in g), and the growth rates of *L. monocytogenes* and LAB (μ_{LM} and μ_{LAB} in h^{-1}), as determined by the functions sfMejlholmDalgaard() and sfMejlholmDalgaardLAB(), respectively.

The following system of differential equations is integrated from time = 0 to time = t,

$$\frac{1}{N_{LM}(t)} \times \frac{dN_{LM}(t)}{dt} = \left(\frac{q_{LM}(t)}{1+q_{LM}(t)}\right) \times \mu_{LM} \times \left(1 - \frac{N_{LM}}{Unit_{Size}\times 10^{MPD_{LM}}}\right) \times \left(1 - \frac{\gamma \times N_{LAB}}{Unit_{Size}\times 10^{MPD_{LAB}}}\right)$$
$$\frac{dq_{LM}(t)}{dt} = \mu_{LM} \times q_{LM}(t)$$
$$\frac{1}{N_{LAB}(t)} \times \frac{dN_{LAB}(t)}{dt} = \left(\frac{q_{LAB}(t)}{1+q_{LAB}(t)}\right) \times \mu_{LAB} \times \left(1 - \frac{N_{LM}}{Unit_{Size}\times 10^{MPD_{LM}}}\right) \times \left(1 - \frac{N_{LAB}}{Unit_{Size}\times 10^{MPD_{LAB}}}\right)$$
$$\frac{dq_{LAB}(t)}{dt} = \mu_{LAB} \times q_{LAB}(t)$$

setting $N_{0\,LM}$, $N_{0\,LAB}$, $q_{0\,LM}$, and $q_{0\,LAB}$ as the initial values. The outputs of the function are the numbers of *L. monocytogenes* and LAB in the food unit at time t ($N_{1\,LM}$, $N_{1\,LAB}$ in CFU) and the natural logarithm of q values at time t for *L. monocytogenes* and LAB ($lnq_{t\,LM}$, $lnq_{t\,LAB}$). In the R script, this function is vectorized for all parameters, meaning that it is prepared to receive and return vectors of simulated data. The function was written in C++.

Appendix A.14. The Function sfCharacteristics

The function sfCharacteristics() samples, at the lot level, the environmental (intrinsic and extrinsic) characteristics of the RTE seafood product, acting therefore as a feeder function for the sfMejlholmDalgaard() and sfMejlholmDalgaardLAB() functions to estimate the specific growth rates of *L. monocytogenes* and LAB, respectively, in the RTE seafood. The sampling of the environmental characteristics of the RTE seafood product assumes that all the units produced in a lot have the same characteristics. Furthermore, since the functions sfMejlholmDalgaard() and sfMejlholmDalgaardLAB() are deterministic, the specific growth rates of *L. monocytogenes* and LAB calculated thereof are the same for all the units produced in a lot.

For every lot i, the RTE seafood characteristics are sampled from multiple Pert distributions defined by minimum (x_{min}), mode (x_{mode}), and maximum (x_{max}) values of the given characteristic x. The lot-specific characteristics are sampled as follows:
$a_{w\,i} \sim Pert\,(a_{w\,min}, a_{w\,mode}, a_{w\,max})$ or $NaCl_i \sim Pert\,(NaCl_{min}, NaCl_{mode}, NaCl_{max})$,
$pH_i \sim Pert\,(pH_{min}, pH_{mode}, pH_{max})$, $LA_{tot\,i} \sim Pert\,(LA_{tot\,min}, LA_{tot\,mode}, LA_{tot\,max})$,
$Phe_i \sim Pert\,(Phe_{min}, Phe_{mode}, Phe_{max})$, $nit_i \sim Pert\,(nit_{min}, nit_{mode}, nit_{max})$,
$CO2_{equi\,i} \sim Pert\left(CO2_{equi\,min}, CO2_{equi\,mode}, CO2_{equi\,max}\right)$,
$AA_{tot\,i} \sim Pert\,(AA_{tot\,min}, AA_{tot\,mode}, AA_{tot\,max})$,
$DA_{tot\,i} \sim Pert\,(DA_{tot\,min}, DA_{tot\,mode}, DA_{tot\,max})$,
$BA_{tot\,i} \sim Pert\,(BA_{tot\,min}, BA_{tot\,mode}, BA_{tot\,max})$,
$CA_{tot\,i} \sim Pert\,(CA_{tot\,min}, CA_{tot\,mode}, CA_{tot\,max})$, and
$SA_{tot\,i} \sim Pert\,(SA_{tot\,min}, SA_{tot\,mode}, SA_{tot\,max})$,

The inputs of the function sfCharacteristics() are the number of lots, r, and the Pert parameters defining the environmental characteristics of RTE seafoods. The values of the Pert parameters used are different for smoked fish and gravad fish, and they are compiled in Table 4. The function's output is a list of vectors of length r for a_w (or $NaCl$), pH, LA_{tot}, Phe, nit, $CO2_{equi}$, AA_{tot}, DA_{tot}, BA_{tot}, CA_{tot}, and SA_{tot}.

Appendix A.15. The Function sfColdChain

The inputs of sfColdChain() are the inputs and parameters of all auxiliary functions (sfMejlholmDalgaard(), sfMejlholmDalgaardLAB(), sfGrowthJameson(), and sfCharacteristics()), whose values are passed, accordingly. Although the cardinal parameters of all environmental factors are pre-established as default in the functions sfMejlholmDalgaard() and sfMejlholmDalgaardLAB() for both *L. monocytogenes* and LAB, respectively, it is still possible to alter any of them by providing the function sfColdChain() with the updated parameter(s).

The function sfCharacteristics() is first applied to sample the lot-specific environmental characteristics of the RTE seafood product for all lots r,

$$\{NaCl_i, pH_i, LA_{tot\,i}, Phe_i, nit_i, CO2_{equi\,i}, AA_{tot\,i}, DA_{tot\,i}, BA_{tot\,i}, CA_{tot\,i}, SA_{tot\,i}\} = sfCharacteristics\\ (r,\ NaCl_{min}, NaCl_{mode}, NaCl_{max}, pH_{min}, pH_{mode}, pH_{max}, LA_{tot\,min}, LA_{tot\,mode}, LA_{tot\,max}, Phe_{min}, Phe_{mode}, \ldots\\ nit_{min}, nit_{mode}, nit_{max}, CO2_{equi\,min}, CO2_{equi\,mode}, CO2_{equi\,max}, AA_{tot\,min}, AA_{tot\,mode}, AA_{tot\,max}, DA_{tot\,min} \cdots\\ DA_{tot\,mode}, DA_{tot\,max}, BA_{tot\,min}, BA_{tot\,mode}, BA_{tot\,max}, CA_{tot\,min}, CA_{tot\,mode}, CA_{tot\,max}, SA_{tot\,min}, SA_{tot\,mode}, SA_{tot\,max})$$

Then, the initial physiological state parameter for *L. monocytogenes*, $h_{0\,LM\,i}$, is sampled for every lot i from a normal distribution with mean μ_{h0} = 2.8 and standard deviation σ_{h0} = 4.6 (assumed from Couvert et al. [71]; Table 4) and subsequently converted to $q_{0\,LM\,i}$ to be used as a parameter of the Baranyi–Roberts growth model.

$$h_{0\,LM\,i} \sim Normal\,(\mu_{h0}, \sigma_{h0})\ h_{0\,LM\,i} > 0;\ i = 1, 2, \ldots, r$$

$$q_{0\,LM\,i} = log\left(\frac{1}{exp(h_{0\,LM\,i})} - 1\right)$$

The log-transformed parameter related to the initial physiological state of LAB cells (ln $q_{0\,LAB\,i}$) is sampled from a Pert distribution, whose parameters $ln\ q_{0\,LABmin} = -12$, $ln\ q_{0\,LABmode}$ = 2.73, and $ln\ q_{0\,LABmax}$ = 1.26 were estimated from Couvert et al. [71] (Table 4). Thus, for every lot i, $q_{0\,LAB\,i}$ is obtained as

$$q_{0\,LAB\,i} \sim exp\{Pert\,(\ln q_{0\,LABmin}, \ln q_{0\,LABmode}, \ln q_{0\,LABmax})\}\ i = 1, 2, \ldots, r$$

The maximum population density of *L. monocytogenes* (MPD_{LM}) and that of LAB MPD_{LAB}) in RTE seafood are sampled for every lot i from Pert distributions, whose parameters $MPD_{LM\,min}$ (6.60 $\log_{10}$ CFU/g), $MPD_{LM\,mode}$ (7.36 $\log_{10}$ CFU/g), $MPD_{LM\,max}$ (8.20 $\log_{10}$

CFU/g), $MPD_{LAB\ min}$ (8.0 $\log_{10}$ CFU/g), $MPD_{LAB\ mode}$ (8.5 $\log_{10}$ CFU/g), and $MPD_{LAB\ max}$ (9.0 $\log_{10}$ CFU/g) were obtained from Pérez-Rodríguez et al. [24] and Mejlholm and Dalgaard [54], respectively (Table 4).

$$MPD_{LM\ i} \sim Pert\ (MPD_{LM\ min}, MPD_{LM\ mode}, MPD_{LM\ max})\ i = 1,\ 2,\ \ldots,\ r$$

$$MPD_{LAB\ i} \sim Pert\ (MPD_{LAB\ min}, MPD_{LAB\ mode}, MPD_{LAB\ max})\ i = 1,\ 2,\ \ldots,\ r$$

The mean concentration of LAB in RTE seafood after packaging ($\overline{C}_{0\ LAB\ i}$, $\log_{10}$ CFU/g) is sampled for every lot i from a Pert distribution with the parameters of minimum ($\overline{C}_{0\ LAB\ min}$), mode ($\overline{C}_{0\ LAB\ mode}$), and maximum ($\overline{C}_{0\ LAB\ max}$). These values were assumed the same for smoked fish and gravad fish and are compiled in Table 4.

$$\overline{C}_{0\ LAB\ i} \sim Pert\ (\overline{C}_{0\ LAB\ min}, \overline{C}_{0\ LAB\ mode}, \overline{C}_{0\ LAB\ max})\ i = 1,\ 2,\ \ldots,\ r$$

If information is provided on the within-lot variability in the concentration of LAB (within-lot standard deviation, $\sigma_{C0\ LAB\ wl}$), the numbers of LAB in the RTE seafood units ($N_{0\ LAB}$) are estimated as

$$N_{0\ LAB\ ij} \sim 10^{Normal\ (\overline{C}_{0\ LAB\ i},\ \sigma_{C0\ LAB\ wl})}\ i = 1,\ 2,\ \ldots,\ r;\ j = 1,\ 2,\ \ldots,\ c_p$$

In the present QRA model, $\sigma_{C0\ LAB\ wl}$ was set to zero. Next, the lot-specific cold chain time ($time_{cc}$) and temperature ($Temp_{cc}$) are sampled. If the correlation between time and temperature during the cold chain ($CorTimeTemp_{cc}$) is different from zero, $time_{cc}$ and $Temp_{cc}$ are sampled for every lot i from their respective Pert distributions, targeting the rank correlation value of $CorTimeTemp_{cc}$ (function cornode from the mc2d package; [78]). Else, $time_{cc}$ and $Temp_{cc}$ are independently sampled.

$$\begin{aligned} time_{cc\ i} &\sim Pert\ (time_{cc\ min}, time_{cc\ mode}, time_{cc\ max})\ i = 1,\ 2,\ \ldots,\ r \\ Temp_{cc\ i} &\sim Pert\ (Temp_{cc\ min}, Temp_{cc\ mode}, Temp_{cc\ max})\ i = 1,\ 2,\ \ldots,\ r \end{aligned}$$

At this point, the cold chain auxiliary microbial kinetic functions can be used. The growth rates of *L. monocytogenes* ($\mu_{LM\ i}$) and LAB ($\mu_{LAB\ i}$) in RTE seafood for every lot i are computed by the functions

$$\mu_{LM\ i} = sfMejlholmDalgaard(Temp_{cc\ i},\ NaCl_i,\ pH_i, LA_{tot\ i}, Phe_i, nit_i, CO2_{equi\ i}, AA_{tot\ i}, DA_{tot\ i}, BA_{tot\ i}, CA_{tot\ i}, SA_{tot\ i})\quad i = 1,\ 2,\ \ldots, r$$

$$\mu_{LAB\ i} = sfMejlholmDalgaardLAB\left(Temp_{cc\ i},\ NaCl_i,\ pH_i, LA_{tot\ i}, Phe_i, nit_i, CO2_{equi\ i}, AA_{tot\ i}, DA_{tot\ i}, BA_{tot\ i}, CA_{tot\ i}, SA_{tot\ i}\right)\quad i = 1,\ 2,\ \ldots, r$$

Finally, the numbers of *L. monocytogenes* and LAB in the RTE seafood units after $time_{CC}$ ($N_{cc\ LM}$, $N_{cc\ LAB}$ in CFU) and the natural logarithm of q values at time t for *L. monocytogenes* and LAB ($lnq_{cc\ LM}$, $lnq_{cc\ LAB}$) are obtained using the function sfGrowthJameson(),

$$\begin{aligned} &\left\{N_{cc\ LM\ ij},\ N_{cc\ LAB\ ij},\ \ln q_{cc\ LM\ ij}, \ln q_{cc\ LAB\ ij}\right\} \\ &= sfGrowthJameson\left(N_{Pack\ ij},\ N_{0\ LAB\ ij},\ time_{cc\ i},\ q_{0\ LM\ i},\ q_{0\ LAB\ i},\ \mu_{LM\ i},\ \mu_{LAB\ i},\ MPD_{LM\ i}, MPD_{LAB\ i}, \gamma, Unit_{SizePack}\right) \\ &i = 1,\ 2,\ \ldots,\ r;\ j = 1,\ 2 \ldots,\ c_p \end{aligned}$$

The lot-specific vectors MPD_{LM} and MPD_{LAB} are also returned.

Appendix A.16. The Function sfPortioning

The number of *L. monocytogenes* ($N_{Portion\ ij}$, [CFU]) in a portion of RTE seafood ($Serv_{size}$ [g]) taken from pack j produced in lot i is sampled from a Beta-binomial distribution,

$$N_{Portion\ ij} \sim Binomial(N_{Home\ LM\ ij}, Beta(b, b(n_{serv} - 1)))$$

where b is the dispersion factor of the Beta distribution and n_{serv} is the rounded number of servings from a pack of RTE seafoods,

$$n_{serv} = \left\lceil \frac{Unit_{SizePack}}{Serv_{size}} \right\rceil$$

The inputs of the function are the *L. monocytogenes* load in the RTE seafood pack unit at the time of consumption, $N_{Home\ LM}$, the net weight of the pack, $Unit_{SizePack}$, and the serving size, $Serv_{size}$ (which was set to be equal to one slice, w_{Slice} = 32.5 g), and the output is the contamination matrix $N_{Portion}$, containing the dose of *L. monocytogenes* in a serving. In addition, the function sfPortioning() returns the mean prevalence of contaminated lots (P_{Tested}, a scalar) and the lot-specific probabilities of contaminated lots ($Prob_{UnitPos\ Tested}$, a vector) unchanged.

References

1. Dass, S.C.; Cummins, E.J.; Abu-Ghannam, N. Prevalence and typing of *Listeria monocytogenes* strains in retail vacuum-packed cold-smoked salmon in the Republic of Ireland. *J. Food Saf.* **2010**, *31*, 21–27. [CrossRef]
2. Domenech, E.; Amoros, J.A.; Martorell, S.; Escriche, I. Safety assessment of smoked fish related to *Listeria monocytogenes* prevalence using risk management metrics. *Food Control* **2012**, *25*, 233–238. [CrossRef]
3. Jami, M.; Ghanbari, M.; Zunabovic, M.; Domig, K.J.; Kneifel, W. *Listeria monocytogenes* in aquatic food products—A review. *Compr. Rev. Food Sci. Food Saf.* **2014**, *13*, 798–813. [CrossRef]
4. Klaeboe, H.; Rosef, O.; Saebo, M. Longitudinal studies on *Listeria monocytogenes* and other *Listeria* species in two salmon processing plants. *Int. J. Environ. Health Res.* **2005**, *15*, 71–77. [CrossRef] [PubMed]
5. Rotariu, O.; Thomas, D.J.I.; Goodburn, K.E.; Hutchison, M.L.; Strachan, N.J.C. Smoked salmon industry practices and their association with *Listeria monocytogenes*. *Food Control* **2014**, *35*, 284–292. [CrossRef]
6. Szymczak, B.; Szymczak, M.; Trafiałek, J. Prevalence of *Listeria* species and *L. monocytogenes* in ready-to-eat foods in the West Pomeranian region of Poland: Correlations between the contamination level, serogroups, ingredients, and producers. *Food Microbiol.* **2020**, *91*, 103532. [CrossRef]
7. Skjerdal, T.; Reitehaug, E.; Eckner, K. Development of performance objectives for *Listeria monocytogenes* contaminated salmon (*Salmo salar*) intended used as sushi and sashimi based on analysis of naturally contaminated samples. *Int. J. Food Microbiol.* **2014**, *184*, 8–13. [CrossRef]
8. Félix, B.; Sevellec, Y.; Palma, F.; Douarre, P.E.; Felten, A.; Radomski, N.; Mallet, L.; Blanchard, Y.; Leroux, A.; Soumet, C.; et al. A European-wide dataset to uncover adaptive traits of *Listeria monocytogenes* to diverse ecological niches. *Sci. Data* **2022**, *9*, 190. [CrossRef]
9. EFSA. The European Union One Health 2022 Zoonoses Report. *EFSA J.* **2023**, *21*, e8442. [CrossRef]
10. Haas, C.N.; Rose, J.B.; Gerba, C.P. *Quantitative Microbial Risk Assessment*; Wiley: New York, NY, USA, 1999.
11. EFSA. The public health risk posed by *Listeria monocytogenes* in frozen fruit and vegetables including herbs, blanched during processing. EFSA Panel of Biological Hazards (BIOHAZ). *EFSA J.* **2020**, *8*, 6092. [CrossRef]
12. ECDC. Surveillance and Disease Data for Listeriosis. *European Centre for Disease Prevention and Control*. Available online: https://www.ecdc.europa.eu/en/all-topics-z/listeriosis/surveillance-and-disease-data/eu-summary-reports (accessed on 15 June 2022).
13. Lachman, R.; Halbedel, S.; Luth, S.; Holzer, A.; Adier, M.; Pietzka, A.; Al Dahouk, S.; Stark, K.; Flieger, A.; Kieta, S.; et al. Invasive listeriosis outbreaks and salmon products: A genomic, epidemiological study. *Emerg. Microbes Infect.* **2022**, *11*, 1308–1315. [CrossRef] [PubMed]
14. ECDC-EFSA. *Prolonged Multi-Country Outbreak of Listeria monocytogenes ST173 Linked to Consumption of Fish Products—19 June 2024*; European Centre for Disease Prevention and Control, European Food Safety Authority: Rome, Italy, 2024; ISBN 978-92-9498-726-6. [CrossRef]
15. IFSAC. IFSAC. Interagency Food Safety Analytics Collaboration. Foodborne Illness Source Attribution Estimates for 2021 for Salmonella, Escherichia coli O157, and Listeria monocytogenes Using Multi-Year Outbreak Surveillance Data, United States. GA and D.C.: U.S. Department of Health and Human Services, Centers for Disease Control and Prevention, Food and Drug Administration, U.S. Department of Agriculture's Food Safety and Inspection. Available online: https://www.cdc.gov/ifsac/media/pdfs/P19-2021-report-TriAgency-508.pdf (accessed on 9 January 2024).

16. EFSA BIOHAZ Panel; Ricci, A.; Allende, A.; Bolton, D.; Chemaly, M.; Davies, R.; Escámez, P.S.F.; Girones, R.; Herman, L.; Koutsoumanis, K. Scientific Opinion on the *Listeria monocytogenes* contamination of ready-to-eat foods and the risk for human health in the EU. *EFSA J.* **2018**, *16*, 5134. [CrossRef]
17. Leclercq, A.; Kooh, P.; Augustin, J.C.; Guillier, L.; Thébault, A.; Cadavez, V.; Gonzales-Barron, U.; Sanaa, M. Risk factors for sporadic listeriosis: A systematic review and meta-analysis. *Microb. Risk Anal.* **2021**, *17*, 100128. [CrossRef]
18. Gonzales-Barron, U.; Cadavez, V.; De Oliveira Mota, J.; Guillier, L.; Sanaa, M. A critical review of risk assessment models for *Listeria monocytogenes* in seafood. *Foods* **2024**, *13*, 716. [CrossRef]
19. Pouillot, R.; Miconnet, N.; Afchain, A.-L.; Delignette-Muller, M.L.; Beaufort, A.; Rosso, L.; Denis, J.-B.; Cornu, M. Quantitative risk assessment of *Listeria monocytogenes* in French cold-smoked salmon: I. Quantitative exposure assessment. *Risk Anal.* **2007**, *27*, 683–700. [CrossRef]
20. Pouillot, R.; Goulet, V.; Delignette-Muller, M.L.; Mahé, A.; Cornu, M. Quantitative risk assessment of *Listeria monocytogenes* in French cold-smoked salmon: II. Risk characterization. *Risk Anal.* **2009**, *29*, 806–819. [CrossRef]
21. Fritsch, L.; Guillier, L.; Augustin, J.C. Next generation quantitative microbiological risk assessment: Refinement of the cold smoked salmon-related listeriosis risk model by integrating genomic data. *Microb. Risk Anal.* **2018**, *10*, 20–27. [CrossRef]
22. Chen, R.; Orsis, R.H.; Guariglia-Oropeza, V.; Wiedmann, M. Development of a modeling tool to assess ad reduce regulatory ad recall risks for cold-smoked salmon due to *Listeria monocytogenes* contamination. *J. Food Prot.* **2022**, *85*, 1335–1354. [CrossRef]
23. FDA-FSIS. *Quantitative Assessment of Relative Risk to Public Health from Foodborne Listeria monocytogenes among Selected Categories of Ready-to-Eat Foods*; FDA-FSIS: Washington, DC, USA, 2003; pp. 1–541.
24. Pérez-Rodríguez, F.; Carrasco, E.; Bover-Cid, S.; Joffré, A.; Valero, A. Closing gaps for performing a risk assessment on *Listeria monocytogenes* in ready-to-eat (RTE) foods: Activity 2, a quantitative risk characterization *on L. monocytogenes* in RTE foods; starting from the retail stage. *EFSA Support. Publ.* **2017**, *14*, 1252E. [CrossRef]
25. Lindqvist, R.; Westöö, A. Quantitative risk assessment for *Listeria monocytogenes* in smoked or gravad salmon and rainbow trout in Sweden. *Int. J. Food Microbiol.* **2000**, *58*, 181–196. [CrossRef]
26. FAO-WHO. *Risk Assessment of Listeria monocytogenes in Ready-to-Eat Foods: Technical Report*; World Health Organization and Food and Agriculture Organization of the United Nations: Geneve, Switzerland, 2004; pp. 1–269.
27. Garrido, V.; García-Jalón, I.; Vitas, A.; Sanaa, M. Listeriosis risk assessment: Simulation modelling and "what if" scenarios applied to consumption of ready-to-eat products in a Spanish population. *Food Control* **2010**, *21*, 231–239. [CrossRef]
28. Gospavic, R.; Haque, M.; Leroi, F.; Popov, V.; Lauzon, H. Quantitative microbial risk assessment for *Listeria monocytogenes* in cold smoked salmon. *WIT Trans. Inf. Commun. Technol.* **2010**, *43*, PI563–PI572.
29. Dass, S. Exposure Assessment of *Listeria monocytogenes* in Vacuum Packed Cold-Smoked Salmon in the Republic of Ireland. Ph.D. Thesis, Technological University Dublin, Dublin, Ireland, 2011.
30. Pasonen, P.; Ranta, J.; Tapanainen, H.; Valsta, L.; Tuominen, P. *Listeria monocytogenes* risk assessment on cold smoked and salt-cured fishery products in Finland—A repeated exposure model. *Int. J. Food Microbiol.* **2019**, *304*, 97–105. [CrossRef] [PubMed]
31. FAO-WHO. *Joint FAO/WHO Expert Meeting on Microbiological Risk Assessment of Listeria monocytogenes in Foods: Summary and Conclusions*; WHO HQ: Geneva, Switzerland, 2023.
32. Nauta, M. The modular process risk model (MPRM): A structural approach to food chain exposure assessment. In *Microbial Risk Analysis of Foods*; Schaffner, D.W., Doyle, M.P., Eds.; ASM Press: Washington, DC, USA, 2008; pp. 99–136.
33. Autio, T.; Hielm, S.; Miettinen, M.; Sjoberg, A.-M.; Aarnisalo, K.; Bjorkroth, J.; Mattila-Sandholm, T.; Korkeala, H. Sources of *Listeria monocytogenes* contamination in a cold-smoked rainbow trout processing plant detected by pulsed-field gel electrophoresis typing. *Appl. Environ. Microbiol.* **1999**, *65*, 150–155. [CrossRef] [PubMed]
34. Cruz, C.D.; Silvestre, F.A.; Kinoshita, E.M.; Landgraf, M.; Franco, B.D.G.M.; Destro, M.T. Epidemiological survey of *Listeria monocytogenes* in a gravlax salmon processing line. *Braz. J. Microbiol.* **2008**, *39*, 375–383. [CrossRef]
35. Di Ciccio, P.; Meloni, D.; Festino, A.R.; Conter, M.; Zanardi, E.; Ghidini, S.; Vergara, A.; Mazzette, R.; Ianieri, A. Longitudinal study on the sources of *Listeria monocytogenes* contamination in cold-smoked salmon and its processing environment in Italy. *Int. J. Food Microbiol.* **2012**, *158*, 79–84. [CrossRef]
36. Markkula, A.; Autio, T.; Lunden, J.; Korkeala, H. Raw and processed fish show identical *Listeria monocytogenes* genotypes with pulsed-field gel electrophoresis. *J. Food Prot.* **2005**, *68*, 1228–1231. [CrossRef]
37. Medrala, D.; Dabrowski, W.; Czekajlo-Kolodziej, U.; Daczkowska-Kozon, E.; Koronkiewcz, A.; Augustynowicz, E.; Manzano, M. Persistence of *Listeria monocytogenes* strains isolated from products in a Polish fish-processing plant over a 1-year period. *Food Microbiol.* **2003**, *20*, 715–724. [CrossRef]
38. Miettinen, H.; Aarnisalo, K.; Salo, S.; Sjoberg, A. Evaluation of surface contamination and the presence of *Listeria monocytogenes* in fish processing factories. *J. Food Prot.* **2001**, *64*, 635–639. [CrossRef]
39. Rorvik, L.M.; Caugant, D.A.; Yndestad, M. Contamination patter of *Listeria monocytogenes* and other *Listeria* spp. in a salmon slaughterhouse and smoked salmon processing plant. *Int. J. Food Microbiol.* **1995**, *25*, 19–27. [CrossRef]
40. Vogel, B.; Huss, H.; Ojeniyi, B.; Ahrens, P.; Gram, L. Elucidation of *Listeria monocytogenes* contamination routes in cold-smoked salmon processing plants detected by DNA-based typing methods. *Appl. Environ. Microbiol.* **2001**, *67*, 2586–2595. [CrossRef] [PubMed]
41. Jarvis, B. *Statistical Aspects of the Microbiological Examination of Foods*, 3rd ed.; Elsevier Science: Amsterdam, The Netherlands, 2016; 336p, ISBN 9780128039731.

42. Jia, Z.; Bai, W.; Li, X.; Fang, T.; Li, C. Assessing the growth of *Listeria monocytogenes* in salmon with or without the competition of background microflora—A one-step kinetic analysis. *Food Control* **2020**, *114*, 107139. [CrossRef]
43. Hoelzer, K.; Pouillot, R.; Gallagher, D.; Silverman, M.B.; Kause, J.; Dennis, S. Estimation of *Listeria monocytogenes* transfer coefficients and efficacy of bacterial removal through cleaning and sanitation. *Int. J. Food Microbiol.* **2012**, *157*, 267–277. [CrossRef]
44. Aarnisalo, K.; Sheen, S.; Raaska, L.; Tamplin, M. Modelling transfer of *Listeria monocytogenes* during slicing of 'gravad' salmon. *Int. J. Food Microbiol.* **2007**, *118*, 69–78. [CrossRef] [PubMed]
45. Gudbjornsdottir, B.; Suihko, M.L.; Gustavsson, P.; Thorkelsson, G.; Salo, S.; Sjoberg, A.M.; Niclasen, O.; Bredholt, S. The incidence of *Listeria monocytogenes* in meat, poultry and seafood plants in the Nordic countries. *Food Microbiol.* **2004**, *21*, 217–225. [CrossRef]
46. Gudmundsdottir, S.; Gudbjornsdottir, B.; Lauzon, H.; Einarsson, H.; Kristinsson, K.; Kristjansson, M. Tracing *Listeria monocytogenes* isolates from cold-smoked salmon and its processing environment in Iceland using pulsed-field gel electrophoresis. *Int. J. Food Microbiol.* **2005**, *101*, 41–51. [CrossRef]
47. Eklund, M.W.; Poysky, F.T.; Paranjpye, R.N.; Lashbrook, L.C.; Peterson, M.E.; Pelory, G.A. Incidence and sources of *Listeria monocytogenes* in cold-smoked fishery products and processing plants. *J. Food Prot.* **1995**, *58*, 502–508. [CrossRef]
48. Porsby, C.H.; Vogel, B.F.; Mohr, M.; Gram, L. Influence of processing steps in cold-smoked salmon production on survival and growth of persistent and presumed non-persistent *Listeria monocytogenes*. *Int. J. Food Microbiol.* **2008**, *122*, 287–295. [CrossRef]
49. Neunlist, M.R.; Ralazamahaleo, M.; Cappelier, J.-M.; Besnard, V.; Federighi, M.; Leroi, F. Effect of salting and cold-smoking process on the culturability, viability and virulence of *Listeria monocytogenes* strain Scott A. *J. Food Prot.* **2005**, *68*, 85–91. [CrossRef]
50. Lopes, S.M.; Carmo da Silva, D.; Tondo, E.C. Survival of *Listeria* monocytogenes in gravlax salmon (*Salmo salar*) recipe. *Int. J. Gastron. Food Sci.* **2023**, *34*, 100836. [CrossRef]
51. Mejlholm, O.; Dalgaard, P. Modeling and predicting the growth boundary of *Listeria monocytogenes* in lightly preserved seafood. *J. Food Prot.* **2007**, *70*, 70–84. [CrossRef] [PubMed]
52. Mejlholm, O.; Dalgaard, P. Modeling and predicting the growth of lactic acid bacteria in lightly preserved seafood and their inhibiting effect on *Listeria monocytogenes*. *J. Food Prot.* **2007**, *70*, 2485–2497. [CrossRef] [PubMed]
53. Mejlholm, O.; Dalgaard, P. Development and validation of an extensive growth and growth boundary model for *Listeria monocytogenes* in lightly preserved and ready-to-eat shrimp. *J. Food Prot.* **2009**, *72*, 2132–2143. [CrossRef] [PubMed]
54. Mejlholm, O.; Dalgaard, P. Development and validation of an extensive growth and growth boundary model for psychrotolerant *Lactobacillus* spp. in seafood and meat products. *Int. J. Food Microbiol.* **2013**, *167*, 244–260. [CrossRef]
55. Mejlholm, O.; Dalgaard, P. Modelling and predicting the simultaneous growth of *Listeria monocytogenes* and psychrotolerand lactic acid bacteria in processed seafood and mayonnaise-based seafood salads. *Food Microbiol.* **2015**, *46*, 1–14. [CrossRef]
56. Mejlholm, O.; Gunvig, A.; Borggaard, C.; Blom-Hanssen, J.; Mellefont, L.; Ross, T.; Leroi, F.; Else, T.; Visser, D.; Dalgaard, P. Predicting growth rates and growth boundary of *Listeria monocytogenes*—An international validation study with focus on processed and ready-to-eat meat and seafood. *Int. J. Food Microbiol.* **2010**, *141*, 137–150. [CrossRef]
57. Gimenez, B.; Dalgaard, P. Modelling and predicting the simultaneous growth of *Listeria monocytogenes* and spoilage microorganisms in cold-smoked salmon. *J. Appl. Microbiol.* **2004**, *96*, 96–109. [CrossRef]
58. Møller, C.O.A.; Ilg, Y.; Aabo, S.; Christensen, B.B.; Dalgaard, P.; Hansen, T.B. Effect of natural microbiota on growth of *Salmonella* spp. in fresh pork—A predictive microbiology approach. *Food Microbiol.* **2013**, *34*, 284–295. [CrossRef]
59. Nauta, M. Microbiological risk assessment models for partitioning and mixing during food handling. *Int. J. Food Microbiol.* **2005**, *100*, 311–322. [CrossRef]
60. Svanevik, C.S.; Lunestad, B.T.; Storesund, J. *Listeria monocytogenes in Salmonid Slaughter Facilities—Screening Program for the Norwegian Food Safety Authority*; Report Series: Rapport fra havforskningen 2021-45; Institute of Marine Research: Oslo, Norway, 2021; IMR: 2021; ISSN 1893-4536. Project No.: 15600.
61. Gonzales-Barron, U.; Cadavez, V.; Thebault, A.; Kooh, P. The Pathogens-in-Foods Database (PIF) (Version 1). Zenodo, 2021. Available online: https://pif.esa.ipb.pt/ (accessed on 10 January 2024). [CrossRef]
62. Daelman, J.; Membré, J.M.; Jacxsens, L.; Vermeulen, A.; Devlieghere, F.; Uyttendaele, M. A quantitative microbiological exposure assessment model for *Bacillus cereus* in REPFEDs. *Int. J. Food Microbiol.* **2013**, *166*, 433–449. [CrossRef]
63. Peiris, I.P.; Lopez-Valladares, G.; Parihar, V.S.; Helmersson, S.; Barbuddhe, S.; Tham, W.; Danielsson-Tham, M.-L. Gravad (Gravlax) and cold-smoked salmon, still a potential source of listeriosis. *J. Foodserv.* **2009**, *20*, 15–20. [CrossRef]
64. FDA. *FDA-iRISK 4.2 Food Safety Modeling Tool: Technical Document*; U.S. Food and Drug Administration, U.S. Department of Agriculture: Washington, USA, 2021.
65. Leistner, L. Basic aspects of food preservation by hurdle technology. *Int. J. Food Microbiol.* **2000**, *55*, 181–186. [CrossRef] [PubMed]
66. Maqsood, S.; Benjakul, S.; Shahidi, F. Emerging role of phenolic compounds as natural food additives in fish and fish products. *Crit. Rev. Food Sci. Nutr.* **2012**, *53*, 162–179. [CrossRef] [PubMed]
67. Niedziela, J.-C.; MacRae, M.; Ogden, I.D.; Nesvadba, P. Control of *Listeria monocytogenes* in salmon; antimicrobial effect of salting, smoking and specific smoke compounds. *LWT—Food Sci. Technol.* **1998**, *31*, 155–161. [CrossRef]
68. Wiernasz, N.; Leroi, F.; Chevalier, F.; Cornet, J.; Cardinal, M.; Rohloff, J.; Passerini, D.; Skırnisdóttir, S.; Pilet, M.F. Salmon gravlax biopreservation with lactic acid bacteria: A polyphasic approach to assessing the impact on organoleptic properties, microbial ecosystem and volatilome composition. *Front. Microbiol.* **2020**, *10*, 3103. [CrossRef]
69. Chen, B.Y.; Pyla, R.; Kim, T.J.; Silva, J.L.; Jung, Y.S. Prevalence and contamination patterns of *Listeria monocytogenes* in catfish processing environment and fresh fillets. *Food Microbiol.* **2010**, *27*, 645–652. [CrossRef]

70. WHO. *Statistical Aspects of Microbiological Criteria Related to Foods: A Risk Managers Guide*; Microbiological Risk Assessment series 24; World Health Organization & Food and Agriculture Organization of the United Nations: Geneve, Switzerland, 2016; p. 120. Available online: https://iris.who.int/bitstream/handle/10665/249531/9789241565318%E2%80%93eng.pdf?sequence=1 (accessed on 20 January 2024).
71. Couvert, O.; Pinon, A.; Bergis, H.; Bourdichon, F.; Carlin, F.; Cornu, M.; Denis, C.; Gnanou, B.; Guillier, L.; Jamet, E.; et al. Validation of a stochastic modelling approach for *Listeria monocytogenes* growth in refrigerated foods. *Int. J. Food Microbiol.* **2010**, *144*, 236–242. [CrossRef]
72. Wiernasz, N.; Gigout, F.; Cardinal, M.; Cornet, J.; Rohloff, J.; Courcoux, P.; Vigneau, E.; Skírnisdottír, S.; Passerini, D.; Pilet, M.-F.; et al. Effect of the manufacturing process on the microbiota, organoleptic properties and volatilome of three salmon-based products. *Foods* **2021**, *10*, 2517. [CrossRef]
73. Hwang, C.-A.; Sheen, S. Modeling the growth characteristics of *Listeria monocytogenes* and native microflora in smoked salmon. *J. Food Sci.* **2009**, *74*, M125–M130. [CrossRef]
74. Orozco, L.N. *The Occurrence of Listeria monocytogenes and Microbiological Quality of Cold Smoked and Gravad Fish on the Iceland Retail Market*; Report, Fisheries Training Programme; The United Nations University: Reykjavic, Iceland, 2000; p. 30.
75. Leblanc, I.; Leroi, F.; Hartke, A.; Auffray, Y. Do stress encountered during the smoked salmon process influence the survival of the spoiling bacterium *Shewanella putrefaciens*? *Lett. Appl. Microbiol.* **2000**, *30*, 437–442. [CrossRef]
76. Endrikat, S.; Gallagher, D.; Pouillot, R.; Hicks Quesenberry, H.; Labarre, D.; Schroeder, C.M.; Kause, J. A comparative risk assessment for *Listeria monocytogenes* in prepackaged versus retail-sliced deli meat. *J. Food Prot.* **2010**, *73*, 612–619. [CrossRef] [PubMed]
77. Marc, L. Développement d'un Modèle Modulaire Décrivant L'effet des Interactions Entre les Facteurs Environnementaux sur les Aptitudes de Croissance de Listeria. Ph.D. Thesis, Université de Bretagne Occidentale, Brest, France, 2001.
78. Pouillot, R.; Delignette-Muller, M.L. Evaluating variability and uncertainty separately in microbial quantitative risk assessment using two R packages. *Int. J. Food Microbiol.* **2010**, *142*, 330–340. [CrossRef] [PubMed]
79. Pouillot, R.; Kiermeier, A.; Guillier, L.; Cadavez, V.; Sanaa, M. Updated Parameters for *Listeria monocytogenes* Dose–Response Model Considering Pathogen Virulence and Age and Sex of Consumer. *Foods* **2024**, *13*, 751. [CrossRef] [PubMed]
80. Pouillot, R.; Gallagher, D.; Tang, J.; Hoelzer, K.; Kause, J.; Dennis, S.B. *Listeria monocytogenes* in retail delicatessens: An interagency risk assessment-model and baseline results. *J. Food Prot.* **2015**, *78*, 134–145. [CrossRef] [PubMed]
81. R Core Team. *R: A Language and Environment for Statistical Computing*; R Foundation for Statistical Computing: Vienna, Austria, 2021; Available online: https://www.R-project.org/ (accessed on 15 September 2023).
82. Hartemink, R.; Georgsson, F. Incidence of *Listeria* species in seafood and seafood salads. *Int. J. Food Microbiol.* **1991**, *12*, 189–195. [CrossRef]
83. Pelroy, G.A.; Peterson, M.E.; Holland, P.J.; Eklund, M.W. Inhibition of *Listeria monocytogenes* in cold-process (smoked) salmon by sodium lactate. *J. Food Prot.* **1994**, *57*, 108–113. [CrossRef]
84. Kang, J.; Stasiewicz, M.J.; Murray, D.; Boor, K.J.; Wiedmann, M.; Bergholz, T.M. Optimization of combinations of bactericidal and bacteriostatic treatments to control *Listeria monocytogenes* on cold-smoked salmon. *Int. J. Food Microbiol.* **2014**, *179*, 1–9. [CrossRef]

foods

MDPI

Article

A Quantitative Risk Assessment Model for *Listeria monocytogenes* in Non-Ready-to-Eat Frozen Vegetables

Ursula Gonzales-Barron [1,2,*], Régis Pouillot [3], Juliana De Oliveira Mota [4], Akio Hasegawa [4], Ana Allende [5], Qingli Dong [6], Matthew J. Stasiewicz [7], Jovana Kovacevic [8], Vasco Cadavez [1,2], Laurent Guillier [9] and Moez Sanaa [4,*]

1 Centro de Investigação de Montanha (CIMO), Instituto Politécnico de Bragança, Campus de Santa Apolónia, 5300-253 Bragança, Portugal; vcadavez@ipb.pt
2 Laboratório Para a Sustentabilidade e Tecnologia em Regiões de Montanha, Instituto Politécnico de Bragança, Campus de Santa Apolónia, 5300-253 Bragança, Portugal
3 Independent Researcher, 18 rue Mohamed Al Ghazi, Rabat 10170, Morocco; rpouillot.work@gmail.com
4 Nutrition and Food Safety Department, World Health Organization, 1202 Geneva, Switzerland; deju@who.int (J.D.O.M.); hasegawaa@who.int (A.H.)
5 Centro de Edafología y Biología Aplicada del Segura, Consejo Superior de Investigaciones Científicas (CEBAS-CSIC), Campus de Espinardo, 25, 30100 Murcia, Spain; aallende@cebas.csic.es
6 School of Health Science and Engineering, University of Shanghai for Science and Technology, Shanghai 200093, China; qdong@usst.edu.cn
7 Department of Food Science and Human Nutrition, University of Illinois at Urbana-Champaign, Champaign, IL 61801, USA; mstasie@illinois.edu
8 Food Innovation Center, Oregon State University, Portland, OR 97209, USA; jovana.kovacevic@oregonstate.edu
9 Risk Assessment Department, French Agency for Food, Environmental and Occupational Health & Safety (Anses), 14 rue Pierre et Marie Curie Maisons-Alfort, 94701 Maisons-Alfort, France; laurent.guillier@anses.fr
* Correspondence: ubarron@ipb.pt (U.G.-B.); sanaam@who.int (M.S.); Tel.: +351-273-303-325 (U.G.-B.)

Abstract: A quantitative risk assessment (QRA) model was developed to evaluate the risk of invasive listeriosis from the consumption of non-ready-to-eat (non-RTE) frozen vegetables. On a lot basis, the QRA model simulates *Listeria monocytogenes* concentration and prevalence in a "Processing module" that comprises blanching, potential recontamination and packaging, any post-packaging inactivation treatment, and within-lot end-product testing and in a subsequent "Consumer's handling module" that encompasses portioning of frozen vegetables, defrosting, and cooking. Based on available published data, the model was coded in nine sequential R functions designed to assess the effectiveness of blanching, the improvement in processing environment hygiene, the implementation of sampling schemes at the end of processing, and improved consumer instructions on the product's package. In a reference scenario, the model estimated that 9.4% of 500 g packages of frozen vegetables would be contaminated, although at mean levels lower than 10 CFU/g, and assuming that 20% of the portions of frozen vegetables would be left to thaw at room temperature for 2 h, the lot-level mean risk of listeriosis in the susceptible population would be 2.935×10^{-14} (median 5.446×10^{-15}) for uncooked 50 g servings and 2.765×10^{-17} (median 5.184×10^{-18}) for cooked 50 g servings. Analysis of selected scenarios suggested that not cooking the non-RTE product contributes to the risk to a greater extent than the level of contamination in the incoming raw vegetables, the latter in turn being more influential than the level of contamination in the processing environment. The QRA model is freely available as an R package with full documentation and can be used as a tool to inform the consideration of strengthened risk management measures in view of the current changes in consumer behavior and new diet trends.

Keywords: frozen produce; blanched vegetables; listeriosis; exposure assessment; simulation

1. Introduction

Recent listeriosis outbreaks have prompted concerns about the risks associated with the consumption of non-RTE frozen vegetables. Between 2013 and 2015, a multi-state outbreak of listeriosis took place in the United States, involving nine cases with three fatalities. Whole-genome sequencing (WGS) analysis demonstrated that an *L. monocytogenes* strain isolated from frozen corn was genetically related to eight bacterial isolates from eight human cases, whereas an isolate from frozen peas from the same manufacturer yielded a different *L. monocytogenes* that was indistinguishable to that from an additional human case. It was concluded that the contamination source of this outbreak was the processing environment [1]. Later on, between 2015 and 2018, an outbreak of invasive *L. monocytogenes* infections occurred across Austria, Denmark, Finland, Sweden, and the UK, involving 53 cases and 10 fatalities. The source of the outbreak was traced to sweet corn grown and quick-frozen in Hungary. WGS identified the outbreak strain (*L. monocytogenes* 4b, multi-locus sequence type ST 6) in various frozen food samples and environmental swabs from a plant where vegetables were processed and frozen. Environmental contamination was indicated as the source of the persistence of the strain causing the outbreak from 2015 to 2018 [2]. The processing of these vegetables included a blanching step before freezing [3,4].

The European Commission [5] and the Codex Alimentarius [6] have established microbiological criteria for *L. monocytogenes* in RTE foods depending on whether the food product permits growth under storage conditions throughout its shelf-life. For RTE foods that support the growth of *L. monocytogenes*, the criterion of absence in 25 g applies to five samples per batch. For RTE foods that do not support the growth of *L. monocytogenes*, the microbiological limit for acceptable batches is 100 CFU/g in five samples at the end of shelf life. While *L. monocytogenes* does not grow in frozen foods and therefore, a threshold of 100 CFU/g would be accepted for these products, such criterion might not be enough to control the production process if defrosted frozen vegetables are stored for a certain period, even at refrigeration temperatures, before consumption [7]. The potential growth of *L. monocytogenes*, initially present in frozen vegetables even at very low concentrations, would then represent a more serious public health risk. On the other hand, non-RTE frozen foods have no regulatory limit when considering the number of *L. monocytogenes* in the product, and these were the kinds of foods causing the outbreaks mentioned above. According to Zoellner et al. [8], outbreak investigations showed that alternative (non-intended) preparations of non-RTE frozen vegetables were the root of the problem, such as defrosting and direct use in salads, or as ingredients in other RTE products subsequently sold to consumers without undergoing any treatment to reduce the level of pathogens.

In this respect, two quantitative risk assessment (QRA) models for non-RTE frozen vegetables have been published: Zoellner et al. [8] and EFSA [7]. EFSA's model [7] focused only on the home preparation module and considered cooking as the most important control measure to reduce risk. They estimated that decreasing the proportion of uncooked servings from 23% (baseline) to 4% (best-case scenario) reduces the predicted listeriosis cases per year from 1.62 to 0.041 in elderly females in the European Union. The EFSA model [7] concluded that blanched frozen vegetables conveyed the lowest risk of listeriosis in comparison to sausage, hot-smoked fish, pate, gravad fish, cooked meat, and cold-smoked fish. The model of Zoellner et al. [8] (made available as the FFLLoRA tool) represented not only the consumer's module but also considered two single features of the processing stage: within-batch testing and batch size. According to the FFLLoRA tool, the median risk of illnesses per batch of frozen vegetables is inferior to 10^{-16} when at least half of the servings are properly cooked. They further stressed the importance of cooking through sensitivity analysis, as it showed that cooking the vegetables serving (Spearman rank coefficient: $r = -0.87$) and log reduction due to proper cooking ($r = -0.48$) were the most important drivers determining the exposure dose of *L. monocytogenes* consumed per serving.

Whereas these two QRA models provided valuable insights into the impact of consumer preparation and handling, the contributions to the final risk from sources and factors inherent to the processing stage remained unknown. This understanding is particularly important since *L. monocytogenes* can contaminate vegetables post-blanching [3,4,9]. Within this context, the main objective of this work was to construct a QRA model of longer scope for non-RTE frozen vegetables, able to represent the inactivation, recontamination, and potential growth of *L. monocytogenes* in non-RTE frozen vegetables from processing to consumption. Widely based on published data, the model was structured to be able to assess the effectiveness of the following treatments/control measures: blanching, any additional inactivation step post-packaging, reducing contamination in the processing facility environment, sampling schemes at the end of processing, better labeling and/or consumer education (represented by instructions on proper storage, defrosting and cooking practices at home), in line with the recent Expert Panel recommendation of the Joint FAO/WHO Expert meeting on microbiological risk assessment [10]. The functionality of the QRA model proposed was illustrated by a reference scenario, supported by literature data, and by the development of what-if scenario analysis.

2. Materials and Methods

2.1. Exposure Assessment

The exposure assessment model of non-RTE frozen vegetables comprises two modules: the "Processing module", which represents blanching, potential recontamination and packaging, post-packaging inactivation treatment, and within-batch testing; and the "Consumer's module", which encompasses portioning, defrosting, and cooking. Each stage was coded as a function estimating stochastically the microbial prevalence and number after a process of microbial growth, death, partitioning, removal, or cross-contamination [11], according to the stage's purpose. Table 1 summarizes the modules, the sequence of stages and processes they consist of, the assumption and data sources employed, and the corresponding programmed functions.

2.1.1. Contaminated Lots of Vegetables Pre-Blanching

Data and assumptions:

The distributions for prevalence and concentration of *L. monocytogenes* in pre-blanched vegetables were modeled from published data. However, since data on pre-blanched vegetables are very scarce, *L. monocytogenes* prevalence in fresh and minimally processed vegetables was instead used. Such data were extracted from the Pathogens-in-Foods database [12]. The lot-to-lot variability in the prevalence of *L. monocytogenes* in pre-blanched vegetables was assumed to follow a Beta distribution with parameters α and β, which were determined using the surveys' data compiled in Table 2. Every observed sampling outcome (s_j positive samples out of n_j total samples) was assumed to originate from the j-th lot of pre-blanched vegetables ($j = 1, 2, \ldots N$), where N is the number of lots sampled ($N = 21$, according to Table 2). The sampling outcomes s_j are considered as realizations of a binomial distribution (n_j, p_j), where p_j is the unobservable prevalence of the lot j. The values of α and β, the beta distribution parameters that describe the between-lot prevalence, were estimated by Markov Chain Monte Carlo (MCMC) simulation as the means of posterior distributions in a Bayesian framework. The model was expressed as

$$\begin{aligned} s_j &\sim Binomial(n_j, p_j) \\ p_j &\sim Beta(\alpha, \beta) \end{aligned} \quad (1)$$

with non-informed priors for α and β assumed to be Gamma distributions with shape and rate parameters both set to 0.01. The parameters were derived from 20,000 iterations and a burn-in period of 5000 iterations. The mean value for α was estimated at 0.5112 (median: 0.4816; standard deviation: 0.1962; 95% credibility interval: 0.220–0.985), and the mean value for β at 9.959 (median: 9.100; standard deviation: 4.915; 95% credibility interval: 2.971–21.92).

Table 1. Sequence of stages, microbial processes represented, data sources, assumptions, and corresponding functions coded in R for the construction of the exposure assessment model of *Listeria monocytogenes* (LM) in non-RTE frozen vegetables.

Module	Stage	Microbial Process	Assumptions	References	Function in R
Processing	Generation of contaminated lots pre-blanching	None	LM prevalence in pre-blanched vegetables is assumed to be comparable to the prevalence found in minimally processed vegetables and fresh whole vegetables sampled at the packinghouse or at retail.	Badosa et al. [13], Cardamone et al. [14], Carp-Carare et al. [15], Cetinkaya et al. [16], De Giusti et al. [17], Gianfranceschi et al. [18], Kokkinakis et al. [19], Lika et al. [20], Losio et al. [21], Magdovitz et al. [22], Moreno et al. [23], Pianetti et al. [24], Vojkovska et al. [25], Wagner et al. [26]	Lot2LotGen()
			LM concentration in pre-blanched vegetables is assumed to be comparable to the concentrations found in freshly harvested vegetables and vegetables sold at retail.	Jeyaletchumi et al. [27], Kuan et al. [28]	
	Blanching	Inactivation	LM during blanching is assumed to follow the kinetics of a cocktail of N-7004 (Scott A, serotype 4b), N-7285 (serotype 1/2a), N-7298 (serotype 1/2b), and N-7017 (Murray B, serotype 4b) LM inoculated in broccoli, mushroom, onions, peas, and pepper.	Mazzotta [29]	fvBlanching()
	Freezing and packaging	Cross-contamination and partitioning	LM can contaminate the bulk of vegetables post-blanching from equipment such as feeders, slicing machines, transporters, conveyor belts, and freezing tunnels.	Truchado et al. [3]	fvPartitioningCC()
			If the contamination event takes place, LM cells are transferred to the bulk according to a transfer coefficient from stainless steel to vegetables.	Hoelzer et al. [30]	
			LM is assumed to be moderately clustered in the bulk of frozen vegetables from a lot.	Nauta [31]	
	Post-packaging treatment	Reduction	Generic inactivation step post-packaging. Assumption of independent, equal probability of inactivation.	Nauta [32]	fvReductionPostpack()
	Within-lot testing	None	At a given probability, a lot of frozen vegetables can be subjected to sampling and testing according to a two-class or three-class microbiological sampling plan.	-	fvTesting()

Table 1. *Cont.*

Module	Stage	Microbial Process	Assumptions	References	Function in R
Consumer's preparation	Portioning	Partitioning	The consumer is assumed to take a portion of frozen vegetables from the pack. LM cells present in a contaminated pack are assumed to be moderately clustered within the package.	Nauta [31]	fvPortioning()
	Defrosting	Growth	The consumer may defrost the frozen vegetables in the fridge, microwave, or at room temperature before cooking or consumption.	Zoellner et al. [8]	fvDefrost()
			LM is assumed to grow according to a log-linear model without a lag phase, based on an exponential growth rate at 5 °C for heat-treated vegetables that distribute as a lognormal distribution.	EFSA [7]	
	Cooking	Reduction	There is a probability that the consumer uses the non-RTE frozen vegetables in a non-intended manner, such as through direct (uncooked) consumption in salads, smoothies, etc.	Willis et al. [33], FSAI [34]	fvCooking()
			For cooked vegetables, different heat treatment intensities can be applied by consumers, varying from very strong heat treatments (i.e., fully cooking) to light heat treatments (i.e., microwave heating). Therefore, there is a variability in the effect of cooking.	EFSA [7]	

Table 2. Retrieved data on *L. monocytogenes* occurrence in minimally processed and fresh whole vegetables used for modeling the lot-to-lot variability in the pathogen's prevalence in pre-blanched vegetables.

Country	Product	Sample Size, *n*	Positive Enrichment, *s*	Prevalence (%)	Source [1]
Minimally processed					
Austria	RTE produce [3]	143	0	0.00	Wagner et al. [26]
Italy	RTE vegetables salads [3]	56	0	0.00	Pianetti et al. [24]
Italy	RTE vegetables [2]	699	2	0.28	De Giusti et al. [17]
Spain	Brocolli fresh-cut [3]	16	1	6.25	Moreno et al. [23]
Italy	Minimally processed pumpkins [3]	33	1	3.03	Cardamone et al. [14]
Italy	RTE packed vegetables [3]	1160	4	0.34	Losio et al. [21]
Fresh whole vegetables					
Italy	Vegetables [3]	738	33	4.47	Gianfranceschi et al. [18]
Greece	Peppers [2]	60	8	13.3	Kokkinakis et al. [19]
	Peppers [2]	60	12	20	
Spain	Raw whole vegetables [3]	141	2	1.41	Badosa et al. [13]
Italy	Whole vegetables [2]	265	3	1.13	De Giusti et al. [17]
Spain	Fresh broccoli [3]	17	2	11.7	Moreno et al. [23]
Romania	Fresh onion [2]	10	0	0	Carp-Carare et al. [15]
Turkey	Raw vegetables [3]	44	6	13.6	Cetinkaya et al. [16]
Albania	Onion [3]	16	0	0	Lika et al. [20]
	Broccoli [3]	21	0	0.00	
Czech Republic	Vegetables [3]	249	16	6.42	Vojkovska et al. [25]
Vegetables arriving at frozen food facility					
USA	Corn	59	8	13.6	Magdovitz et al. [22]
	Carrots	54	0	0.00	
	Green beans	72	3	4.17	
	Peas	96	6	6.25	

[1] Data extracted from the Pathogens-in-Foods database [12], except for [22]. In all sources, a subsample weight of 25 g was used for enrichment. [2] Sampled at the packinghouse. [3] Sampled at retail.

Thus, the prevalence of *L. monocytogenes* in a lot of pre-blanched vegetables was sampled from a Beta (0.5112, 9.959) distribution. For reference, such a distribution has a mean of 5.13% and a 95% variability interval of 0.06–28.48%.

Available data on the concentration of *L. monocytogenes* in pre-blanched (or fresh) vegetables were very scarce. In Malaysia, Jeyaletchumi et al. [27] found 3.0 MPN/g in a sample of freshly harvested carrots and 3.6 MPN/g in a sample of freshly harvested yardlong beans, whereas Kuan et al. [28] determined 3.0 MPN/g in one sample of carrots at retail. Non-published data of *L. monocytogenes* in green beans sold in markets from our laboratory were added to the data set: 21.0, 36.0, and 75.0 MPN/g. Because the six observations available were obtained from the MPN based on serial dilution analysis, the statistical model of the MPN was exploited in a Bayesian framework to model the distribution of the total variability in the $\log_{10}$ concentration of *L. monocytogenes*.

A conditional probability of observing the tube counts given the unknown *L. monocytogenes* concentration in a contaminated vegetable unit was built from the MPN triplets of every vegetable sample tested. According to a three-dilution (0.1, 0.01, 0.001) MPN table, the tube result triplets (x_1-x_2-x_3) of the contaminated vegetable samples were 1-0-0 (3.6 MPN/g); 2-2-0 (21.0 MPN/g); 2-3-1 (36.0 MPN/g); 3-1-1 (75.0 MPN/g); and for the two incidences of 3 MPN/g, the triplets 0-0-1 and 0-1-0 were assumed.

The MPN technique assumes that the microorganisms from a sample at concentration λ (cells per gram) are well mixed in the homogenate and that the aliquots taken from it contain a Poisson ($\lambda \times df_i$) distributed number of microorganisms, where df_i is the dilution factor or the amount of sample in g per ml in the tube corresponding to the dilution i. The probability p_i of having a positive test tube (at least one cell present in the tube) belonging to the dilution i arises from the Poisson process, whereas out of n_i serial dilution analysis tubes (or trials), the numbers of positive tubes (x_i or successes) follow a binomial distribution. For our data, the number of dilutions is three (i = 3), and the number of analysis tubes per dilution is three ($n_1 = n_2 = n_3 = 3$). Thus, the mean μ_{C_0} and the standard deviation σ_{C_0} of the normal distribution characterizing the total variability in the $\log_{10}$ concentration of *L. monocytogenes* in pre-blanched vegetables (C_0) were estimated in a Bayesian framework through MCMC simulation by adjusting the MPN triplets ((x_{1j}, x_{2j}, x_{3j}) for j = 6 observations) to the model

$$\begin{aligned} x_{1j} &\sim Binomial(3, p_{1j}) \\ x_{2j} &\sim Binomial(3, p_{2j}) \\ x_{3j} &\sim Binomial(3, p_{3j}) \\ p_{1j} &= 1 - exp(-\lambda_j \times df_1) \\ p_{2j} &= 1 - exp(-\lambda_j \times df_2) \\ p_{3j} &= 1 - exp(-\lambda_j \times df_3) \\ C_0 = log_{10}\lambda_j &\sim Normal(\mu_{C0}, \sigma_{C0}) \end{aligned} \quad (2)$$

where df_1, df_2, and df_3 were given the values of 0.1, 0.01, and 0.001, respectively. The prior for μ_{C_0} was defined as a non-informed Normal distribution with a mean of 1.0 and a standard deviation of $1/\sqrt{0.001}$; and the prior for σ_{C_0} was set as a non-informed $1/\sqrt{Gamma(0.0001, 0.0001)}$. Following a simulation of 20,000 iterations and a burn-in period of 5000 iterations, the mean value for μ_{C_0} was estimated at 1.038 $\log_{10}$ CFU/g (median: 1.058; standard deviation: 0.3061; 95% credibility interval: 0.379–1.590), and the mean value for σ_{C_0} at 0.5436 $\log_{10}$ CFU/g (median: 0.4966; standard deviation: 0.3795; 95% credibility interval: 0.010–1.446).

The R function:

The function Lot2LotGen() generates a contamination matrix N of dimensions $r \times c$, whose number of rows r represents the number of lots and can therefore be understood as a number of iterations that will correspond to between-lot variability and the number of columns c represents the number of units (frozen vegetable packs) of net weight $Unit_{size}$ grams to be produced in the lot, considered as fixed and equal for all lots. For every lot, $i = \{1, 2, \ldots, r\}$, a value of prevalence P_i is sampled from the between-lot prevalence distribution $Beta\ (\hat{\alpha},\ \hat{\beta})$. The number of contaminated units s_i in a given lot i is determined as

$$s_i \sim Binomial(c, P_i)$$

and assigned to random locations, flagged with ones within the matrix N. The probability that the lot i is contaminated $Prob_{UnitPos}$ is estimated as

$$Prob_{UnitPos\ i} = 1 - P(s_i = 0)$$

considering that a lot is contaminated if at least one unit is contaminated. The mean prevalence of contaminated lots (*P*) is calculated as the complement of the beta-binomial probability evaluated at $s = 0$

$$P = 1 - \frac{\Gamma\ (c + \beta)}{\Gamma(c + \alpha + \beta)}$$

where $\Gamma(x)$ is the gamma function.

For the contaminated units, the numbers of *L. monocytogenes* are produced as follows. First, the mean concentration C_{0bi} for every lot i is sampled from,

$$C_{0bi}\ \sim\ Normal\left(\mu_{C0}, \sqrt{prop_{Var} \times \sigma^2_{C0}}\right)$$

where $prop_{Var}$ is the proportion of the total variance ${\sigma_{C0}}^2$ attributed to between-lot variance ($prop_{Var}$ is assumed to be 0.7); and, secondly, the numbers of *L. monocytogenes* (N_{ij}) in the contaminated units s_i of the lot i is determined as

$$N_{ij} = \left\lceil Unit_{size} \times 10^{\ \max(Normal(C_{0bi}, \sqrt{(1-prop_{Var}) \times \sigma^2_{C0}}),\ \ log_{10}0.501/Unit_{size})} \right\rceil\ j = 1,\ 2 \ldots,\ s_i$$

The value of 0.501 is set to ensure that at least one bacterial cell is present in a contaminated unit. The non-contaminated units (c—s_i) of every lot are assigned zero (CFU) in the matrix N. Apart from the parameters defining the between-lot prevalence (α, β) and concentration (μ_{C_0}, σ_{C_0}) of *L. monocytogenes*, the function Lot2LotGen() requires other inputs, namely *nLots*, *sizeLot*, $Unit_{size}$, and $prop_{Var}$. The outputs of the function are the contamination matrix (*N*), the mean prevalence of contaminated lots (*P*, a scalar), and the probability that the iterated lot is contaminated ($Prob_{UnitPos}$, a vector).

2.1.2. Blanching of Vegetables

Data and assumptions:

Blanching is a technological heat treatment applied to vegetables before freezing, whose main objective is to inactivate enzymes that cause quality degradation. The level of reduction of *L. monocytogenes* during blanching is variable, as it depends on the vegetables and the process parameters applied (time/temperature); the latter are, however, neither regulated nor strictly defined [35]. Ceylan et al. [36] assessed the thermal inactivation of a cocktail of *L. monocytogenes* on peas, spinach, broccoli, potatoes, and carrots that were exposed to hot water and steam under specific time and temperature combinations. Although these authors presented many results, their reductions were all expressed in a "higher than x log" form, which made them unusable for the present QRA model. Data from Ceylan et al. (2017) could be used instead to compare the simulation outputs. Instead, the results from Mazzotta [29] were employed to model the inactivation kinetics of *L. monocytogenes* during blanching. Mazzotta [29] carried out challenge studies mimicking blanching to determine the heat resistance parameters of a cocktail of stationary-phase or starved *L. monocytogenes* cells (serotypes 1/2a, 1/2b, and 4b) in onions, broccoli florets, green peppers, mushrooms, and peas. The D values they determined at 56, 60, and 62 °C are shown in Figure 1.

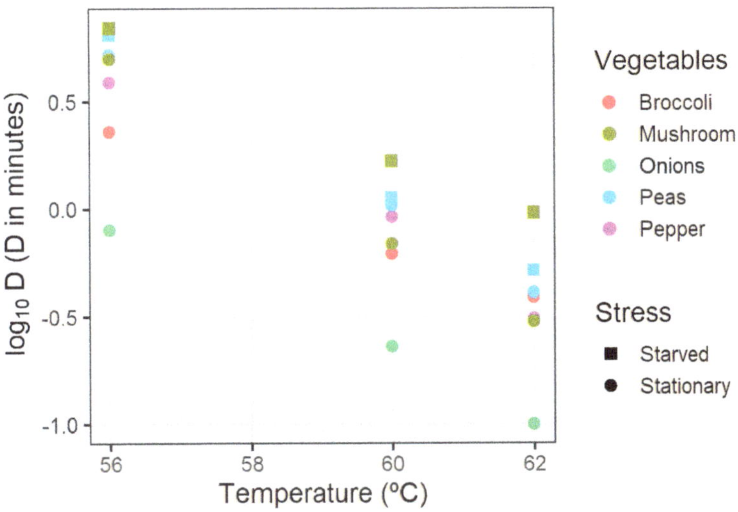

Figure 1. Heat resistance data of stationary-phase or starved *L. monocytogenes* inoculated in various vegetables, taken from Mazzotta [29], used to fit the Bigelow model for blanching.

The *L. monocytogenes* heat resistance data extracted from Mazzotta [29] in the context of evaluating the blanching process were modeled by adjusting a Bigelow equation of the form [37]

$$\log_{10} D = \log_{10} D_{ref} - \frac{\left(T - T_{ref}\right)}{z_T}$$

where D is the decimal reduction time in min, T is the treatment temperature (°C), D_{ref} is the D (min) at the reference temperature (T_{ref} = 70 °C), and z_T is the z value (°C) or change in temperature required for a one-$\log_{10}$ reduction in D. Due to the inability to differentiate between stress effects and specific vegetable effects in the experimental setup, it was decided to incorporate all available data for the analysis. The parameter estimates were as follows: $log_{10}\ D_{ref} = -1.7840$ (standard error = 0.2518) and z_T = 6.061 (standard error = 0.7852), with a residual sum of squares 1.6815. For the QRA model, the variability in $log_{10}\ D_{ref}$ is defined as Normal (−1.784, 0.2518), and z_T as a single point estimate (6.061). The mix of raw vegetables is assumed to be blanched at a temperature ($Temp_{Blanch}$) of 83 °C for a duration ($time_{Blanch}$) of 1.0 min.

The R function:

The function fvBlanching() simulates the effect of blanching on the numbers of *L. monocytogenes* present in the vegetables' portions according to the Bigelow model explained above; and updates the probabilities of lots being contaminated (after blanching) and the mean prevalence of contaminated lots (after blanching). fvBlanching() is fed by the outputs of the function Lot2LotGen(), and by the blanching time and temperature ($time_{Blanch}$ [min], $Temp_{Blanch}$ [°C]). For every lot i, a value of $log_{10}\ D_{ref\,i}$ is sampled from the normal distribution, and the probability of survival of a microbial cell $p_{SurviveBlanching\ i}$ for that lot is determined as

$$p_{SurviveBlanchingi} = 10^{time_{Blanch} \times \left(\log_{10} D_{refi} - \frac{(Temp_{Blanch} - 70)}{6.061}\right)}$$

Understanding $p_{SurviveBlanching\ i}$ as a binomial probability, the number of surviving cells after blanching $N_{blanched\ ij}$ can be estimated as,

$$N_{Blanched\ ij} \sim Binomial\left(N_{ij}, p_{SurviveBlanching\ i}\right) \qquad j = 1, 2 \ldots, c$$

It is noteworthy mentioning that $p_{SurviveBlanching}$ does not include the come-up time in its calculation. To update the prevalence estimates after blanching, the probability that at least one cell survives in a lot ($atLeastOne_i$) is calculated:

$$atLeastOne_i = 1 - \left(1 - p_{SurviveBlanching\ i}\right)^{\sum_{j=1}^{c} N_{ij}}$$

The probability vector of contaminated lots after blanching $Prob_{UnitPos\ blanched\ i}$ is then updated,

$$Prob_{UnitPosblanchedi} = Prob_{UnitPosi} \times atLeastOne_i$$

Likewise, the mean prevalence of contaminated lots $P_{Blanched}$.

$$P_{Blanched} = P \times \frac{\sum_{i=1}^{r} atLeastOne_i}{r}$$

The outputs of the function fvBlanching() are the contamination matrix (numbers of *L. monocytogenes* per unit) after blanching ($N_{Blanched}$), the mean prevalence of contaminated lots after blanching ($P_{Blanched}$, a scalar), and the probabilities that the sampled lots are contaminated ($Prob_{UnitPos\ blanched}$, a vector).

2.1.3. Freezing and Packaging

Data and assumptions:

Studies have shown that there is a probability for cross-contamination to occur after blanching from equipment in contact with food, such as post-blanching conveyor belts, freezing tunnels, and packaging machines [3,38]. Furthermore, non-food contact surfaces and harboring sites play a role in the spread of *L. monocytogenes* to the final product [3,22,39]. Since freezing tunnels and packaging machines have been associated with high recoveries of *L. monocytogenes* in processing facilities (12.5% in Truchado et al. [3]; 36.4% in Magdovitz et al. [22]; 41.3% in Pappelbaum et al. [38]), the present QRA model was designed to accommodate the possibility of bacterial transfer from surfaces such as conveyor belts, freezing tunnels and/or packaging machines. Thus, for the model, the overall risk of post-blanching recontamination is considered as one process. The probability of bacterial transfer occurring after blanching, P_{cc}, was set to the value provided by the most recent source, Truchado et al. [3] (0.125). The distribution for the transfer coefficient of *L. monocytogenes* from stainless steel to vegetables [Normal (−0.44, 0.40) in $\log_{10}$] reported by Hoelzer et al. [30] was employed in the QRA model. Following the potential recontamination event, the frozen bulk vegetables are packaged into bag units, which, in modeling terms, involve the representation of the microbial process of partitioning. In a contaminated lot, *L. monocytogenes* is assumed to be moderately clustered in the bulk of frozen vegetables; therefore, a dispersion factor $b = 1$ is assumed [31].

The R function:

The function fvPartitioningCC() simulates two microbial processes: firstly, microbial transfer by the potential recontamination of blanched vegetables when in direct contact with contaminated surfaces between transport by conveyors and packaging and subsequently, partitioning of the bulk of frozen vegetables into packed units. fvPartitioningCC() is fed by the outputs of the function fvBlanching(), and by the probability of recontamination (P_{cc}), the dispersion factor (b_{CCFV}), the parameters of the normal distribution about the $\log_{10}$ transfer rate from stainless steel to vegetables ($TR_{equip\text{-}veg}$), and the load of *L. monocytogenes* cells on the surface of conveyors, freezer and/or packaging machine ready to be transferred (N_{equip}).

The algorithm starts by calculating the total numbers of *L. monocytogenes* in every lot *i* before the potential recontamination event (N_{tot}),

$$N_{tot\ i} = \sum_{j=1}^{c} N_{ij}$$

If the recontamination event takes place, at a probability P_{cc}, a fraction ($TR_{equip\text{-}veg}$) of the numbers of cells on the contaminated equipment surface(s) (N_{equip}) is transferred to the bulk of vegetables ($N_{from\text{-}equip}$). The algorithm considers that if the bulk of vegetables is not contaminated with *L. monocytogenes* (actually not represented in the contamination matrix), it may become contaminated from equipment; and if the lot is contaminated, its load will increase or remain the same depending on the recontamination event happening or not, respectively. This raises three possible scenarios for contaminated lots [40]: (1) no recontamination occurring in lots already contaminated ($Bulk_{Pos}\_CC_{Neg}$); (2) recontamination occurring in lots already contaminated ($Bulk_{Pos}\_CC_{Pos}$); and (3) recontamination occurring in lots that were not contaminated ($Bulk_{Neg}\_CC_{Pos}$). The fourth case, represented by the status of the non-contaminated lot ($Bulk_{Neg}\_CC_{Neg}$), is not taken into account in the upcoming calculations since it does not affect the microbial numbers.

The probability of each of the three contaminated lots taking place is calculated as

$$P\left(Bulk_{Pos}\_CC_{Neg\ i}\right) = Prob_{UnitPosblanchedi} \times (1 - P_{cc})$$

$$P\left(Bulk_{Pos}\_CC_{Pos\ i}\right) = Prob_{UnitPosblanchedi} \times P_{cc}$$

$$P\left(Bulk_{Neg}\_CC_{Pos\ i}\right) = \left(1 - Prob_{UnitPosblanchedi}\right) \times P_{cc}$$

The status of every lot ($i = 1, 2, \ldots, c$) is then randomly sampled from the probabilities $\{P\left(Bulk_{Pos}\_CC_{Neg\ i}\right),\ P\left(Bulk_{Pos}\_CC_{Pos\ i}\right),\ P\left(Bulk_{Neg}\_CC_{Pos\ i}\right)\}$. This strategy is adopted to maintain the dimension $[r, c]$ of the input contamination matrix $N_{Blanched}$. The extent of recontamination in CFU per lot *i* is computed as

$$TR_{equip-veg\ i} \sim Normal\ (-0.44,\ 0.40), \quad TR_i \leq 0$$

$$N_{from-equip\ i} \sim Binomial\left(N_{equip},\ 10^{TR_{equip-veg\ i}}\right), \quad N_{from-equip\ i} > 0$$

where N_{equip} can be understood as the total load of *L. monocytogenes* cells on the surface of conveyors, freezer and/or packaging machine in contact with the bulk of vegetables. Next, depending on the iterated lot status, the value of $N_{from\text{-}equip\ i}$ is added to $N_{tot\ i}$ (or not), in order to determine the numbers of *L. monocytogenes* in the lot *i* after the potential recontamination event ($N_{tot\ postCC\ i}$), according to

$$N_{tot\ postCC\ i} = \begin{cases} N_{tot\ i} & i \in Bulk_{Pos}\_CC_{Neg} \\ N_{tot\ i} + N_{from-equip\ i} & i \in Bulk_{Pos}\_CC_{Pos} \\ N_{from-equip\ i} & i \in Bulk_{Neg}\_CC_{Pos} \end{cases}$$

Next, the partitioning algorithm randomly distributes the total numbers of cells $N_{tot\ postCC\ i}$ from a contaminated lot into *c* packed units, following a Dirichlet-multinomial process:

$$\begin{bmatrix} p_{dist\ 1} & p_{dist\ 2} & \cdots & p_{dist\ c} \end{bmatrix}_i \sim Dirichlet\left(\begin{bmatrix} b_{j=1} & b_{j=2} & \cdots & b_{j=c} \end{bmatrix}_i\right) \quad i = 1,\ 2,\ \ldots,\ r$$

$$\begin{bmatrix} N_{pack\ 1} & N_{pack\ 2} & \cdots & N_{pack\ c} \end{bmatrix}_i \sim Multinomial\left(N_{tot\ postCC\ i},\ [p_{dist\ 1}\ \ p_{dist\ 2}\ \ \cdots\ \ p_{dist\ c}]_i\right) \quad i = 1,\ 2,\ \ldots,\ r$$

A fixed value for the parameter b is used in order to avoid increasing the uncertainty in the model's outputs, given the lack of data about the dispersion of *L. monocytogenes* in the bulk of frozen vegetables.

The resulting matrix N_{pack} contains the numbers of *L. monocytogenes* cells in *c* packs of frozen vegetables (horizontal dimension) in a sample of *r* contaminated lots (vertical dimension). Finally, the probability vector of contaminated lots after packaging $Prob_{UnitPos\ pack\ i}$ and the mean prevalence of contaminated lots after packaging P_{Pack} are updated.

$$Prob_{UnitPospacki} = 1 - (1 - Prob_{UnitPos\ blanchedi}) \times (1 - P_{cc})$$

$$P_{Pack} = 1 - (1 - P_{Blanched}) \times (1 - P_{cc})$$

The outputs of the function fvPartitioningCC() are the matrix N_{Pack}, the vector $Prob_{UnitPos\ pack}$ and the scalar P_{Pack}.

2.1.4. Post-Packaging Treatment

Data and assumptions:

Although the application of a post-packaging inactivation treatment for frozen foods is challenging [41], a generic post-processing inactivation treatment was considered to represent any alternative approach to reduce *L. monocytogenes* risk during processing [42]. The $\log_{10}$ reduction attained by the generic treatment is assumed to follow a triangular distribution with a minimum value of $Interv_{min}$, a mode of $Interv_{mode}$, and a maximum value of $Interv_{max}$ representing the variability in *L. monocytogenes* inactivation. In the current model, such values were set to zero to assume there is no post-packaging treatment.

The R function:

The fvReductionPostpack() function receives the outputs of the function fvPartitioningCC() and the values of $Interv_{min}$, $Interv_{mode}$, and $Interv_{max}$ set by the user. For every lot *i*, the probability of survival of a microbial cell $p_{SurvivePostpack\ i}$ is sampled as

$$p_{SurvivePostpack\ i} = 10^{-Triang\ (Interv_{min},\ Interv_{mode}, Interv_{max})}$$

and the number of surviving cells after the inactivation treatment N_{Interv} is obtained, under a binomial process assumption, by

$$N_{Interv\ ij} \sim Binomial\left(N_{pack\ ij}, p_{SurvivePostpack\ i}\right) \qquad j = 1, 2\ldots, c$$

To update the prevalence estimates after the inactivation treatment, the probability of at least one cell surviving in the lot *i* ($atLeastOne_i$) is calculated:

$$atLeastOne_i = 1 - \left(1 - p_{SurvivePostpack\ i}\right)^{\sum_{j=1}^{c} N_{ij}}$$

The probability vector of contaminated lots after the inactivation treatment $Prob_{UnitPos\ interv\ i}$ and the mean prevalence of contaminated lots P_{Interv} are then updated:

$$Prob_{UnitPosintervi} = Prob_{UnitPos\ packi} \times atLeastOne_i$$

$$P_{Interv} = P_{Pack} \times \frac{\sum_{i=1}^{r} atLeastOne_i}{r}$$

2.1.5. Within-Lot Testing

Data and assumptions:

The QRA model enables the microbiological testing of food unit samples taken from a lot, according to a two-class or a three-class mixed sampling plan [43]. In a two-class plan (the most commonly used for *L. monocytogenes*), n units are randomly sampled and analyzed per lot. For each sampled unit, a quantity of g grams is used in the enrichment essay and tested. The lot is rejected if more than c sampled units are positive in detection. In a three-class mixed sampling plan, sampled units are also enumerated, and the lot is rejected if more than c sampled units are positive in detection or if at least one sampled unit has an estimated concentration greater than a predefined concentration M [44]. It is assumed that the enumeration test is conducted only on positive samples in detection by direct plating of $g_{TestedEnum}$ g taken from the same sample.

The R function:

The function fvTesting() receives the outputs of the function fvReductionPostpack(), n, the number of tested units, g the sub-sample weight in grams used for detection, c the number of positive samples accepted (two-class or three-class mixed plan), M the maximum limit concentration, $p_{lot\ tested}$, the proportion of tested lots, Se, the probability of the test (enumeration or detection) to detect, independently, each bacteria present in a sample, and $g_{TestedEnum}$, the sub-sample weight in grams used for the enumeration essay.

Assuming perfect homogenization of the sample, each of the bacteria present in each of the $r \times c$ units of $Unit_{size}$ weight has a probability of being present in the g grams of the sub-sample and detected equal to $Se \times g/Unit_{size}$. The number of bacteria detected in the detection essay is then

$$N_{detected\ ij} \sim Binomial\left(N_{Pack\ ij}, Se \times g/Unit_{size}\right)$$

and the detection test is positive if $N_{detected\ ij} > 0$.

If the sampling plan is a 3-class plan, an enumeration test (direct plating) is performed. The number of bacteria enumerated in the sample is

$$N_{enumerated\ ij} \sim Binomial\left(N_{Pack\ ij}, Se \times g_{TestedEnum}/Unit_{size}\right)$$

and the estimated concentration is $N_{enumerated\ ij}/g_{TestedEnum}$ CFU/g. The algorithm assumes that the enumeration is performed only on samples positive in detection.

In order to evaluate the probability of each of the lot to be rejected, 1000 (by default) Monte Carlo samples of n samples are constituted for each lot and, for each of these Monte Carlo samples, the microbiological criteria is applied (i.e., in a 2-class plan, the test is positive if $>c$ samples among n are detected; whereas in a 3-class plan, the test is positive if $>c$ samples among n are detected or if at least one sample has an estimated concentration $>M$ CFU/g). The mean number of times the lot is rejected among the Monte Carlo samples multiplied by the probability for the lot to be tested is an estimate of $P_{pos\ i}$, the probability of the lot i to be rejected.

The function updates the prevalence outputs using the Bayes' theorem. Given the probability of the lot i being contaminated pre-testing ($Prob_{UnitPos\ interv\ i}$) and the probability for the lot i to be rejected ($P_{pos\ i}$), the prevalence of contaminated lots given that they were not detected after testing ($Prob_{UnitPos\ tested\ i}$) is

$$Prob_{UnitPos\ tested\ i} = \frac{\left(1 - P_{pos\ i}\right) \times Prob_{UnitPos\ interv\ i}}{\left(1 - P_{pos\ i}\right) \times Prob_{UnitPos\ interv\ i} + \left(1 - Prob_{UnitPos\ interv\ i}\right)}$$

The mean prevalence of contaminated lots after within-lot testing (P_{Tested}, a scalar) is therefore,

$$P_{Tested} = \frac{\left(1 - \frac{\sum_{i=1}^{r} P_{pos\ i}}{r}\right) \times P_{Interv}}{\left(1 - \frac{\sum_{i=1}^{r} P_{pos\ i}}{r}\right) \times P_{Interv} + \left(1 - P_{Interv}\right)}$$

Contaminated lots detected after testing are not removed from the matrix, and therefore the input matrix N_{Interv} is returned unchanged. The function fvTesting() only updates the prevalence outputs.

2.1.6. Portioning

Data and assumptions:

The product is expected to be kept frozen during distribution and retail, and therefore, no growth of *L. monocytogenes* in these stages is modeled. During handling at home, it is assumed that the consumer removes a portion $Serv_{size}$ (g) from the pack of frozen vegetables of net weight $Unit_{size}$ g. Such a portion later becomes the serving size. The portion $Serv_{size}$ is assumed to be 50 g, as in the model of EFSA [7]. Then, the number of *L. monocytogenes* cells in this small unit can be considered to be a sample from a Beta-binomial distribution because only one portion per pack is retained (as opposed to simulating all portions that can be obtained from a pack). A moderate clustering of *L. monocytogenes* cells in the frozen vegetables contained in the pack is assumed (dispersion $b = 1$ [31]). The dispersion parameter and the number of portions that can be obtained from a pack are assumed to be independent of the microbial numbers. The likelihood that a non-contaminated portion of frozen vegetables can be obtained from a contaminated pack of frozen vegetables is taken into account.

The R function:

The function fvPortioning() mimics the microbial process of partitioning; yet, differently from the function fvPartitioningCC(). The numbers of *L. monocytogenes* in a portion ($N_{Portion\ ij}$) taken from the pack j produced in the lot i is sampled from a beta-binomial distribution

$$N_{Portion\ ij} \sim Binomial\left(N_{Interv\ ij}, Beta(b, b(n_{serv} - 1)\ )\right)$$

where n_{serv} is the rounded number of servings from a pack of frozen vegetables,

$$n_{serv} = \left\lceil \frac{Unit_{size}}{Serv_{size}} \right\rceil$$

Next, the probability that a contaminated pack ij produces a non-contaminated portion ($\pi_{0\ ij}$) is calculated as the beta-binomial probability of $N_{portion\ ij} = 0$.

$$\pi_{0\ ij} = \frac{\Gamma(b \times n_{serv}) \times \Gamma\left(N_{Packij} + b \times (n_{serv} - 1)\right)}{\Gamma(b \times (n_{serv} - 1)) \times \Gamma\left(b \times n_{serv} + N_{Packij}\right)}$$

The mean probability of non-contaminated portions from a lot i ($\overline{\pi}_{0\ i}$) is computed as,

$$\overline{\pi}_{0\ i} = \frac{\sum_{j=1}^{c} \pi_{0\ ij}}{c}$$

and is used to estimate the vector of lot-specific probabilities of contaminated portions $Prob_{UnitPos\ portion}$ and the overall prevalence of contaminated portions ($P_{Portion}$, a scalar).

$$Prob_{UnitPosportioni} = Prob_{UnitPos\ testedi} \times \overline{\pi}_{0\ i}$$

$$P_{Portion} = P_{Tested} \times \frac{\sum_{i=1}^{r} \overline{\pi}_{0i}}{r}$$

Such probabilities and the contamination matrix of frozen vegetable portions ($N_{Portion}$) are the outputs of the function fvPortioning(). The size of the contamination matrix remains as $[r, c]$.

2.1.7. Defrosting

Data and assumptions:

L. monocytogenes can survive for long periods of time in frozen foods [38] and resume growth when defrosted [35]. To this respect, Kataoka et al. [45] have shown that the lag phase time required for *L. monocytogenes* to start growth depends on the defrost/storage temperature and the type of vegetable. For instance, these authors found that, at ambient temperature (20 °C), the lag phase duration of a cocktail of *L. monocytogenes* (1/2a, 1/2b, 1/2c, 4b) inoculated in corn (0.53 h) was significantly lower than that of peas (3.40 h). On the other hand, EFSA [7] has indicated that the blanching treatment would lead to a product more supportive of *L. monocytogenes* growth (if re-contaminated) during subsequent storage after thawing. This is likely due to the absence of an antagonistic effect by the background microflora, which have been inactivated during heat treatment.

The present QRA model disregards the occurrence of a lag phase and assumes a log-linear growth of *L. monocytogenes* in the defrosted portion taken from the pack of frozen vegetables. The lognormal distribution of the exponential growth rate of *L. monocytogenes* in heat-treated vegetables at a reference temperature of 5 °C (EGR_{5C}; $\log_{10}$/h) was taken from EFSA [7], which was built using growth data collected from the scientific literature on *L. monocytogenes* in blanched vegetables including corn, green peas, carrots, broccoli, beans and asparagus stored in air, later converted to EGR_{5C} using the square-root secondary model [46]. Thus, it is considered that such a distribution involves the variability between vegetables, between *L. monocytogenes* strains, and between effects of blanching treatments. The mean μ_{EGR5C} = 0.0117 and standard deviation σ_{EGR5C} = 0.00816 in $\log_{10}$/h of the lognormal distribution for EGR_{5C}, proposed in EFSA [7], were converted to the $\log_e$ scale for use as parameters of the lognormal distribution (−4.646, 0.6296, respectively). Whereas the *L. monocytogenes* minimum temperature for growth (T_{min}) was regarded as −1.18 °C [7,46], the maximum population density (MPD) in thawed vegetables was assumed to be 8.0 $\log_{10}$ CFU/g [45].

There is limited information on the consumers' way of preparing non-RTE frozen vegetables, except for the surveys conducted in the UK [33] and later in Ireland [34]. In Ireland, the survey participants were asked about their preparation habits, and a high proportion (84–89%) recognized that vegetables such as sweet corn, peas, mushrooms, and mixed vegetables needed to be cooked prior to consumption. Those who consumed non-RTE products without cooking (~20%) indicated that their common ways of preparation were in salads and as garnish [34]. In this way, the present QRA model considered that the probability of defrosting the portion of vegetables ($p_{Defrost}$) is 0.20 since it can be assumed that 80% of the time, the consumers cook the vegetables directly from the frozen state. If defrosted, the user-selected time-temperature ($Time_{Defrost}$ [h], $Temp_{Defrost}$ [°C]) combinations for defrosting/storage may reflect several consumer practices such as placing the frozen vegetables on the counter-top at room temperature, in the refrigerator for overnight defrost, or using the microwave for quick defrost. It will be assumed that consumers leave frozen vegetables to defrost for a couple of hours on the countertop 20% of the time ($Temp_{Defrost}$ = 25 °C, $Time_{Defrost}$ = 2 h, and $p_{Defrost}$ = 0.20).

The R function:

The fvDefrost() function simulates the growth of *L. monocytogenes* in the vegetables during defrosting and is fed by the outputs of the function fvPortioning() along with $Time_{Defrost}$, $Temp_{Defrost}$, $p_{Defrost}$, and $Unit_{size}$. The algorithm begins by randomly allocating the portions of frozen vegetables that will be defrosted by the consumer at a probability $p_{Defrost}$, which is independent of the lot. According to a Bernoulli distribution with parameter $p_{Defrost}$, the defrost status matrix *Defrost* is created, where $Defrost_{ij}$ = 1 represents the defrosted portions and $Defrost_{ij}$ = 0 the non-defrosted ones.

$$Defrost_{ij} \sim Bernoulli\left(p_{Defrost}\right) \quad i = 1, 2 \ldots, r; \; j = 1, 2 \ldots, c$$

For the portions $Defrost_{ij}$ = 1, the *L. monocytogenes* growth in $\log_{10}$ CFU/g is estimated as,

$$Delta_{ij} \sim lognormal(\mu_{EGR5C}, \sigma_{EGR5C}) \left(\frac{Temp_{Defrost} - T_{min}}{5 - T_{min}} \right)^2 \times Time_{Defrost}$$

and the matrix $N_{Defrost}$ of the numbers of *L. monocytogenes* after thawing is finally computed taking into account the MPD restriction:

$$N_{Defrost\ ij} = \begin{cases} Unit_{size} \times 10^{\left(\log_{10} \frac{N_{Portion\ ij}}{Unit_{size}} + Delta_{ij}\right)}, & if\ Unit_{size} \times 10^{\left(\log_{10} \frac{N_{Portion\ ij}}{Unit_{size}} + Delta_{ij}\right)} < Unit_{size} \times 10^{MPD} \\ Unit_{size} \times 10^{MPD}, & if\ Unit_{size} \times 10^{\left(\log_{10} \frac{N_{Portion\ ij}}{Unit_{size}} + Delta_{ij}\right)} \geq Unit_{size} \times 10^{MPD} \end{cases}$$

The function acts only on the contaminated portions ($N_{Portion\ ij} > 0$), producing the contamination matrix $N_{Defrost}$, and returning unaffected the vector of lot-specific probabilities of contaminated portions ($Prob_{UnitPos\ portion}$) and the overall prevalence of contaminated portions ($P_{Portion}$, a scalar). It is noteworthy to mention that the growth model does not allow for the come-up time.

2.1.8. Cooking

Data and assumptions:

According to a survey at retail in the UK [33], 77% of the frozen vegetable packages recommended cooking before consumption, 4% specified being RTE, and on 19% of the packages, there was no indication for cooking. In Ireland, 9.3% of the frozen vegetable samples did not have any non-RTE label or cooking instructions printed on the packaging indicating that frozen vegetables must be fully cooked. Of public health impact concern is tackling those non-intended uses of non-RTE frozen vegetables as alternatives of consumption. The present QRA model assumed in its scenarios a probability of cooking a serving (p_{Cook}) of 1 (always cooked) or 0 (never cooked).

It is expected that the extent of inactivation of *L. monocytogenes* depends on the time of cooking and type of cooking. In EFSA's generic model [7], the reduction by cooking was assumed to follow a Pert distribution with a minimum 1.0 $\log_{10}$, mode 3.0 $\log_{10}$, and maximum 9.0 $\log_{10}$. The present QRA model represented such variability as a triangular distribution ($Cook_{min}$, $Cook_{mode}$, $Cook_{max}$) with the same parameters.

The R function:

The function fvCooking() simulates the microbial process of reduction of *L. monocytogenes* in the (defrosted) vegetables during cooking. It takes the outputs of the function fvDefrost() and the parameters p_{Cook}, $Cook_{min}$, $Cook_{mode}$, $Cook_{max}$. The algorithm starts by randomly flagging the portions of defrosted vegetables that will be cooked, regardless of the lot or their contamination status. According to Bernoulli distribution with parameter p_{Cook}, the cooked status matrix *Cook* is produced, where $Cook_{ij}$ = 1 represents the portions to be cooked, and $Cook_{ij}$ = 0 the ones that will not.

$$Cook_{ij} \sim Bernoulli(p_{Cook}) \quad i = 1, 2 \ldots, r; \ j = 1, 2 \ldots, c$$

For the portions $Cook_{ij}$ = 1, the probability of survival of a microbial cell $p_{SurviveCooking\ ij}$ is sampled as,

$$p_{SurviveCooking\ ij} = 10^{-Triang\ (Cook_{min},\ Cook_{mode}, Cook_{max})}$$

and the number of surviving cells after cooking N_{Cook} is obtained by,

$$N_{Cook\ ij} \sim Binomial\left(N_{Defrost\ ij}, p_{SurviveCooking\ ij}\right) \qquad N_{Cook\ ij} > 0$$

To update the prevalence estimates after cooking, the probability of at least one cell surviving in a cooked portion ($atLeastOne_{ij}$) is calculated:

$$atLeastOne_{ij} = 1 - \left(1 - p_{SurviveCooking\ ij}\right)^{N_{Defrost\ ij}}$$

The probability vector of contaminated lots after cooking $Prob_{UnitPos\ cooked\ i}$ and the mean prevalence of contaminated portions (after the potential cooking event) P_{Cooked} are then updated,

$$Prob_{UnitPoscookedi} = Prob_{UnitPos\ portioni} \times \left\{ (1 - p_{Cook}) + p_{Cook} \times \frac{\sum_{i=1}^{c} atLeastOne_{ij}}{c} \right\}$$

$$P_{Cooked} = P_{Portion} \times \left\{ (1 - p_{Cook}) + p_{Cook} \times \frac{\sum_{i=1}^{r} \sum_{j=1}^{c} atLeastOne_{ij}}{r \times c} \right\}$$

The function fvCooking() returns the contamination matrix N_{Cooked} containing the load of *L. monocytogenes* in the servings originated from the contaminated lots, the lot-specific probability of contaminated servings to be consumed ($Prob_{UnitPos\ cooked}$, a vector) and the mean prevalence of contaminated servings to be consumed (P_{Cooked}, a scalar).

2.2. Risk Characterization

Several dose-response relationships are available for *L. monocytogenes* [47]. For purposes of comparison with the model of Zoellner et al. [8], the dose-response model chosen to estimate the risk of listeriosis per serving in the susceptible population was that of FAO-WHO [48]. According to this exponential model, each ingested *L. monocytogenes* cell has an independent probability r of causing invasive listeriosis, which is assumed constant within a given population. FAO-WHO [48] inferred a median value for r of 1.06×10^{-12} for the "population with increased susceptibility". The function DRQuick() from the doseresponsemodels R package [47] was employed to calculate the marginal probabilities of invasive listeriosis in the susceptible population $RiskServing_{ij}$ from the input matrix $N_{Cooked\ ij}$, containing the *L. monocytogenes* doses (CFU) in servings j from contaminated lots i. The mean risk for every lot i ($RiskLot_i$) was calculated as a risk averaged across servings j, and weighed by the lot-specific probability $Prob_{UnitPos\ cooked\ i}$:

$$RiskLot_i = \frac{\sum_{j=1}^{c} RiskServing_{ij} \times Prob_{UnitPos\ cooked\ i}}{c}$$

2.3. QRA Model's Ouputs

The model's outputs were summarised at three stages: end of processing, end of exposure assessment (point of consumption), and risk characterization. The descriptors at the *end of processing* were calculated from the outputs of the function fvTesting(), and included descriptive statistics (mean, median, and 95% confidence interval) of the mean concentration of *L. monocytogenes* (CFU/g) in the fraction of contaminated lots, the prevalence of contaminated packs, the probability that a contaminated pack contains more than 10 CFU *L. monocytogenes* per g frozen vegetables, and the probability that a contaminated pack contains more than 100 CFU *L. monocytogenes* per g frozen vegetables. The model's descriptors at the *point of consumption* were computed from the outputs of the function fvCooking(), and encompassed descriptive statistics (mean, median, and 95% confidence interval) of the concentration of *L. monocytogenes* in any serving, the prevalence of contaminated servings, the probability that a contaminated serving contains more than 10 CFU *L. monocytogenes* per g vegetables, and the probability that a contaminated serving contains more than 100 CFU *L. monocytogenes* per g vegetables. The descriptors for risk characterization include the summary statistics, mean, median, and 2.5 and 97.5 percentiles of the lot-level mean risk per serving *RiskLot*.

2.4. QRA Model's Functionality: Reference and What-If Scenarios

To illustrate the utility of the QRA model, a reference scenario was run, in addition to three processing-related scenarios, which simulated changes in the initial *L. monocytogenes* contamination of the vegetables, in the blanching temperature, and in the contamination level of the processing environment.

(a) Reference, which is the baseline scenario constituted of parameters whose values represent as much as possible the current situation supported by actual data and, in their absence, reasonable assumptions.
(b) Initial contamination, assessed by scenarios representing mean initial *L. monocytogenes* concentrations in pre-blanched vegetables that are lower ($\mu_{C0} - 1\ \log_{10}$ CFU/g) or higher ($\mu_{C0} + 1\ \log_{10}$ CFU/g) than the reference scenario ($\mu_{C0} = 1.038\ \log_{10}$ CFU/g).
(c) Blanching temperature, assessed by scenarios with temperatures lower than the reference scenario ($Temp_{Blanch}$ = 71 and 77 °C).
(d) Recontamination after blanching, evaluated by scenarios representing total loads of *L. monocytogenes* on equipment surfaces after blanching in a lot, lower ($N_{equip}/100$) and higher (N_{equip}*100) than the reference scenario (N_{equip} = 10000 CFU).

To assess the impact of the most important consumer's handling practice (i.e., cooking the non-RTE product), all of the above scenarios were run in two modalities, assuming that vegetables are not cooked ($p_{Cook} = 0$) and that vegetables are always cooked ($p_{Cook} = 1$). Models were run for 2000 lots (*r*) and 5000 units per lot (*c*).

2.5. QRA Model's Implementation

All the functions described in Section 2.1 were programmed in R version 4.4.1 [49] and compiled in the package qraLm, which can be installed from the Github repository: https://github.com/WorldHealthOrganization/qraLm, accessed on 10 September 2024. The reference manual can be found at: https://WorldHealthOrganization.github.io/qraLm/reference/, accessed on 10 September 2024.

3. Results and Discussion

3.1. Reference Scenario and Comparison with Other QRA Models

According to the reference scenario, the mean lot concentration of *L. monocytogenes*, determined after simulating 2000 lots, was found to be very low (95% credibility interval: 8.0×10^{-7}–2.5×10^{-4}) with an overall prevalence of 9.40% of contaminated 500 g packs at the end of processing. Furthermore, virtually all of these contaminated packs would have *L. monocytogenes* concentrations lower than 10 CFU/g (Table 3). The mean concentration in a contaminated pack at the end of processing would be 3.35×10^{-4} CFU/g (95% credibility interval: 0–0.004 CFU/g).

Surveys' data on the prevalence of *L. monocytogenes* in non-RTE frozen vegetables sampled at the end of processing and retail displayed variable estimates (Table 4), ranging from 0% [25,38,50–52] to 45.6% [38], yet a random-effects meta-analysis of the ten primary sources (Table 4) produced a pooled estimate (5.66%; 95% CI: 2.28–13.3%) that was not distant from the prevalence estimated by the QRA model (9.40%). In terms of enumeration analysis, Moravkova et al. [53], Willis et al. [33], and FSAI [34] found that all *L. monocytogenes*-positive samples presented concentrations below 100 CFU/g, which is also in agreement with the reference scenario of the QRA model.

The simulation for the reference scenario predicts that the *L. monocytogenes* prevalence in uncooked 50 g servings is 1.37%, although such contaminated servings would have a very low probability of exceeding the concentration of 10 CFU/g (0.0046%; Table 5). In comparison with the QRA model of Zoellner et al. [8], our model estimated at least a five-fold lower number of *L. monocytogenes* in a contaminated serving, whereas Zoellner et al. [8] predicted a median of 32 CFU in a 140 g uncooked serving, our model estimated a median of 2 CFU (95% CI: 1.0–6.0 CFU) in a 50 g uncooked contaminated serving (results not tabulated).

Table 3. Simulation outcomes of the exposure assessment model of *L. monocytogenes* in 500 g packs of non-RTE frozen vegetables at the end of processing for the reference and selected what-if scenarios.

Scenario	Mean Counts (CFU/g) in Contaminated Lots (Mean; Median; [95% CI])	Prevalence of Contaminated Packs	P(N > 10 CFU/g in a Contaminated Pack)	P(N > 100 CFU/g in a Contaminated Pack)
Reference (μ_{C0} = 1.038 $\log_{10}$ CFU/g; $Temp_{Blanch}$ = 83 °C; N_{equip} = 10^4 CFU)	0.0003; 5.72 × 10^{-5} [8.00 × 10^{-7}–0.0025]	0.094	0	0
Initial mean contamination				
Lower (μ_{C0} = 0.038 $\log_{10}$ CFU/g)	2.12 × 10^{-4}; 6.80 × 10^{-6} [4.00 × 10^{-7}–0.0024]	0.050	0	0
Higher (μ_{C0} = 2.038 $\log_{10}$ CFU/g)	0.0017; 0.0005 [3.20 × 10^{-6}–0.0120]	0.267	0	0
Blanching temperature				
Very low ($Temp_{Blanch}$ = 71 °C)	0.0148; 0.0038 [2.76 × 10^{-5}–0.1129]	0.560	0	0
Low ($Temp_{Blanch}$ = 77 °C)	0.0016; 0.0005 [3.20 × 10^{-6}–0.0117]	0.264	0	0
Recontamination				
Lower (N_{equip} = 10^2 CFU)	1.60 × 10^{-4}; 4.14 × 10^{-5} [8.00 × 10^{-7}–0.0012]	0.054	0	0
Higher (N_{equip} = 10^6 CFU)	0.0178; 5.72 × 10^{-5} [8.00 × 10^{-7}–0.2345]	0.158	0	0
Worst-case (μ_{C0} = 2.038 $\log_{10}$ CFU/g; $Temp_{Blanch}$ = 71 °C; N_{equip} = 10^6 CFU)	0.1641; 0.0513 [0.0003–1.1550]	0.818	0.0004	0

Table 4. *L. monocytogenes* prevalence in minimally processed and fresh whole vegetables sampled at packinghouse or at retail.

Country	Product	Sample Size	Positive Enrichment	Prevalence (%)	Source [1]
End of processing					
Spain	Frozen vegetables	906	11	1.21	Aguado et al. [54]
Poland	Frozen mixed vegetables	248	113	45.6	Pappelbaum et al. [38]
	Frozen leeks	29	0	0.00	
	Frozen onions	45	0	0.00	
	Frozen vegetables	73	17	23.3	
	Frozen corn	12	1	8.33	
	Frozen green peas	110	22	20.0	
Retail					
Turkey	Frozen pepper	216	0	0.00	Lee et al. [52]
Poland	Frozen vegetable mix (broccoli, carrot, green beans, peas, corn, red beans, onions, pepper, potatoes)	9100	504	5.54	Skowron et al. [55]
Spain	Frozen vegetables	1750	31	1.77	Vitas et al. [56]
Portugal	Frozen sliced green peppers	31	7	22.6	Mena et al. [51]
	Frozen sliced red peppers	33	0	0.00	
	Frozen peas	27	4	14.8	
Czech Republic	Frozen vegetables (carrots, broccoli, peas, mix, sprout)	66	0	0.00	Vojkovska et al. [25]
Multiple	Frozen vegetables (peas, carrots, corn)	43	9	20.9	Moravkova et al. [53]
	Frozen vegetables	673	69	10.3	Willis et al. [33]
Ireland	Frozen vegetables	366	21	5.73	FSAI [34]

[1] Data extracted from the Pathogens-in-Foods database [12], except FSAI [34]. In all sources, a subsample weight of 25 g was used for enrichment.

Table 5. Simulation outcomes of the exposure assessment model of *L. monocytogenes* in 50 g servings of non-RTE frozen vegetables for the reference and selected what-if scenarios, as solved by the intended use (cooking) and the non-intended use (not cooking).

Scenario	Cooking	Counts (CFU/g) in Any Serving (Mean; Median; [95% CI])	Prevalence of Contaminated Servings	P(N > 10 CFU/g in a Contaminated Serving)	P(N > 100 CFU/g in a Contaminated Serving)
Reference (μ_{C0} = 1.038 $\log_{10}$ CFU/g; $Temp_{Blanch}$ = 83 °C; N_{equip} = 10^4 CFU)	No	0.0006; 6.59×10^{-5} [2.41×10^{-7}–0.0030]	0.0137	4.59×10^{-5}	0
	Yes	5.21×10^{-7}; 9.80×10^{-8} [3.91×10^{-10}–3.95×10^{-6}]	2.61×10^{-5}	0	0
Initial contamination					
Lower (μ_{C0} = 0.038 $\log_{10}$ CFU/g)	No	3.15×10^{-4}; 8.79×10^{-6} [1.24×10^{-7}–0.0021]	0.0080	4.67×10^{-5}	0
	Yes	2.96×10^{-7}; 1.16×10^{-8} [1.95×10^{-10}–3.44×10^{-6}]	1.48×10^{-5}	0	0
Higher (μ_{C0} = 2.038 $\log_{10}$ CFU/g)	No	0.0027; 0.0005 [1.30×10^{-6}–0.0139]	0.0545	0.0001	0
	Yes	2.54×10^{-6}; 8.41×10^{-7} [2.00×10^{-9}–175×10^{-5}]	0.0001	0	0
Blanching temperature					
Very low ($Temp_{Blanch}$ = 71 °C)	No	0.0228; 0.0035 [1.32×10^{-5}–0.1244]	0.2011	0.0005	2.98×10^{-5}
	Yes	1.75×10^{-5}; 5.57×10^{-6} [2.14×10^{-8}–0.0001]	0.0008	0	0
Low ($Temp_{Blanch}$ = 77 °C)	No	0.0027; 0.0005 [1.30×10^{-6}–0.0136]	0.0537	0.0001	0
	Yes	2.46×10^{-6}; 8.24×10^{-7} [2.00×10^{-9}–1.67×10^{-5}]	0.0001	0	0
Recontamination					
Lower (N_{equip} = 10^2 CFU)	No	2.63×10^{-4}; 4.12×10^{-5} [2.41×10^{-7}–0.0015]	0.0060	9.50×10^{-6}	0
	Yes	2.52×10^{-7}; 6.15×10^{-8} [3.92×10^{-10}–1.91×10^{-6}]	1.26×10^{-5}	0	0
Higher (N_{equip} = 10^6 CFU)	No	0.0267; 6.70×10^{-5} [2.41×10^{-7}–0.1882]	0.0876	0.0017	0.0001
	Yes	1.72×10^{-5}; 9.79×10^{-8} [3.91×10^{-10}–0.0002]	0.0008	0	0
Worst-case (μ_{C0} = 2.038 $\log_{10}$ CFU/g; $Temp_{Blanch}$ = 71 °C; N_{equip} = 10^6 CFU)	No	0.2712; 0.0212 [0.0001–1.5584]	0.5110	0.0050	0.0002
	Yes	0.0001; 5.62×10^{-5} [2.04×10^{-7}–0.0007]	0.0062	0	0

According to our model, cooking the non-RTE vegetables decreases the prevalence of *L. monocytogenes* in 50 g servings to 0.0026%, yielding for the mean count in any serving a negligible value of 5.21×10^{-7} CFU/g (Table 5). In our model, if all servings were uncooked, the mean risk of listeriosis per serving in the susceptible population would be 2.93×10^{-14} (median 3.50×10^{-15}; 95% CI: 1.28×10^{-17}–1.61×10^{-13}); which reveals a risk level far lower than the one predicted by the model of Zoellner et al. [8] for uncooked servings (median 7.8×10^{-12} in the susceptible population).

The higher levels of risk obtained in the model of Zoellner et al. [8] can be explained at least by three facts: (1) the higher probability of thawing (0.5, as opposed to 0.2 in our model); (2) the possibility of longer storage at room temperature (from 0.25 to 10 h, as opposed to a fixed value of 2 h in our model); and (3) the absence of the processing stage of blanching in their model, which is a step that largely reduces the initial contamination.

To this respect, in our model, blanching decreased the mean *L. monocytogenes* load in contaminated lots from 1.417 CFU/g in pre-blanched vegetables to 0.00020 CFU/g in post-blanch vegetables. In turn, the QRA model of Zoellner et al. [8] begins from the

L. monocytogenes concentration in the lot after manufacturing (post-blanch vegetables), assuming a mean of 0.30 CFU/g, which is 1500-fold higher than our post-blanching mean concentration. Moreover, through sensitivity analysis, these authors concluded that it was the high initial *L. monocytogenes* concentration in the manufactured lot that was the most important factor driving the high risk of listeriosis in the uncooked servings.

In the present QRA model, cooking the non-RTE frozen vegetables decreases the mean risk of listeriosis per serving in a lot by 3.0 $\log_{10}$, from 2.935×10^{-14} (uncooked servings) to 2.765×10^{-17} (cooked servings; Table 6); which is an outcome in line with the findings of Zoellner et al. model [8], who found risk levels lower than 10^{-16} linked to the cooked servings (actual mean risk value not presented by these authors given the floating point limitation of the simulation software used).

Table 6. Statistics of the mean risk of invasive listeriosis per lot of non-RTE frozen vegetables in the susceptible population for the reference and selected what-if scenarios, as solved by the intended use (cooking) and the non-intended use (not cooking). The logarithm base 10 of the mean risk reduction due to cooking is shown ($\log_{10}$RR).

Scenario	Cooking	Mean	Median	2.5 pct	97.5 pct	$\log_{10}$ RR
Reference (μ_{C0} = 1.038 $\log_{10}$ CFU/g; $Temp_{Blanch}$ = 83 °C; N_{equip} = 10^4 CFU)	No	2.935×10^{-14}	5.446×10^{-15}	2.183×10^{-17}	2.187×10^{-13}	
	Yes	2.765×10^{-17}	5.184×10^{-18}	2.077×10^{-20}	2.086×10^{-16}	3.03
Initial contamination						
Lower (μ_{C0} = 0.038 $\log_{10}$ CFU/g)	No	1.673×10^{-14}	6.497×10^{-16}	1.088×10^{-17}	1.882×10^{-13}	
	Yes	1.569×10^{-17}	6.183×10^{-19}	1.035×10^{-20}	1.803×10^{-16}	3.03
Higher (μ_{C0} = 2.038 $\log_{10}$ CFU/g)	No	1.452×10^{-13}	4.625×10^{-14}	1.114×10^{-16}	1.032×10^{-12}	
	Yes	1.347×10^{-16}	4.444×10^{-17}	1.059×10^{-19}	9.267×10^{-16}	3.03
Blanching temperature						
Very low ($Temp_{Blanch}$ = 71 °C)	No	1.209×10^{-12}	3.171×10^{-13}	1.195×10^{-15}	9.023×10^{-12}	
	Yes	9.275×10^{-16}	2.950×10^{-16}	1.137×10^{-18}	6.263×10^{-15}	3.12
Low ($Temp_{Blanch}$ = 77 °C)	No	1.426×10^{-13}	4.556×10^{-14}	1.112×10^{-16}	9.503×10^{-13}	
	Yes	1.304×10^{-16}	4.363×10^{-17}	1.058×10^{-19}	8.818×10^{-16}	3.04
Recontamination						
Lower (N_{equip} = 10^2 CFU)	No	1.400×10^{-14}	3.422×10^{-15}	2.183×10^{-17}	1.035×10^{-13}	
	Yes	1.334×10^{-17}	3.257×10^{-18}	2.077×10^{-20}	9.769×10^{-17}	3.02
Higher (N_{equip} = 10^6 CFU)	No	1.418×10^{-12}	5.466×10^{-15}	2.183×10^{-17}	1.809×10^{-11}	
	Yes	9.106×10^{-16}	5.194×10^{-18}	2.076×10^{-20}	1.148×10^{-14}	3.19
Worst-case (μ_{C0} = 2.038 $\log_{10}$ CFU/g; $Temp_{Blanch}$ = 71 °C; N_{equip} = 10^6 CFU)	No	1.437×10^{-11}	3.725×10^{-12}	1.143×10^{-14}	1.077×10^{-10}	
	Yes	6.866×10^{-15}	2.957×10^{-15}	1.080×10^{-17}	3.883×10^{-14}	3.32

The model of EFSA [7] also assessed consumer's handling scenarios, and in the case of cooking, they predicted that in contrast to consuming uncooked servings, the consumption of cooked servings reduces the mean risk of listeriosis by 3.2 $\log_{10}$ in the female susceptible population and by 3.5 $\log_{10}$ in the male susceptible population, which are both reduction levels of comparable magnitude to our estimation. Nonetheless, in absolute terms, EFSA's model [7] predicted higher mean risk of illness at 4.0×10^{-10} and 1.9×10^{-9} for uncooked servings and 2.3×10^{-13} and 5.3×10^{-13} for cooked servings, in the female and male susceptible populations, respectively.

The higher level of listeriosis risk per serving estimated by EFSA [7]—in contrast with our results—arises as a consequence of their more conservative assumptions, namely (1) a higher initial mean concentration (which can reach up to 5 $\log_{10}$ CFU/g); (2) the longer time between vegetables being defrosted and consumed for the uncooked servings' risk estimation (mean of 9 h); and (3) the longer time between vegetables being defrosted and cooked for the cooked servings' risk estimation (mean of 4.6 h). By contrast, in our model, not all portions are assumed to be thawed, but only 20% of them, meaning that only in such a fraction *L. monocytogenes* growth is simulated.

3.2. What-If Scenarios

According to the present QRA model, decreasing the *L. monocytogenes* mean concentration in pre-blanched vegetables by 1.0 $\log_{10}$ (Scenario: μ_{C0} = 0.038 $\log_{10}$ CFU/g) would generate practically the same effect as reducing the total *L. monocytogenes* cells on equipment by 2.0 $\log_{10}$ CFU in a processing lot (Scenario: $N_{equip} = 10^2$ CFU). These two scenarios would produce means of *L. monocytogenes* prevalence in packs at the end of processing of 5.0% and 5.4%—yet in very low levels of contamination (97.5th percentile of concentration of contaminated lots of 0.0024 and 0.0012 CFU/g, respectively; Table 3); which would in turn yield mean proportions of contaminated uncooked servings of 0.8% and 0.6%, respectively, and of contaminated cooked servings of 0.00148% and 0.00126%, respectively (Table 5). In both scenarios, the model estimated negligible probabilities of finding contaminated servings at the time of consumption with more than 10 CFU/g *L. monocytogenes*, even if non-RTE frozen vegetables are consumed uncooked (0.00467% and 0.00095%; Table 5).

Increasing the total *L. monocytogenes* cells on equipment in a processing lot (Scenario: $N_{equip} = 10^6$ CFU) was found to have a stronger impact on exposure than increasing the *L. monocytogenes* mean concentration in pre-blanched vegetables by 1.0 $\log_{10}$ (Scenario: μ_{C0} = 2.038 $\log_{10}$ CFU/g); as attested by the higher prevalence of contaminated uncooked servings (8.76% versus 5.45%) and cooked servings (0.08% versus 0.01%; Table 5). Higher contamination levels on equipment produce a greater spread of the microbial cells in the lot being processed, causing the microbial concentration to be not only higher on average (0.0178 CFU/g versus 0.0017 CFU/g; Table 3) but also more variable from lot to lot (97.5 percentile of 0.2345 CFU/g versus 0.0120 CFU/g; Table 3) and more variable between servings, particularly noticeable in the uncooked ones (97.5 percentile of 0.1882 CFU/g versus 0.0139 CFU/g; Table 5).

According to AFFI [50] and PROFEL [35], blanching by immersion in hot water or steam can be carried out at temperatures as low as 65 °C, subject to the type of commodity, size of pre-cut pieces, and seasonal variability. Thus, what-if scenarios with blanching temperatures (71 °C, 77 °C) lower than the one assumed for the reference scenario (83 °C) were conducted. Lowering the blanching temperature to 71 °C would produce a six-fold higher prevalence of contaminated packs at the end of processing (56.0%) in comparison with the reference scenario, with higher *L. monocytogenes* concentrations in contaminated lots (mean 0.0148 CFU/g; 97.5 percentile 0.113 CFU/g; Table 3). Correspondingly, its respective uncooked servings would have a high prevalence of contamination (20.1%), with increased concentrations of *L. monocytogenes* per serving of up to 0.124 CFU/g (97.5 percentile; Table 5). Nonetheless, it should be noticed that even though a low blanching temperature produced the highest contamination prevalence estimates among the processing-related scenarios evaluated, the numbers of *L. monocytogenes* still remained low: at the end of processing, no pack exceeded the *L. monocytogenes* concentration in frozen vegetables of 10 CFU/g; whereas, at consumption, on average 0.05% of the contaminated servings would exceed 10 CFU/g (Table 5).

The worst-case scenario illustrates a scenario of multiple co-occurring risk factors during processing—namely, a high *L. monocytogenes* concentration in pre-blanched vegetables, greater contamination on equipment post-blanching, and a milder blanching treatment. It should be kept in mind, however, that the worst-case scenario of multiple co-occurring risk factors evaluated in this study is rare because (1) the concentrations of *L. monocytogenes* in raw produce are very likely to be low and (2) normally, the cross-contamination of food from the processing environment will not result in high numbers, since *L. monocytogenes* numbers on surfaces are generally low (Zoellner et al. [8]).

In the worst-case scenario, the prevalence of contaminated packs at the end of processing was greatly increased (81.8%; Table 3), with contaminated lots attaining higher concentrations at a mean of 0.1641 CFU/g (equivalent to a mean of 82 CFU in 500 g packs from a contaminated lot). In such a rare scenario, not cooking the vegetables would be linked to a high mean risk per lot of 1.437×10^{-11} (Table 6), exposing the consumers to contaminated uncooked servings at a probability of 51.1%, and when contaminated,

0.50% of them would have concentrations higher than 10 CFU/g (Table 5). Yet, proper (intended) consumer handling at home (i.e., cooking) would decrease the prevalence of contaminated servings to 0.62% with very low concentration levels that hardly ever would exceed 10 CFU/g.

Comparing the risk per lot estimates between the scenarios evaluated, it was found that, opposite to the mean concentration of *L. monocytogenes* in pre-blanched vegetables (Figure 2) and blanching temperature (Figure 3), changes in the microbial load on post-blanching equipment produced a very low effect on the average of the mean risk per lot and a higher effect on its between-lot variability, in particular, at the high contamination level on equipment of 10^6 CFU (Figure 4).

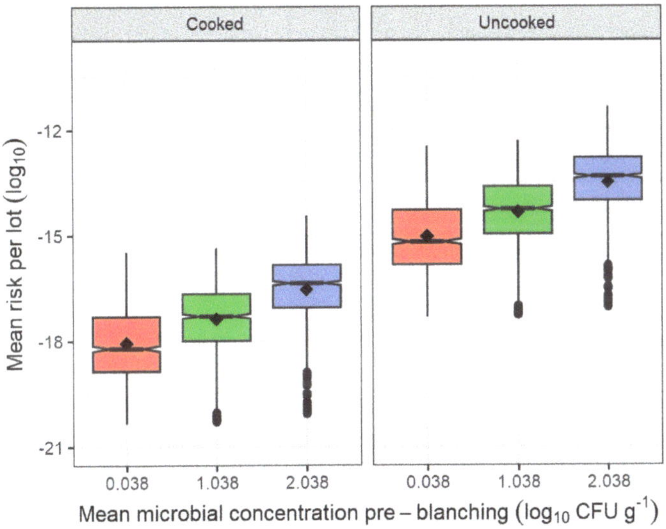

Figure 2. Effect of mean concentration of *L. monocytogenes* in pre-blanched vegetables and home cooking status on the mean risk of listeriosis in a serving per lot.

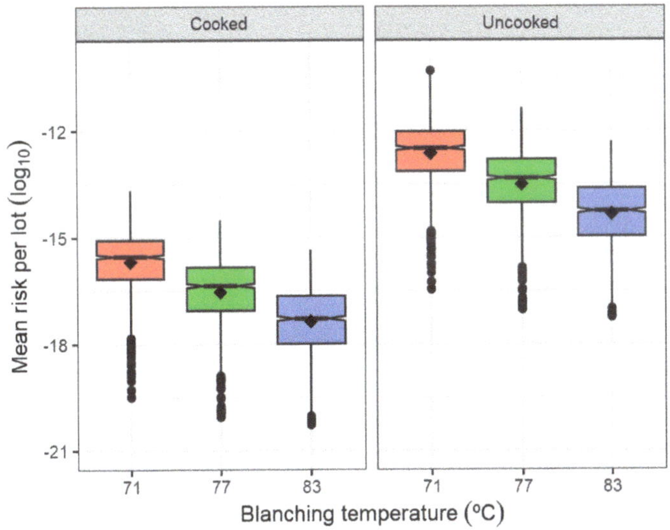

Figure 3. Effect of blanching temperature and home cooking status on the mean risk of listeriosis in a serving per lot.

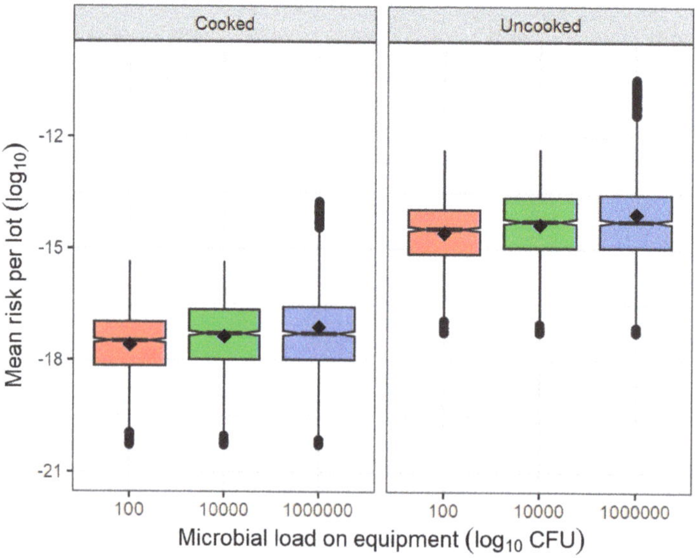

Figure 4. Effect of microbial load on equipment processing vegetables after blanching and home cooking status on the mean risk of listeriosis in a serving per lot.

Therefore, according to the present QRA model, it can be deduced that, on a batch basis, the contribution of the incoming raw produce to the final risk is greater than the contribution of re-contamination post-blanching. In environments that have been proven to harbor *L. monocytogenes* niches and lead to frequent cross-contamination, such as those of deli meats retail establishments, a QRA model [57] demonstrated that minimizing the incoming *L. monocytogenes* cells and the transfers from the environment or niches directly decreases the predicted risk of illness; and that more frequent environmental contamination events have more impact on risk than a greater number of *L. monocytogenes* cells per cross-contamination event.

Simulation results have shown the impact of the initial contamination of raw materials on the prevalence of *L. monocytogenes* in frozen vegetables at the end of processing and, consequently, the final listeriosis risk, which underlines the need for applying preventive measures of good agricultural practices. In addition, in the processing facilities, the maintenance of good hygiene conditions, including both food contact surfaces and non-food contact surfaces, cannot be sufficiently overemphasized and constitutes a strong factor that increases the likelihood of contaminated frozen vegetables, especially if cross-contamination events occur in the post-blanching processing environment such as conveyor belts and freezing equipment. The most critical steps should be considered when establishing a routine monitoring program for *L. monocytogenes* in the food processing environment of frozen vegetables. Further, sanitation standard operating procedures and corrective actions should be implemented to reduce *L. monocytogenes* contamination in the facilities. EFSA [7] has conducted an in-depth assessment of the pre-requisite program activities in the production process of blanched frozen vegetables, mainly including raw materials, infrastructure, cleaning and disinfection, water and air control, and personnel. Furthermore, EFSA [7] has provided recommendations on routine monitoring, processing environment monitoring, and end-product monitoring for processors of blanched frozen vegetables. In the same year, the European Association of Food and Vegetable Processors [35] published hygiene guidelines for the control of *L. monocytogenes* in non-RTE frozen vegetables, aiming to prevent the pathogen from colonizing and persisting in complex biofilm formation and to prevent contamination with the pathogen post thermal treatment during further handling and before packaging. PROFEL [35] details the environmental controls that need to be established to corroborate the effectiveness of the implemented pre-requisite programs and HACCP plan and to assess the potential accumulation of *L. monocytogenes* in the broader production environment.

The comparison of what-if scenarios demonstrated that the final risk was most sensitive to whether the serving was cooked or not (Figures 2–4). Across all scenarios, the effect of cooking on the final risk proved to be consistent, in the order of reducing the mean risk per lot by 3.02–3.32 $\log_{10}$ (Table 6). This quantification is very pertinent in light of the increased popularity of smoothies, dips, and similar uncooked blended preparations that can be potentially elaborated from non-RTE frozen vegetables [58]. The absence of a pre-cooking step can have serious food safety consequences, as demonstrated by the *L. monocytogenes* outbreak associated with frozen corn that occurred in Europe between 2015 and 2018 and caused 53 cases and 10 deaths [7]. Therefore, frozen raw vegetables used in uncooked food preparations by consumers should not be overlooked as a source of potential foodborne outbreaks. The present model demonstrated that such an improper practice can turn a low-risk food (2.765×10^{-17}) into a medium-risk food (2.935×10^{-14}). Therefore, in addition to processing environment monitoring activities and end-product testing at the processing stage, consumers should be clearly informed through proper package labeling that these products are of a non-RTE nature and that thorough cooking should be ensured prior to consumption [7]. Further indications should also be given on how these products should be stored and the maximum time to be kept unfrozen in the refrigerator. Likewise, if the non-RTE frozen vegetable is intended for catering, the caterer should be instructed to ensure proper handling and adequate heat treatment during preparation. According to the reference scenario, the mean risk per serving is very low; and would consequently be the probability of observing more than one case per lot. However, this model is not contradictory with the reported outbreaks linked to these products since the implicated frozen vegetables covered long production periods and possibly involved multiple lots caused by a persistent strain belonging to the group of the most virulent strains. If we were to consider an extreme scenario of highly contaminated incoming vegetables (μ_{C0} = 2.038 $\log_{10}$ CFU/g), frequent failure to disinfect processing equipment ($N_{equip} = 10^6$ CFU, $P_{cc} = 0.30$), milder blanching process ($Temp_{Blanch}$ = 68 °C, $Time_{Blanch}$ = 0.75 min) and unintended use in salads or smoothies ($p_{Cook} = 0$), the lot-level mean risk per serving would be as high as 5.335×10^{-8} (median: 4.727×10^{-9}; 95% CI: 1.244×10^{-11}–4.002×10^{-7}) in the elderly population (as per the dose-response model of Pouillot et al. [59]). It is, therefore, very likely that the outbreaks that occurred in the USA and Europe would have arisen from an unusual combination of failures at both processing and consumer levels that turned a regularly negligible-risk product such as frozen vegetables into a high-risk product.

In the present study, in order to illustrate the QRA model's functionality, only a few what-if scenarios (i.e., parameters) have been evaluated. However, it is possible to obtain an in-depth assessment of the effect of multiple parameters (e.g., prevalence of *L. monocytogenes* in pre-blanch vegetables, blanching time, probability of recontamination, probability of defrosting, defrosting time/temperature, etc.) on the final risk of listeriosis, through the installation of the R package qraLm (https://github.com/WorldHealthOrganization/qraLm, accessed on 10 September 2024) containing all model's functions and an accompanying shiny interactive dashboard.

Author Contributions: Conceptualization, U.G.-B., R.P. and M.S.; methodology, U.G.-B., R.P. and M.S.; software, R.P., U.G.-B., L.G., M.S. and V.C.; validation, U.G.-B. and R.P.; formal analysis, R.P., U.G.-B., L.G., V.C. and M.S.; investigation, U.G.-B., R.P., J.D.O.M., A.H., V.C., L.G. and M.S.; resources, M.S.; data curation, U.G.-B., L.G., J.D.O.M. and A.H.; writing—original draft preparation, U.G.-B. and R.P.; writing—review and editing, U.G.-B., R.P., L.G., V.C. and M.S.; visualization, U.G.-B., R.P., L.G., V.C., M.S., J.D.O.M., A.H., A.A., Q.D., M.J.S. and J.K.; supervision, M.S.; project administration, M.S.; funding acquisition, M.S. All authors have read and agreed to the published version of the manuscript.

Funding: This research received no external funding.

Institutional Review Board Statement: Not applicable.

Informed Consent Statement: Not applicable.

Data Availability Statement: The original contributions presented in the study are included in the article, further inquiries can be directed to the corresponding authors.

Acknowledgments: The FAO/WHO Joint Expert Meeting on Microbiological Risk Assessment (JEMRA) on *Listeria monocytogenes* in foods (Part 1), held in Rome, Italy, from 24 to 28 October 2022, recommended the development of a quantitative risk assessment model for *Listeria monocytogenes* in non-ready-to-eat frozen vegetables that considers processing and consumer's handling. The model was finalized during the JEMRA meeting on *Listeria monocytogenes* in foods Part 2: Risk Assessment Models, which was held in Geneva, Switzerland, from 29 May to 2 June 2023. The authors thank the JEMRA experts for Part 1 (https://www.who.int/publications/m/item/jemra-of-listeria-monocytogenes-in-foods, accessed on 30 July 2024) and Part 2 (https://www.who.int/publications/m/item/jemra-of-listeria-monocytogenes-in-foods-part-2-risk-assessment-models, accessed on 30 July 2024) as well as FAO/WHO secretariat for their suggestions and feedback.

Conflicts of Interest: Author R.P. is self-employed. He participated in this work in his own personal capacity. The remaining authors declare that the research was conducted in the absence of any commercial or financial relationships that could be construed as potential conflicts of interest.

References

1. CDC. *Multistate Outbreak of Listeriosis Linked to Frozen Vegetables (Final Update)*; Centers for Disease Control and Prevention: Atlanta, GA, USA, 2016. Available online: https://archive.cdc.gov/#/details?url=https://www.cdc.gov/listeria/outbreaks/frozen-vegetables-05-16/index.html (accessed on 25 July 2022).
2. EFSA (European Food Safety Authority); ECDC (European Centre for Disease Prevention and Control). *Multi-Country Outbreak of Listeria monocytogenes Serogroup IVb, Multi-Locus Sequence Type 6, Infections Linked to Frozen Corn and Possibly to Other Frozen Vegetables—First Update*; EFSA Supporting Publication: Oxford, UK, 2018; Volume 15, p. 1448E. [CrossRef]
3. Truchado, P.; Gil, M.I.; Querido-Ferreira, A.P.; López Capón, C.; Álvarez-Ordoñez, A.; Allende, A. Frozen vegetable processing plants can harbour diverse *Listeria monocytogenes* populations: Identification of critical operations by WGS. *Foods* **2022**, *11*, 1546. [CrossRef] [PubMed]
4. EFSA (European Food Safety Authority). *Urgent Scientific and Technical Assistance to Provide Recommendations for Sampling and Testing in the Processing Plants of Frozen Vegetables Aiming at Detecting Listeria monocytogenes*; EFSA Supporting Publication: Oxford, UK, 2018; Volume 15, p. 1445E. [CrossRef]
5. Commission Regulation (EC). No 2073/2005 of 15 November 2005 on microbiological criteria for foodstuffs. *Off. J. Eur. Union.* **2005**, *338*, 1–26.
6. Codex Alimentarius Commission. *Guidelines on the Application of General Principles of Food Hygiene to the Control of Listeria monocytogenes in Ready-to-Eat Foods*; CAC/GL 61-2007; Codex Alimentarius Commission: Rome, Italy, 2007.
7. Panel, E.B. The public health risk posed by *Listeria monocytogenes* in frozen fruit and vegetables including herbs, blanched during processing. EFSA Panel of Biological Hazards (BIOHAZ). *EFSA J.* **2020**, *8*, e06092. [CrossRef]
8. Zoellner, C.; Wiedmann, M.; Ivanek, R. An assessment of listeriosis risk associated with a contaminated production lot of frozen vegetables consumed under alternative consumer handling scenarios. *J. Food Prot.* **2019**, *82*, 2174–2193. [CrossRef] [PubMed]
9. McLauchlin, J.; Aird, H.; Amar, C.; Barker, C.; Dallman, T.; Lai, S.; Painset, A.; Willis, C. An outbreak of human listeriosis associated with frozen sweet corn consumption: Investigations in the UK. *Int. J. Food Microbiol.* **2021**, *338*, 108994. [CrossRef]
10. FAO/WHO. *Joint FAO/WHO Expert Meeting on Microbiological Risk Assessment of Listeria monocytogenes in Foods: Summary and Conclusions*; WHO: Geneva, Switzerland, 2023.
11. Nauta, M.J. Modelling bacterial growth in quantitative microbiological risk assessment: Is it possible? *Int. J. Food Microbiol.* **2002**, *73*, 297–304. [CrossRef]
12. Gonzales-Barron, U.; Cadavez, V.; Thebault, A.; Kooh, P. The Pathogens-in-Foods Database (PIF) (Version 1). 2021. Available online: https://pif.esa.ipb.pt/ (accessed on 18 April 2023). [CrossRef]
13. Badosa, E.; Trias, R.; Pares, D.; Pla, M.; Montesinos, E. Microbiological quality of fresh fruit and vegetable products in Catalonia (Spain) using normalised plate-counting methods and real time polymerase chain reaction (QPCR). *J. Sci. Food Agric.* **2008**, *88*, 605–611. [CrossRef]
14. Cardamone, C.; Aleo, A.; Mammina, C.; Oliveri, G.; Di Noto, A.M. Assessment of the microbiological quality of fresh produce on sale in Sicily, Italy: Preliminary results. *J. Biol. Res.* **2015**, *22*, 3. [CrossRef]
15. Carp-Carare, C.; Vlad-Sabie, A.; Floristean, V.-C. Detection and serotyping of *Listeria monocytogenes* in some food products from North-East of Romania. *Rev. Romana Med. Lab.* **2013**, *21*, 285–292. [CrossRef]
16. Cetinkaya, F.; Mus, T.E.; Yibar, A.; Guclu, N.; Tavsanli, H.; Cibik, R. Prevalence, serotype identification by multiplex polymerase chain reaction and antimicrobial resistance patterns of *Listeria monocytogenes* isolated from retail foods. *J. Food Safety* **2014**, *34*, 42–49. [CrossRef]

17. De Giusti, M.; Aurigemma, C.; Marinelli, L.; Tufi, D.; De Medici, D.; Di Pasquale, S.; De Vito, C.; Boccia, A. The evaluation of the microbial safety of fresh ready-to-eat vegetables produced by different technologies in Italy. *J. Appl. Microbiol.* **2010**, *109*, 996–1006. [CrossRef] [PubMed]
18. Gianfranceschi, M.; Gattuso, A.; Tartaro, S.; Aureli, P.B. Incidence of *Listeria monocytogenes* in food and environmental samples in Italy between 1990 and 1999: Serotype distribution in food, environmental and clinical samples. *Eur. J. Epidemiol.* **2003**, *18*, 1001–1006. [CrossRef] [PubMed]
19. Kokkinakis, E.; Boskou, G.; Fragkiadakis, G.A.; Kokkinaki, A.; Lapidakis, N. Microbiological quality of tomatoes and peppers produced under the good agricultural practices protocol AGRO 2-1 & 2-2 in Crete, Greece. *Food Control* **2007**, *18*, 1538–1546.
20. Lika, M.; Malollari, I.; Lajqi, V. Microbiological contamination of vegetable food as risk factor for human population. *J. Environ. Prot. Ecol.* **2014**, *15*, 764–770.
21. Losio, M.; Pavoni, E.; Bilei, S.; Bertasi, B.; Bove, D.; Capuano, F.; Farneti, S.; Blasi, G.; Comin, D.; Cardamone, C.; et al. Microbiological survey of raw and ready-to-eat leafy green vegetables marketed in Italy. *Int. J. Food Microbiol.* **2015**, *210*, 88–91. [CrossRef]
22. Magdovitz, B.F.; Gummalla, S.; Garren, D.; Thippareddi, H.; Berrang, M.E.; Harrison, M.A. Prevalence of Listeria species and *Listeria monocytogenes* on raw produce arriving at frozen food manufacturing facilities. *J. Food Prot.* **2021**, *84*, 1898–1903. [CrossRef]
23. Moreno, Y.; Sanchez-Contreras, J.; Montes, R.M.; Garcia-Hernandez, J.; Ballesteros, L.; Antonia Ferrus, M. Detection and enumeration of viable *Listeria monocytogenes* cells from ready-to-eat and processed vegetable foods by culture and DVC-FISH. *Food Control* **2012**, *27*, 374–379. [CrossRef]
24. Pianetti, A.; Sabatini, L.; Citterio, B.; Pierfelici, L.; Ninfali, P.; Bruscolini, E. Changes in microbial populations in ready-to-eat vegetable salads during shelf-life. *Ital. J. Food Sci.* **2008**, *20*, 245–254.
25. Vojkovská, H.; Myšková, P.; Gelbíčová, T.; Skočková, A.; Koláčková, I.; Karpíšková, R. Occurrence and characterization of food-borne pathogens isolated from fruit, vegetables and sprouts retailed in the Czech Republic. *Food Microbiol.* **2017**, *63*, 147–152. [CrossRef]
26. Wagner, M.; Auer, B.; Trittremmel, C.; Hein, I.; Schoder, D. Survey on the *Listeria* contamination of ready-to-eat food products and household environments in Vienna, Austria. *Zoonoses Public Health* **2007**, *54*, 16–22. [CrossRef]
27. Jeyaletchumi, P.; Tunung, R.; Selina, P.M.; Chai, L.C.; Radu, S.; Farinazleen, M.G.; Cheah, Y.K.; Mitsuaki, N.; Yoshitsugu, N.; Kumar, M.P. Evaluation of *Listeria* spp. and *Listeria monocytogenes* in selected vegetable farms. *J. Trop. Agric. Food Sci.* **2011**, *39*, 255–266.
28. Kuan, C.-H.; Rukayadi, Y.; Ahmad, S.H.; Wan, C.; Radzi, M.; Thung, T.-Y.; Premarathne, J.; Chang, W.-S.; Loo, Y.-Y.; Tan, C.-W.; et al. Comparison of the microbiological quality and safety between conventional and organic vegetables sold in Malaysia. *Front. Microbiol.* **2017**, *8*, 1433. [CrossRef] [PubMed]
29. Mazzota, A.S. Heat resistance of *Listeria monocytogenes* in vegetables: Evaluation of blanching process. *J. Food Prot.* **2001**, *64*, 385–387. [CrossRef] [PubMed]
30. Hoelzer, K.; Pouillot, R.; Gallagher, D.; Silverman, M.B.; Kause, J.; Dennis, S. Estimation of *Listeria monocytogenes* transfer coefficients and efficacy of bacterial removal through cleaning and sanitation. *Int. J. Food Microbiol.* **2012**, *157*, 267–277. [CrossRef] [PubMed]
31. Nauta, M. Microbiological risk assessment models for partitioning and mixing during food handling. *Int. J. Food Microbiol.* **2005**, *100*, 311–322. [CrossRef]
32. Nauta, M. The modular process risk model (MPRM): A structural approach to food chain exposure assessment. In *Microbial Risk Analysis of Foods*; Schaffner, D.W., Doyle, M.P., Eds.; ASM Press: Washington, DC, USA, 2008; pp. 99–136.
33. Willis, C.; McLauchlin, J.; Aird, H.; Amar, C.; Barker, C.; Dallman, T.; Elviss, N.; Lai, S.; Sadler-Reeves, L. Occurrence of *Listeria* and *Escherichia coli* in frozen fruit and vegetables collected from retail and catering premises in England 2018–2019. *Int. J. Food Microbiol.* **2020**, *334*, 108849. [CrossRef]
34. FSAI. *National Microbiological Survey and Consumer Habits in Relation to Frozen Vegetables, Fruits and Herbs (19NS6)*; Monitoring & Surveillance Series; Food Safety Authority of Ireland: Dublin, Ireland, 2022; 62p, ISBN 978-1-910348-54-3.
35. PROFEL. *Hygiene Guidelines for the Control of Listeria monocytogenes in the Production of Quick Frozen Vegetables*; European Association of Fruit and Vegetable Processors: Brussels, Belgium, 2020; 58p.
36. Ceylan, E.; McMahon, W.; Garren, D. Thermal inactivation of *Listeria monocytogenes* during water and steam blanching of vegetables. *J. Food Prot.* **2017**, *80*, 1550–1556. [CrossRef]
37. Bigelow, W. The logarithmic nature of thermal death time curves. *J. Infect. Dis.* **1921**, *29*, 528–536. [CrossRef]
38. Pappelbaum, K.; Grif, K.; Heller, I.; Wüirzner, R.; Hein, I.; Ellerbroek, L.; Wagner, M. Monitoring hygiene on- and at-line is critical for controlling *Listeria monocytogenes* during produce processing. *J. Food Prot.* **2008**, *71*, 735–741. [CrossRef]
39. Fagerlund, A.; Moretro, T.; Heir, E.; Briandet, R.; Langsrud, S. Cleaning and disinfection of biofilms composed of *Listeria monocytogenes* and background microbiota from meat processing surfaces. *Appl. Environ. Microbiol.* **2017**, *83*, e01046-17. [CrossRef]
40. FDA. *FDA-iRISK 4.2 Food Safety Modeling Tool: Technical Document*; U.S. Food and Drug Administration. U.S. Department of Agriculture: Rockwell, MD, USA, 2021.
41. van der Sman, R.G.M. Impact of Processing Factors on Quality of Frozen Vegetables and Fruits. *Food Eng. Rev.* **2020**, *12*, 399–420. [CrossRef]

42. Gómez-Galindo, M.; Truchado, P.; Allende, A.; Gil, M.I. Optimization of the use of a commercial phage-based product as a control strategy of *Listeria monocytogenes* in the fresh-cut Industry. *Foods* **2023**, *12*, 3171. [CrossRef] [PubMed]
43. Pouillot, R.; Chen, Y.; Van Doren, J.M. Elucidating the influence of the lower and upper microbiological limits: When is a 3-class sampling plan useful to test for pathogens in food? *Food Control* **2024**, *163*, 110544. [CrossRef]
44. WHO. *Statistical Aspects of Microbiological Criteria Related to Foods: A Risk Managers Guide*; Microbiological Risk Assessment Series 24; World Health Organization & Food and Agriculture Organization of the United Nations: Geneva, Switzerland, 2016; 120p. Available online: https://iris.who.int/bitstream/handle/10665/249531/9789241565318%E2%80%93eng.pdf?sequence=1 (accessed on 10 April 2024).
45. Kataoka, A.; Wang, H.; Elliot, P.H.; Whiting, R.C.; Hayman, M. Growth of *Listeria monocytogenes* in thawed frozen foods. *J. Food Prot.* **2017**, *80*, 447–453. [CrossRef] [PubMed]
46. FDA-FSIS. *Quantitative Assessment of Relative Risk to Public Health from Foodborne Listeria monocytogenes Among Selected Categories of Ready-to-Eat Foods*; FDA-FSIS: Washington, DC, USA, 2003; pp. 1–541.
47. Pouillot, R.; Kiermeier, A.; Guillier, L.; Cadavez, V.; Sanaa, M. Updated Parameters for *Listeria monocytogenes* Dose–Response Model Considering Pathogen Virulence and Age and Sex of Consumer. *Foods* **2024**, *13*, 751. [CrossRef]
48. FAO-WHO. *Risk Assessment of Listeria monocytogenes in Ready-to-Eat Foods: Technical Report*; World Health Organization and Food and Agriculture Organization of the United Nations: Geneva, Switzerland, 2004; pp. 1–269.
49. R Foundation for Statistical Computing. *R Core Team R: A Language and Environment for Statistical Computing*; R Foundation for Statistical Computing: Vienna, Austria, 2021; Available online: https://www.R-project.org/ (accessed on 1 July 2024).
50. AFFI. *How to Validate Your Blancher*; American Frozen Food Institute: Arlington, VA, USA, 2018. Available online: https://affi.org/listeria/how-to-validate-your-blancher/ (accessed on 15 September 2023).
51. Mena, C.; Almeda, G.; Carneiro, L.; Teixeira, P.; Hogg, T.; Gibbs, P.A. Incidence of *Listeria monocytogenes* in different food products commercialized in Portugal. *Food Microbiol.* **2004**, *21*, 213–216. [CrossRef]
52. Lee, S.; Cetinkaya, F.; Soyutemiz, G.E. Occurrence of *Listeria* species in the processing stages of frozen pepper. *J. Food Saf.* **2007**, *27*, 134–147. [CrossRef]
53. Moravkova, M.; Verbikova, V.; Michna, V.; Babak, V.; Cahlikova, H.; Karpiskova, R.; Kralik, P. Detection and quantification of *Listeria monocytogenes* in ready-to-eat vegetables, frozen vegetables and sprouts examined by culture methods and real-time PCR. *J. Food. Nutr. Res.* **2017**, *5*, 832–837. [CrossRef]
54. Aguado, V.; Vitas, A.; Garcia-Jalon, I. Characterization of *Listeria monocytogenes* and *Listeria innocua* from a vegetable processing plant by RAPD and REA. *Int. J. Food Microbiol.* **2004**, *90*, 341–347. [CrossRef]
55. Skowron, K.; Grudlewska, K.; Lewandowski, D.; Gajewski, P.; Reśliński, A.; Gospodarek-Komkowska, E. Antibiotic susceptibility and ability to form biofilm of *Listeria monocytogenes* strains isolated from frozen vegetables. *Acta Aliment.* **2019**, *48*, 65–75. [CrossRef]
56. Vitas, A.; Aguado, V.; Garcia-Jalon, I. Occurrence of *Listeria monocytogenes* in fresh and processed foods in Navarra (Spain). *Int. J. Food Microbiol.* **2004**, *90*, 349–356. [CrossRef] [PubMed]
57. Pouillot, R.; Gallagher, D.; Tang, J.; Hoelzer, K.; Kause, J.; Dennis, S.B. *Listeria monocytogenes* in retail delicatessens: An interagency risk assessment-model and baseline results. *J. Food Prot.* **2015**, *78*, 134–145. [CrossRef] [PubMed]
58. Perchec, A.-M.; Esgurra, C. *Microbiological Risks Associated with Frozen Raw Produce Used in Uncooked Food Preparations*; New Zealand Food Safety Technical Paper No: 2023/14; New Zealand Food Safety, Ministry for Primary Industries: Wellington, New Zealand, 2023; 33p. Available online: https://www.mpi.govt.nz/dmsdocument/57904%E2%80%93Microbiological-risks-associated-with-frozen-produce-used-in-uncooked-food-preparations (accessed on 26 June 2024).
59. Pouillot, R.; Hoelzer, K.; Chen, Y.; Dennis, S.B. *Listeria monocytogenes* dose response revisited—Incorporating adjustments for variability in strain virulence and host susceptibility. *Risk Anal.* **2015**, *35*, 90–108. [CrossRef] [PubMed]

foods

MDPI

Article

Updated Parameters for *Listeria monocytogenes* Dose–Response Model Considering Pathogen Virulence and Age and Sex of Consumer

Régis Pouillot [1,*], Andreas Kiermeier [2], Laurent Guillier [3], Vasco Cadavez [4,5] and Moez Sanaa [6,*]

1 Independent Researcher, 18 rue Mohamed Al Ghazi, Rabat 10170, Morocco
2 Statistical Process Improvement Consulting and Training Pty Ltd., Gumeracha 5233, Australia; andreas.kiermeier@gmail.com
3 Risk Assessment Department, French Agency for Food, Environmental and Occupational Health & Safety (Anses), 14 rue Pierre et Marie Curie, 94701 Maisons-Alfort, France
4 Centro de Investigação de Montanha (CIMO), Campus de Santa Apolónia, Instituto Politécnico de Bragança, 5300-253 Bragança, Portugal; vcadavez@ipb.pt
5 Laboratório para a Sustentabilidade e Tecnologia em Regiões de Montanha, Campus de Santa Apolónia, Instituto Politécnico de Bragança, 5300-253 Bragança, Portugal
6 Nutrition and Food Safety Department, World Health Organization, 1211 Geneva, Switzerland
* Correspondence: rpouillot.work@gmail.com (R.P.); sanaam@who.int (M.S.); Tel.: +212-6-10-99-22-80 (R.P.)

Abstract: Better knowledge regarding the *Listeria monocytogenes* dose–response (DR) model is needed to refine the assessment of the risk of foodborne listeriosis. In 2018, the European Food Safety Agency (EFSA) derived a lognormal Poisson DR model for 14 different age–sex sub-groups, marginally to strain virulence. In the present study, new sets of parameters are developed by integrating the EFSA model for these sub-groups together with three classes of strain virulence characteristics ("less virulent", "virulent", and "more virulent"). Considering classes of virulence leads to estimated relative risks (RRs) of listeriosis following the ingestion of 1000 bacteria of "less virulent" vs. "more virulent" strains ranging from 21.6 to 24.1, depending on the sub-group. These relatively low RRs when compared with RRs linked to comorbidities described in the literature suggest that the influence of comorbidity on the occurrence of invasive listeriosis for a given exposure is much more important than the influence of the virulence of the strains. The updated model parameters allow better prediction of the risk of invasive listeriosis across a population of interest, provided the necessary data on population demographics and the proportional contribution of strain virulence classes in food products of interest are available. An R package is made available to facilitate the use of these dose–response models.

Keywords: *Listeria monocytogenes*; hazard characterization; dose–response; virulence; clonal complex

1. Introduction

Listeria monocytogenes poses a significant threat to public health due to its ability to cause severe infections, particularly in vulnerable populations. In 2022, 27 European Union (EU) member states reported 2738 confirmed invasive human cases of listeriosis, a notification rate of 0.62 cases per 100,000 population, with a fatality rate of 18.1% and a total of 286 deaths [1]. In the realm of food safety, understanding the dose–response (DR) relationship of *L. monocytogenes* is pivotal for accurate and specific risk assessment and risk management.

In the context of microbiological food safety risk assessment, the DR model for any pathogen is the mathematical description of the relationship between the number of ingested pathogenic organisms and the probability of infection or illness. The DR relationship is poorly understood for human listeriosis. However, we know that this relationship is affected by the three aspects of the infectious disease triangle [2–4], namely, (i) the food matrix, (ii) host susceptibility, and (iii) pathogen characteristics/virulence. In practice, most DR models are developed from limited data and, to date, cannot fully incorporate all these effects. The DR model for *L. monocytogenes* is no exception in this respect [4].

The DR models developed by the FAO and WHO in 2004 [5] for *L. monocytogenes* were of the exponential form. In these DR models, the probability that the exposure to a single *L. monocytogenes* cell results in illness, the "r-value", is considered independent of the presence of other cells, and is considered constant for all cells and all individuals of a given sub-population [2,6]. The response of individuals exposed to *L. monocytogenes* is highly variable [4]. However, in the original FAO and WHO [5] risk assessment for invasive listeriosis, the only variability considered in the DR model was that between the general and susceptible populations and this was accommodated by fitting different r-values for the two populations. This group of two models had been developed for *L. monocytogenes* in ready-to-eat (RTE) foods, based on the method described by Buchanan et al. [7]. The method involved fitting the DR model to surveillance data detailing *L. monocytogenes* contamination observed in the food supply and matching the expected versus actual total number of listeriosis cases annually per sub-population as closely as possible.

More recent DR models for *L. monocytogenes* were developed to better incorporate host susceptibility. Pouillot et al. [8] proposed a DR model (called the "lognormal Poisson model") that considers the variability in the probability for a *L. monocytogenes* cell causing illness within a sub-population. Host susceptibility was considered using epidemiological surveillance data from France for 11 sub-populations that were defined according to underlying health conditions [8]. A similar approach was used by the EFSA and considered 14 age–sex sub-groups (age classes: 1–4, 5–14, 15–24, 25–44, 45–64, 65–74, and ≥75 years old and sex classes: male and female) as surrogates for underlying health conditions [9].

In contrast to the work on host susceptibility, Fritsch et al. [10] focused on the pathogen and differences in virulence. These researchers derived specific parameters of a lognormal Poisson model by incorporating parameters specific to three distinct categories of virulence, characterized using clonal complexes (CCs) and sequence types (STs).

Finally, with respect to the food matrix, the FAO and WHO [11] concluded that not enough information was available to consider these aspects as part of the DR model.

While the Fritsch et al. [10] model evaluated the DR for three different classes of *Listeria* virulence marginally to the various sub-populations, the EFSA model [9] evaluated the DR model for 14 different sub-populations marginally to strain virulence. The objective of this study was to develop new sets of parameters for the lognormal Poisson model by integrating the EFSA model for age–sex sub-groups together with three classes of virulence characteristics as defined by Fritsch et al. [10], giving a total of 42 DR curves, one for each different age–sex and virulence class combination.

2. Materials and Methods

2.1. Dose–Response Model

The DR models we considered here link the ingested number of bacteria (as an actual number or as the mean of a Poisson distribution) with the marginal probability (over strains and over individuals of a given population) of developing an invasive case of listeriosis.

The DR model used by FAO and WHO [5] assumes that each ingested *L. monocytogenes* cell has a given and independent probability (r) of triggering invasive listeriosis. In this model, r is assumed to be constant within a given sub-population. Hence, the probability of developing invasive listeriosis following the ingestion of exactly n bacteria is

$$\mathrm{Prob}(\mathrm{ill}|r,n) = 1 - (1-r)^n \tag{1}$$

which is a "binomial DR model" as it follows a binomial process [12]. In contrast, if the number of bacteria is expressed as the mean of a Poisson distribution of parameter d, then the dose–response, integrated over the (serving-to-serving) variability for an average dose d, is written as

$$\mathrm{Prob}(\mathrm{ill}|r,d) = 1 - \exp(-r \times d), \tag{2}$$

which is the "exponential DR model" [6]. FAO and WHO [5] inferred some sets of r parameters for two sub-populations and used, for the food examples described in their risk assessment, a median value for r of 1.06×10^{-12}, which equals $10^{-11.975}$, for their defined "population with increased susceptibility" and 2.37×10^{-14}, which equals $10^{-13.625}$, for their "healthy population" ([5], page 60).

Pouillot et al. [8] revisited this DR model and considered a model where r is not constant but follows a lognormal (base 10) distribution. The marginal (over strains and individuals from a given sub-population) model can then be written as follows:

$$\text{Prob}(\text{ill}|n,\theta) = 1 - \int_0^1 (1-r)^n f(r,\theta)dr, \tag{3}$$

with $f(r,\theta = \{\mu,\sigma\})$ being the density of the lognormal (x follows the lognormal (base 10) distribution with parameter μ and σ if $\log_{10}(x) \sim \text{Normal}(\mu,\sigma)$) distribution with parameters μ (mean) and σ (standard deviation) and n being the exact number of bacteria ingested, giving the "lognormal binomial DR model". Alternatively, if the dose d is expressed as the mean of a serving-to-serving Poisson distribution, then the result is the "lognormal Poisson DR model", which is given by

$$\text{Prob}(\text{ill}|n,\theta) = 1 - \int_0^1 \exp(-r \times d) f(r,\theta)dr \tag{4}$$

Note that the exponential DR model is a special case of the lognormal Poisson model with $\sigma = 0$. In Pouillot et al. [8], the parameters of the lognormal distributions for various sub-populations with underlying health conditions were estimated using Goulet et al.'s [13] data regarding relative risk of listeriosis in 11 sub-populations in France (pregnancy, non-hematological cancer, other cancers, etc.) and exposure data estimated from the United States. Note that the inference process considered no post-retail growth of *L. monocytogenes* and a maximal population density (MPD) of *L. monocytogenes* in products equals to 6.1 $\log_{10}$ cfu/g.

2.2. Inference Process

In the current work, Pouillot et al.'s [8] lognormal Poisson framework is applied to data originating from the EU, considering various classes of *L. monocytogenes* virulence. As stated above, the lognormal Poisson model has two parameters, i.e., the mean (μ) and the standard deviation (σ) of the normal distribution of the $\log_{10}(r)$ parameter. Here, estimates of μ and σ are made for each combination of the exposed sub-population and the class of virulence that is being considered. The estimation process described by Pouillot et al. [8] involves estimating σ for each class of virulence, and then scaling μ to exposure and epidemiological data.

The following information is needed:

1. A classification of *L. monocytogenes* strains as a function of their virulence;
2. Estimates of the inter-individual variability in susceptibility within each sub-population and estimates of the inter-strain variability in virulence within each class of virulence;
3. Estimates of the exposure of the considered sub-populations to the various classes of virulence;
4. Estimates of the corresponding number of cases of invasive listeriosis for each combination of sub-populations/class of virulence.

The following sub-sections will develop how those estimates were obtained.

2.2.1. Classification of Strain Virulence

We used the classes of virulence defined by Fritsch et al. ([10], their Figure 3). These categories were named "hypovirulence", "medium virulence", and "hypervirulence". The concept of a virulence class was initially introduced for *L. monocytogenes* by Roche et al. in 2001 [14]. These authors initially categorized virulence into three classes, named "virulent", "hypovirulent", and "avirulent". However, we contend that no strains of *L. monocytogenes* are completely avirulent. Therefore, we suggest renaming these classes as "less virulent", "virulent", and "more virulent"." The term "hypovirulent" literally means "less virulent", but we have chosen not to use it to prevent confusion with the terminology used by Roche et al. [14], who applied "hypovirulent" to what we consider the medium virulence category.

Fritsch et al. [10] used the clinical frequency data compiled by Maury et al. [15], who characterized 6633 *L. monocytogenes* strains isolated from clinical samples and food. A clinical frequency was calculated by dividing the number of clinical isolates of a particular CC by the total number of clinical and food isolates of that CC. Three groups of CCs or STs were then created using a dendrogram from those clinical frequencies. For CCs/STs not present in that list, Fritsch et al. [10] recommended classifying them in the "virulent" group (Table 1).

Table 1. Classification of strains in three classes of virulence (from [10]).

Less Virulent Strains	Virulent Strains	More Virulent Strains
CC121, CC204, CC31, CC9, CC193, CC19, ST214	CC14, CC155, CC177, CC18, CC20, CC21, CC26, CC3, CC37, CC379, C388, CC398, CC5, CC59, CC8, CC403 and all others	CC1, CC101, CC2, CC220, CC224, CC4, CC451, CC54, CC6, CC7, CC87

2.2.2. Estimate of the Standard Deviation of $\log_{10}$(r) within a Sub-Population/Class of Virulence

Pouillot et al. [8] showed that the standard deviation of $\log_{10}(r)$ for a given sub-population could be estimated from the inter-individual variability in the susceptibility, σ_p for sub-population p, and the inter-strain variability of virulence, σ_v for virulence class v.

Similarly to Fritsch et al. [10] and EFSA [9], the estimate ($\sigma_p = 0.55 \log_{10}$ cfu) of the inter-individual variability in the susceptibility for all sub-populations was used, as suggested by Pouillot et al. [8]. This estimate came from USFDA and FSIS assessments in which 90% of the individual variability within the population group with medium variability in susceptibility may be contained within a range of 1.8 $\log_{10}$ ([15], their Table IV-8). If $Q_{90,p}$ is the $\log_{10}$ difference between the 5th and the 95th percentile of a lognormal distribution,

$$\sigma_p = \frac{Q_{90,p}}{2 \cdot \Phi^{-1}(0.95)} = 0.55, \quad (5)$$

where Φ^{-1} denotes the inverse of the standard normal cumulative density function. For the inter-strain variability of virulence, the experimental data on mice reported in Fritsch et al. [10], collected from USFDA and FSIS ([15], their Table IV-3), were used. The median lethal doses (LD_{50}) obtained by intraperitoneal infectious route in normal mice for 26 strains were collected from this table. The values were clustered in 3 different groups ([10], their Figure 2). From these, the estimates of $\sigma_{v=lv}$ = 1.12, $\sigma_{v=v}$ = 0.63, and $\sigma_{v=mv}$ = 0.52 $\log_{10}$ cfu were estimated for the "less virulent" (LD_{50} ranging from 6.80 to 9.70 $\log_{10}$ cfu), "virulent" (LD_{50} ranging from 4.49 to 6.23 $\log_{10}$ cfu), and "more virulent" (LD_{50} ranging from 2.57 to 3.67 $\log_{10}$ cfu) *L. monocytogenes* strains, respectively. Note that the larger standard deviation for the "less virulent" strains is related to the choice of distance cutoff for $\log_{10}$(r) used and incorporates two small but potentially heterogeneous virulence groups; these two groups were more variable than any observed in the other virulence classes.

These estimates of the inter-individual variability in the susceptibility and of the inter-strain variability of virulence can then be combined [8], assuming independence, e.g., for a sub-population consuming strains of the "more virulent" class:

$$\sigma_{v=mv,p} = \sqrt{\sigma_p^2 + \sigma_{v=mv}^2} = \sqrt{0.55^2 + 0.52^2} = 0.756. \quad (6)$$

2.2.3. Proportion of Various Classes of Strain Virulence in EU RTE and Clinical Cases

Møller Nielsen et al. [16] selected a total of 1143 *L. monocytogenes* isolates, including 333 human clinical isolates and 810 isolates from the food chain. The food strains originated partially from the EU-wide baseline survey conducted in 2010 and 2011 [17]. The baseline survey collected representative samples of three types of RTE food: seafood (smoked and gravad fish), meats (packaged heat-treated meat products), and cheese and dairy (soft and semi-soft cheeses). Due to the lack of strains isolated from meat and cheeses, additional isolates from RTE meat products and cheeses were obtained from as many different EU member states as possible.

Among the 576 food strains available in Annexes 1, 2, and 3 of that report, 290 strains from the baseline survey (Annex 1 in [16]) were selected for the seafood category (out of 294, 3 were excluded due to a lack of information on the sampling stage and 1 was excluded because the allelic type was not specified), 176 were retained for the meat product category, and 89 for the dairy products categories (i.e., all the strains from Annexes 1 and 2 in [16] under the categories "meat and meat products" and "milk and milk products"). Five strains isolated from vegetables (Annex 3 in [16]) were not considered due to the small number of strains. A total of 555 food strains from this study were eventually retained. Clinical isolates from assumed sporadic human cases collected during the baseline survey period, 2010–2011, were included in the study (262 strains).

The isolates were whole genome sequenced and their CC characterized. In Table 2, the proportion of each of these virulence classes estimated in seafood products, meat products, and cheese, as well as in sporadic cases of invasive listeriosis are presented. Note that the "others/unknown" category will be considered as "virulent".

Table 2. Proportion of the different classes of virulence in seafood, meats, and cheese and dairy, and in sporadic cases in the EU from [16].

	Less Virulent	Virulent	More Virulent	Others/Unknown	n
RTE Seafood	149 (51.4%)	102 (35.2%)	36 (12.4%)	3 (1.0%)	290
RTE Meats	104 (59.1%)	35 (19.9%)	35 (19.9%)	2 (1.1%)	176
RTE cheese and dairy	11 (12.4%)	42 (47.2%)	29 (32.6%)	7 (7.9%)	89
Sporadic	22 (8.4%)	76 (29.0%)	156 (59.5%)	8 (3.1%)	262

2.2.4. Exposure of the EU Population to Various Classes of Virulence

To estimate the exposure of the EU population to the various classes of virulence of *L. monocytogenes*, the exposure part of the generic quantitative microbial risk assessment (gQMRA) model of EFSA [9] was adapted. The EFSA model was based on that developed by Pérez-Rodriguez et al. [18].

In this model, the exposure to *L. monocytogenes* in the EU from various RTE food categories—including heat-treated meat, smoked and gravad fish, and soft and semi-soft cheeses—was estimated. Exposure was assessed for each of 14 sub-populations, i.e., male and female sub-population of individuals aged 1–4, 5–14, 15–24, 25–44, 45–64, 65–74, and ≥75 years of age. Note that no underlying health conditions, such as pregnancy or illnesses, were considered in this study. The process started at the retail stage and ended at consumption. For prevalence and concentration at retail, data from the EU-wide baseline survey was complemented with EU monitoring data and US data [19]. Food serving size and the number of servings per year were estimated from the EU food consumption database. Growth of *L. monocytogenes* was modeled from retail to consumption, using temperature–time profiles during transport and storage. The packaging conditions, i.e., reduced oxygen packaging (ROP, including both vacuum and modified atmosphere packaging) vs. normal packaging, were considered as a factor modifying the growth potential of *L. monocytogenes* in all RTE food sub-categories (except in soft and semi-soft cheese, for which no ROP was considered). MPD of 6.23 $\log_{10}$ cfu/g for cooked meat and sausages, 7.28 $\log_{10}$ cfu/g for cheeses, 7.29 $\log_{10}$ cfu/g for fish products, or 7.53 $\log_{10}$ cfu/g for pâté were used. The exposure was calculated at the end of the simulation model by multiplying the concentration by the serving size. The overall exposure to *L. monocytogenes* was then the sum of exposure from the consumption of 13 RTE foods (cold smoked fish, hot smoked fish, gravid fish, cooked meat, sausage, and pâté for normal packaging and ROP and soft and semi-soft cheese for normal packaging). The original EFSA model derived an empirical cumulative distribution function (ecdf) of the overall exposure to *L. monocytogenes* from consumption of RTE for each of the 14 sub-populations. See [9] for details and availability of the original code and data.

The R [20] code from EFSA [9] was adapted in this study to separate the estimated exposure to the various classes of strain virulence. For that purpose, the model was scaled to the frequencies of the various virulence classes as observed in Table 2. For example, the prevalence considered for cheese (13/3114) was multiplied by 12.4%, 47.2% + 7.9% = 55.1%, and 32.6% (see Table 2) to estimate the specific exposure to "less virulent", "virulent", and "more virulent" *L. monocytogenes* strains from cheese, respectively. Similar proportions were applied for all food categories. The model resulted in three empirical distribution functions (one per virulence class) for each of the 14 age–sex sub-populations. Note that this evaluation considered, as underlying assumptions, that the initial concentration of *L. monocytogenes* in contaminated food is independent of the virulence class, and that the general behavior, and growth in particular, of *L. monocytogenes* in all those products is also independent of the class of virulence.

In addition, the total number of eating occasions per year/sub-populations/considered RTE estimated by EFSA [9] from the EFSA consumption database was used.

2.2.5. Prevalence of Invasive Listeriosis in the EU Population according to Classes of Virulence

To estimate the number of invasive human listeriosis cases per sub-population for each class of virulence, the proportion of each class of virulence observed in sporadic cases in EU (Table 2) was applied to the number of invasive listeriosis cases in the EU during the 2008–2015 period for each of the 14 sub-population group, as estimated by EFSA ([9], their Table 1, page 28).

2.3. *Scaling of the Model to Epidemiological Data*

These inferences led to:

- Estimates of the number of invasive human listeriosis cases in the EU from 2008 to 2015 within 14 sub-populations for 3 different classes of virulence (42 categories);
- An estimate of the prevalence of contaminated RTE and the ecdf of the exposure to *L. monocytogenes* from contaminated servings of RTE for each of these 42 categories;
- Estimates of the standard deviation of the lognormal (base 10) distribution of r for 3 classes of virulence, considered as similar in each sub-population for a given class of virulence.

With these data, the mean of the $\log_{10}(r)$ population can be estimated to match the model outputs with the epidemiological surveillance data for the EU. The number of cases of invasive listeriosis linked to *L. monocytogenes* of the class of virulence v in population $C_{p,v}$ is actually [8]:

$$C_{p,v} = TEO_p \times \pi_{p,v} \times \left(1 - \int_{d=0}^{\infty} \int_{lr=-\infty}^{\infty} \exp\left(-10^{lr} \times d\right) f_{p,v}(d) \varphi\left(lr; \mu_{p,v}, \sigma_v\right) dlr\, dd\right), \quad (7)$$

where p is the index for the sub-population, v is the index for the virulence class, TEO_p is the total eating occasions of RTE for population p during the 2008–2015 period (estimated as the annual total eating occasions for population p multiplied by 8 (years)), $\pi_{p,v}$ is the prevalence of contaminated RTE for population p and virulence class v, lr is the $\log_{10}$ of r, $f_{p,v}$ is the density function of the expected dose for population p and virulence class v for contaminated servings, and $\varphi(x; \mu, \sigma)$ is the probability density function of the normal distribution. In practice, the integration over $f_{p,v}$ is obtained by discretizing the ecdf by steps of 0.1 $\log_{10}$ cfu.

2.4. Implementation and Diffusion

Equation (7) has a single unknown, $\mu_{p,v}$, which can be estimated numerically using the integrate function from the R stats package. To transfer the DR models to users, a dedicated R package was developed. The code and data used to derive the model and the dedicated R package are available at www.github.com/rpouillot/FoodsDR (accessed on 1 February 2024).

3. Results

The combined standard deviation for the consuming sub-population and strain virulence were estimated to be 1.247 for the "less virulent", 0.836 for the "virulent", and 0.756 $\log_{10}$ cfu for the "more virulent" strains. The full set of parameters, namely, the mean and standard deviation of the lognormal distribution for r for each combination sub-population/class of virulence, is provided in Table 3.

Table 3. Mean (μ) and standard deviation (σ) of the lognormal (base 10) distributions of r for the 14 sub-populations and the 3 classes of virulence. The arithmetic mean of r is given in parenthesis.

Population	Less Virulent (σ = 1.247)	Virulent (σ = 0.836)	More Virulent (σ = 0.756)
Female 1–4 yo	−14.166 (4.22×10^{-13})	−12.296 (3.22×10^{-12})	−11.671 (9.71×10^{-12})
Male 1–4 yo	−14.124 (4.65×10^{-13})	−12.256 (3.54×10^{-12})	−11.625 (1.08×10^{-11})
Female 5–14 yo	−14.516 (1.89×10^{-13})	−12.582 (1.67×10^{-12})	−12.046 (4.09×10^{-12})
Male 5–14 yo	−14.633 (1.44×10^{-13})	−12.690 (1.30×10^{-12})	−12.165 (3.11×10^{-12})
Female 15–24 yo	−14.002 (6.45×10^{-13})	−12.123 (4.80×10^{-12})	−11.521 (1.37×10^{-11})
Male 15–24 yo	−14.668 (1.33×10^{-13})	−12.730 (1.19×10^{-12})	−12.200 (2.87×10^{-12})
Female 25–44 yo	−13.708 (1.21×10^{-12})	−11.815 (9.76×10^{-12})	−11.239 (2.63×10^{-11})
Male 25–44 yo	−14.444 (2.22×10^{-13})	−12.522 (1.92×10^{-12})	−11.973 (4.84×10^{-12})
Female 45–64 yo	−13.755 (1.09×10^{-12})	−11.890 (8.21×10^{-12})	−11.272 (2.43×10^{-11})
Male 45–64 yo	−13.753 (1.09×10^{-12})	−11.869 (8.61×10^{-12})	−11.274 (2.42×10^{-11})
Female 65–74 yo	−13.418 (2.36×10^{-12})	−11.594 (1.62×10^{-11})	−10.916 (5.52×10^{-11})
Male 65–74 yo	−13.283 (3.23×10^{-12})	−11.447 (2.28×10^{-11})	−10.785 (7.45×10^{-11})
Female >75 yo	−13.234 (3.61×10^{-12})	−11.437 (2.33×10^{-11})	−10.718 (8.70×10^{-11})
Male >75 yo	−13.255 (3.44×10^{-12})	−11.468 (2.17×10^{-11})	−10.734 (8.38×10^{-11})

Figure 1 illustrates the DR curves for doses ranging from 1 to 10^{12} bacteria (as a mean of the Poisson dose) for the EFSA model [9] that are to be compared with the DR obtained in this study. The FAO and WHO [5] DR for the healthy population (exponential dose–response with parameter 1.06×10^{-12}) and for the increased susceptibility population (exponential dose–response with parameter 2.37×10^{-14}) are provided for comparison. As expected, the "more virulent" strains lead to a DR model that is shifted to the left when compared to the "marginal" model (EFSA, [9]) or, even more, the "less virulent strains".

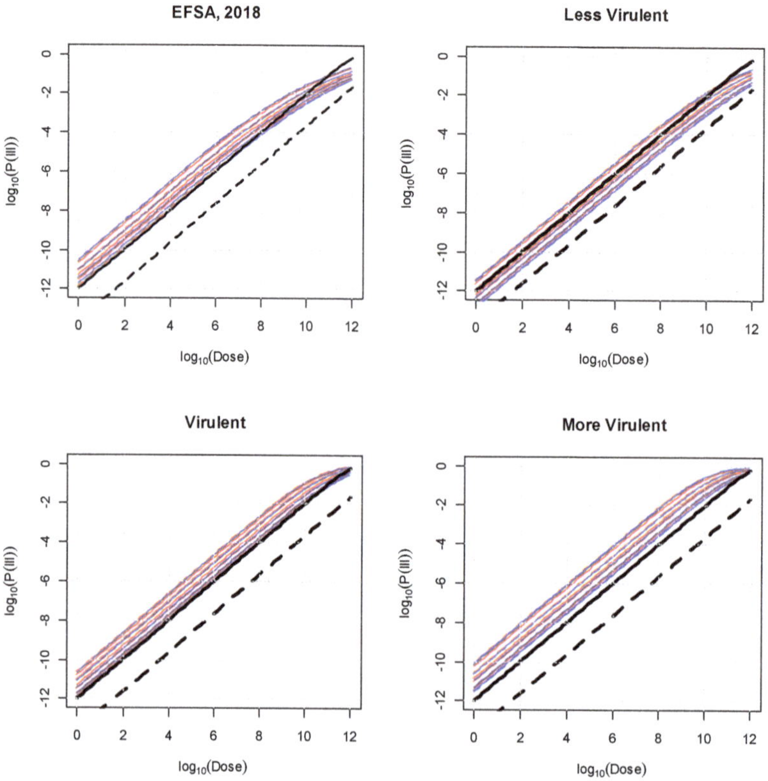

Figure 1. Dose–response curves for 14 age–sex groups (females = red lines and males = blue lines) from EFSA [9] (top, left) and this study ("less virulent", "virulent", and "more virulent" strains). FAO and WHO [5] models for the general (black dashed line) and susceptible (black solid line) populations.

The risk of invasive listeriosis for selected doses ranging from 100 bacteria to 1,000,000,000 bacteria using the FAO and WHO [5], the Pouillot et al. [8], the EFSA [9] and the model from this study are presented in Table 4. This table also reports, for an easier comparison, the relative risk (RR) from an exposure to 10^3 bacteria when compared to the FAO and WHO [5] model, healthy population. Note that, from this table it is possible to obtain the RR of any model vs. another one by calculating the ratio of their respective RRs. As an example, the RR "this study, female > 75 years old—more virulent strains" vs. "this study, female > 75 years old—less virulent strains" is estimated to be 3669.7/152.3 = 24.1.

Table 4. Risk of invasive listeriosis for selected doses according to the dose–response model and the population. Relative risk from a dose of 10^3 bacteria (reference: FAO and WHO model, "healthy population" [5]).

Model	Population	Virulence	Dose (Bacteria)				Relative Risk
			10^2	10^3	10^6	10^9	
FAO and WHO [5]	Healthy Population	All strains	2.37×10^{-12}	2.37×10^{-11}	2.37×10^{-08}	2.37×10^{-05}	Reference
	Increased susceptibility		1.06×10^{-10}	1.06×10^{-09}	1.06×10^{-06}	1.06×10^{-03}	44.7
Pouillot et al. [8]	Less than 65 yo, no underlying condition	All strains	8.15×10^{-10}	8.14×10^{-09}	7.33×10^{-06}	2.74×10^{-03}	343.5
	More than 65 yo, no underlying condition		1.55×10^{-08}	1.53×10^{-07}	1.10×10^{-04}	1.94×10^{-02}	6473.8
	Pregnancy		2.06×10^{-07}	1.98×10^{-06}	9.73×10^{-04}	7.70×10^{-02}	83,675.8
	Non-hematological cancer		8.09×10^{-08}	7.89×10^{-07}	4.54×10^{-04}	4.85×10^{-02}	33,270.7
	Hematological Cancer		9.66×10^{-07}	8.93×10^{-06}	3.18×10^{-03}	1.50 × 10-01	376,694.8
	Renal or Liver Failure		2.84×10^{-07}	2.71×10^{-06}	1.25×10^{-03}	8.93×10^{-02}	114,387.0
	Solid Organ Transplant		3.18×10^{-07}	3.03×10^{-06}	1.37×10^{-03}	9.40×10^{-02}	127,857.9
	Inflammatory diseases		8.66×10^{-08}	8.44×10^{-07}	4.81×10^{-04}	5.03×10^{-02}	35,604.0
	HIV/AIDS		6.73×10^{-08}	6.58×10^{-07}	3.90×10^{-04}	4.41×10^{-02}	27,763.9
	Diabetes		7.78×10^{-09}	7.72×10^{-08}	5.96×10^{-05}	1.27×10^{-02}	3259.4
	Heart Diseases		5.26×10^{-09}	5.23×10^{-08}	4.18×10^{-05}	9.90×10^{-03}	2207.8
EFSA [9]	Female 1–4 yo	All strains	2.80×10^{-10}	2.80×10^{-09}	2.63×10^{-06}	1.22×10^{-03}	118.1
	Male 1–4 yo		3.58×10^{-10}	3.58×10^{-09}	3.33×10^{-06}	1.48×10^{-03}	151.1
	Female 5–14 yo		1.27×10^{-10}	1.27×10^{-09}	1.22×10^{-06}	6.54×10^{-04}	53.8
	Male 5–14 yo		1.04×10^{-10}	1.04×10^{-09}	9.99×10^{-07}	5.54×10^{-04}	43.8
	Female 15–24 yo		4.97×10^{-10}	4.97×10^{-09}	4.57×10^{-06}	1.89×10^{-03}	209.5
	Male 15–24 yo		9.67×10^{-11}	9.67×10^{-10}	9.32×10^{-07}	5.23×10^{-04}	40.8
	Female 25–44 yo		9.92×10^{-10}	9.90×10^{-09}	8.83×10^{-06}	3.15×10^{-03}	417.7
	Male 25–44 yo		1.81×10^{-10}	1.81×10^{-09}	1.72×10^{-06}	8.66×10^{-04}	76.3
	Female 45–64 yo		8.72×10^{-10}	8.70×10^{-09}	7.81×10^{-06}	2.87×10^{-03}	367.2
	Male 45–64 yo		9.47×10^{-10}	9.45×10^{-09}	8.45×10^{-06}	3.05×10^{-03}	398.9
	Female 65–74 yo		2.09×10^{-09}	2.08×10^{-08}	1.78×10^{-05}	5.33×10^{-03}	877.5
	Male 65–74 yo		2.89×10^{-09}	2.88×10^{-08}	2.41×10^{-05}	6.66×10^{-03}	1215.8
	Female > 75 yo		3.06×10^{-09}	3.04×10^{-08}	2.54×10^{-05}	6.92×10^{-03}	1284.7
	Male > 75 yo		3.06×10^{-09}	3.04×10^{-08}	2.54×10^{-05}	6.92×10^{-03}	1284.7
This Study	Female 1–4 yo	Less virulent	4.22×10^{-11}	4.22×10^{-10}	4.22×10^{-07}	3.77×10^{-04}	17.8
	Male 1–4 yo		4.65×10^{-11}	4.65×10^{-10}	4.65×10^{-07}	4.12×10^{-04}	19.6
	Female 5–14 yo		1.89×10^{-11}	1.89×10^{-10}	1.88×10^{-07}	1.75×10^{-04}	8.0
	Male 5–14 yo		1.44×10^{-11}	1.44×10^{-10}	1.44×10^{-07}	1.35×10^{-04}	6.1
	Female 15–24 yo		6.15×10^{-11}	6.15×10^{-10}	6.14×10^{-07}	5.35×10^{-04}	26.0
	Male 15–24 yo		1.33×10^{-11}	1.33×10^{-10}	1.33×10^{-07}	1.25×10^{-04}	5.6
	Female 25–44 yo		1.21×10^{-10}	1.21×10^{-09}	1.21×10^{-06}	9.96×10^{-04}	51.2
	Male 25–44 yo		2.22×10^{-11}	2.22×10^{-10}	2.22×10^{-07}	2.05×10^{-04}	9.4
	Female 45–64 yo		1.09×10^{-10}	1.09×10^{-09}	1.09×10^{-06}	9.02×10^{-04}	45.9
	Male 45–64 yo		1.09×10^{-10}	1.09×10^{-09}	1.09×10^{-06}	9.05×10^{-04}	46.0
	Female 65–74 yo		2.36×10^{-10}	2.36×10^{-09}	2.36×10^{-06}	1.80×10^{-03}	99.7
	Male 65–74 yo		3.23×10^{-10}	3.23×10^{-09}	3.21×10^{-06}	2.37×10^{-03}	136.1
	Female > 75 yo		3.61×10^{-10}	3.61×10^{-09}	3.59×10^{-06}	2.61×10^{-03}	152.3
	Male > 75 yo		3.44×10^{-10}	3.44×10^{-09}	3.42×10^{-06}	2.50×10^{-03}	144.9
	Female 1–4 yo	Virulent	3.22×10^{-10}	3.22×10^{-09}	3.22×10^{-06}	3.08×10^{-03}	136.0
	Male 1–4 yo		3.54×10^{-10}	3.54×10^{-09}	3.54×10^{-06}	3.37×10^{-03}	149.2
	Female 5–14 yo		1.67×10^{-10}	1.67×10^{-09}	1.67×10^{-06}	1.63×10^{-03}	70.4
	Male 5–14 yo		1.30×10^{-10}	1.30×10^{-09}	1.30×10^{-06}	1.27×10^{-03}	54.9
	Female 15–24 yo		4.80×10^{-10}	4.80×10^{-09}	4.80×10^{-06}	4.52×10^{-03}	202.6
	Male 15–24 yo		1.19×10^{-10}	1.19×10^{-09}	1.19×10^{-06}	1.17×10^{-03}	50.1
	Female 25–44 yo		9.76×10^{-10}	9.76×10^{-09}	9.76×10^{-06}	8.80×10^{-03}	412.0
	Male 25–44 yo		1.92×10^{-10}	1.92×10^{-09}	1.92×10^{-06}	1.86×10^{-03}	80.9
	Female 45–64 yo		8.21×10^{-10}	8.21×10^{-09}	8.20×10^{-06}	7.48×10^{-03}	346.2
	Male 45–64 yo		8.61×10^{-10}	8.61×10^{-09}	8.61×10^{-06}	7.83×10^{-03}	363.3
	Female 65–74 yo		1.62×10^{-09}	1.62×10^{-08}	1.62×10^{-05}	1.40×10^{-02}	685.3
	Male 65–74 yo		2.28×10^{-09}	2.28×10^{-08}	2.28×10^{-05}	1.89×10^{-02}	960.5
	Female > 75 yo		2.33×10^{-09}	2.33×10^{-08}	2.33×10^{-05}	1.93×10^{-02}	982.4
	Male > 75 yo		2.17×10^{-09}	2.17×10^{-08}	2.17×10^{-05}	1.81×10^{-02}	916.1
	Female 1–4 yo	More Virulent	9.71×10^{-10}	9.71×10^{-09}	9.71×10^{-06}	9.06×10^{-03}	409.6
	Male 1–4 yo		1.08×10^{-09}	1.08×10^{-08}	1.08×10^{-05}	9.99×10^{-03}	454.5
	Female 5–14 yo		4.09×10^{-10}	4.09×10^{-09}	4.09×10^{-06}	3.95×10^{-03}	172.4
	Male 5–14 yo		3.11×10^{-10}	3.11×10^{-09}	3.11×10^{-06}	3.03×10^{-03}	131.2
	Female 15–24 yo		1.37×10^{-09}	1.37×10^{-08}	1.37×10^{-05}	1.25×10^{-02}	578.3
	Male 15–24 yo		2.87×10^{-10}	2.87×10^{-09}	2.87×10^{-06}	2.80×10^{-03}	121.2
	Female 25–44 yo		2.63×10^{-09}	2.63×10^{-08}	2.62×10^{-05}	2.27×10^{-02}	1107.6
	Male 25–44 yo		4.84×10^{-10}	4.84×10^{-09}	4.84×10^{-06}	4.65×10^{-03}	204.0
	Female 45–64 yo		2.43×10^{-09}	2.43×10^{-08}	2.43×10^{-05}	2.12×10^{-02}	1026.1
	Male 45–64 yo		2.42×10^{-09}	2.42×10^{-08}	2.42×10^{-05}	2.10×10^{-02}	1020.0
	Female 65–74 yo		5.52×10^{-09}	5.52×10^{-08}	5.52×10^{-05}	4.33×10^{-02}	2330.5
	Male 65–74 yo		7.45×10^{-09}	7.45×10^{-08}	7.45×10^{-05}	5.55×10^{-02}	3145.5
	Female > 75 yo		8.70×10^{-09}	8.70×10^{-08}	8.69×10^{-05}	6.28×10^{-02}	3669.7
	Male > 75 yo		8.38×10^{-09}	8.38×10^{-08}	8.37×10^{-05}	6.10×10^{-02}	3536.8

From Figure 1, and considering a dose of 10^3 bacteria, which lies within the linear part of the DR curves, for calculating the RR, the following observations can be made: for the FAO and WHO [5] model, the RR between the general and susceptible populations is 44.7, with the latter resulting in the higher probability of illness of 1.06×10^{-9}.

The RRs between the Pouillot et al. [8] and the FAO and WHO [5] DR display some differences. Indeed, for the less than 65-year-old individual with no underlying health conditions, the estimated risk is 344 times higher using the Pouillot et al. (2015) model than when using the FAO and WHO healthy population model and is still 7.68 times higher than the FAO and WHO increased susceptibility population model. When compared with the EFSA model [9], the Pouillot et al. [8] model estimates a risk for the less than 65-year-old individual with no underlying conditions, for a given dose of 10^3 bacteria, that is 8.41 times higher than the risk estimated from the EFSA model for the least susceptible population (male, 15–24 years old). The Pouillot et al. [8] model is more "conservative" than the FAO and WHO [5] model or the EFSA one [9].

The RR between the least (males between 15 and 24 years old) and the most (males and females > 75 years old) susceptible EFSA populations was 31.5, with the latter resulting in the higher probability of illness of 3.04×10^{-8} from a dose of 10^3 bacteria. The RRs between the least (males between 15 and 24 years old) and the most (males and females > 75 years old) susceptible EFSA populations and the general population of the FAO and WHO model were 40.8 and 1285, respectively. In contrast, RRs between the least (males between 15 and 24 years old) and the most (males and females > 75 years old) susceptible EFSA populations and the susceptible population of the FAO and WHO model were 0.91 and 28.7, respectively. These observations hold for the linear part of the DR model, i.e., doses up to about 10^8. At greater doses (>10^{11}), the RRs reduce, and may reverse, i.e., the FAO and WHO susceptible population model predicts higher probability of illness compared with the EFSA model.

As explained above, the new DR model takes the three classes of strain virulence into account, in addition to the age–sex groups of the EFSA model, resulting in a total of 42 DR curves. Instead of displaying all of these, the least (males between 15 and 24 years old) and the most (females > 75 years old) susceptible EFSA populations were selected and displayed for each of the three virulence classes: "less virulent", "virulent", and "more virulent". It should be noted that when integrating the virulence class information with the EFSA model, females over 75 years of age are marginally more susceptible than males of the same age. The resulting plot is shown in Figure 2.

As before, considering a dose of 10^3 bacteria, which lies within the linear part of the DR curves, for calculating the RR, the following observations can be made from studying Figure 2 and Table 4. The RRs between the "less virulent" class and the "more virulent" class of *L. monocytogenes* strains depend on the population of interest, i.e., they are not constant. The RRs are 21.6 and 24.1 for the least susceptible population (males between 15 and 24 years old) and the most susceptible population (females > 75 years old), respectively. Comparing the general and susceptible population FAO/WHO model with the new DR model, and using the virulent class of strains, the following RRs can be calculated:

(i) Susceptible FAO/WHO population versus the most susceptible population (females > 75 years old) yields an RR of 22.0;
(ii) General FAO/WHO population versus least susceptible population (males between 15 and 24 years old) yields an RR of 50.1.
(iii) The RR between the least susceptible group (males between 15 and 24 years old) exposed to the "less virulent" class of strains versus the most susceptible group (females > 75 years old) exposed to the most virulent class of strains is 655.1.
(iv) Taking class of strain virulence and age–sex sub-populations into account results in a wider range of possible probability of illness at a specific dose (RR = 655) compared with the range obtained from the previous FAO and WHO model (RR = 45).

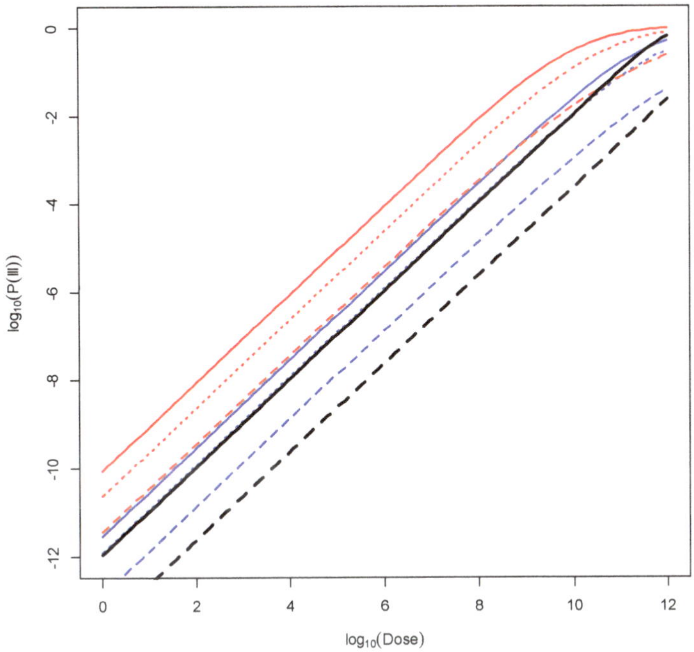

Figure 2. DR model curves for the least susceptible population (blue, males between 15 and 24 years old) and the most susceptible population (red, females > 75 years old); "less virulent" class of strains (dashed red/blue lines), "virulent" (dotted red/blue lines), and "more virulent" (solid red/blue lines); and FAO and WHO [5] susceptible population (black solid line) and general population (black dashed line).

Overall, to compare the marginal DR over all sub-populations, we weighted the risk of invasive listeriosis for a 10^3 dose according to the proportion of each sub-population as provided by the FAO and WHO ([5], 83% of healthy population and 17% of increased susceptibility population), Pouillot et al. ([8], their Table I, based on frequency of each sub-population in France), and the EFSA ([9], their Table 22, based on the number of eating occasions per sub-population). Using as a reference the FAO and WHO model [5], the overall RR for Pouillot et al.'s [8] model at a 10^3 bacteria dose is estimated to be 576, the overall RR for the EFSA model [9] is 53, and the overall RR for this study model is 6 for the "less virulent" strains, 45 for the "virulent" strains, and 144 for the "more virulent" strains.

4. Discussion

DR models are a critical component of the hazard characterization part of any food safety risk assessment, especially when they are quantitative [2]. For quantitative microbial risk assessments, they link the number of pathogenic organisms ingested by a consumer, as estimated at the end of the exposure assessment, with the probability of the health outcome of interest, usually infection or illness. Different mathematical forms for DR models are available, each with their own assumptions [6]. Nevertheless, so-called "single-hit" models, which assume that each pathogenic cell ingested has a probability of resulting in the health outcome of interests—the *r*-value—which is independent of all other pathogenic cells ingested, are the most common DR models. By incorporating additional assumptions and parameterizations for the *r*-value, increasingly complex models can be developed with the aim of better accounting for the food matrix, pathogen, and host characteristics of the infectious disease triangle [2].

In the case of *L. monocytogenes*, the health outcome of interest is generally invasive disease due to the associated sequalae and relatively high mortality rate [5]. The 2004 model used by the FAO and WHO [5] was of the exponential form, and broad differentiation of the host's susceptibility—healthy versus susceptible—was incorporated through the use of two different *r*-values. However, FAO and WHO experts at the time recommended that "*More complete investigation of outbreaks and determination of the virulence characteristics of L. monocytogenes will make the dose-response relationships more accurate and precise*" [5].

Considerable effort has been expended over the last two decades in that regard and several different data sources can contribute to this [11]. Advances in refining the DR model for invasive listeriosis in humans have been as follows:

- Pouillot et al. [8] modeled the *r*-value as a $\log_{10}$-normal distribution. The host's underlying health conditions, such as pregnancy, cancer, heart disease, etc., were incorporated using RRs for various sub-populations and a proportion of each sub-population estimated from the French population [13]. Exposure to *L. monocytogenes* was estimated from US data [19]. No growth from retail to consumption was considered. For the MPD, an important parameter in *L. monocytogenes* risk assessment [21], a value of 6.1 $\log_{10}$ cfu/g was used;
- Fritsch et al. [10] estimated *r*-values for 26 strains of *L. monocytogenes* and then grouped these strains into three classes of virulence according to the similarity of the *r*-value, namely, "hypovirulent" (called here "less virulent"), "medium virulent" (called here "virulent"), and "hypervirulent" (called here "more virulent");
- The EFSA [9] estimated *r*-values for 14 age–sex combinations. Age and sex were used as surrogates for underlying health conditions because the demographic stratification into the 14 sub-populations was readily available across the EU, unlike actual information on underlying health conditions. The EFSA considered a bacterial growth from retail to consumption and MPDs ranging from 6.23 to 7.53 $\log_{10}$ cfu/g according to the considered RTE [9].
- In 2022, an FAO and WHO expert group [11] considered that the most appropriate DR model approaches would include variability in sub-population susceptibility *and* variability in strain virulence. The availability of an exposure model to *L. monocytogenes* in the EU [9] and of CC data from representative samples of food and clinical cases in the same population [16] made this derivation possible.
- The novelty of the current work is to conflate these three works to develop a DR model for invasive listeriosis that could consider differences in strain virulence class as well as better account for host susceptibility through the inclusion of surrogate information—age and sex. Consequently, this new model allows for more specific estimation of the risk of invasive listeriosis at the cost of two additional pieces of information, namely, population demographics and proportional contribution of strain virulence classes in food products of interest.

The assumption underlying the inference of the number of invasive human listeriosis cases per sub-population for each class of virulence was that the proportion of cases linked to "less virulent", "virulent", or "more virulent" strains is independent of age, sex, EU country of residence, underlying health conditions, etc. This assumption could be refined in the future if additional data regarding the cases of listeriosis based on age, sex, and country of residence become available.

It should be noted that the EFSA sub-groups, based on sex and age, were used as a surrogate for underlying health conditions [9]. This was linked to the fact that the EFSA data are more broadly representative geographically and that information about underlying health conditions was not available. Ideally, however, representative data on specific underlying health conditions, similar to those obtained from France [13], would be preferable; this would also allow the DR model to be more globally relevant. As an example, the difference between males and females for the classes of age "15–24 years old" and "25–44 years old" is most probably linked to pregnancy [9]. The DR model derived here may help to refine risk assessment but will not be able to really quantify the increased RRs of listeriosis for the most susceptible population, as was done by Pouillot et al. [8].

The major assumption regarding the derivation of this DR model is the assumption of a single lognormal distribution underlying the distribution of r parameters in each of the considered sub-populations/classes of virulence. As described above, the use of highly heterogeneous sub-populations (pregnant and non-pregnant women for the female 15–24 and 25–44-year-old groups, individuals with severe underlying health conditions, and healthy individuals for groups of aged individuals) may impair this assumption. The other assumptions used in this derivation are all the various assumptions used in the EFSA generic quantitative microbial risk assessment model [9], notably the fact that all listeriosis cases were assumed to be linked to the considered limited number of RTE foods. This study makes the additional assumption of a similar behavior of the strains in this generic quantitative microbial risk assessment irrespective of their class of virulence.

Our estimates for μ and σ are simple point estimates. The uncertainty surrounding this DR model combines all the uncertainty linked to the exposure model, the estimated number of eating occasions, the reported number of clinical cases per sub-population, and the estimated proportions of classes of virulence in food and clinical cases. Regarding the exposure model and the number of eating occasions, the original publications ([18], their Table 25 and [9], their table J.1) characterized, documented, and explained all types of uncertainty, following the EFSA's recommendations [22]. All the assumptions and limitations identified in these reports apply here. However, no quantification of the uncertainty was performed in these studies. The EFSA [9] performed an importance analysis for some parameters that they considered as the most influential, i.e., regarding duration and temperature of storage in the consumer's refrigerator and MPD. Those parameters are considered to have the largest effect in risk assessments linked to *L. monocytogenes* in RTE foods [15,21,23]. The EFSA [9] found that the risk assessment model was very sensitive to a shift in the *L. monocytogenes* MPD, as expected [21]. We re-ran the model using a shift of +0.5 $\log_{10}$ of the MPD of *L. monocytogenes* for all RTE foods and observed that the estimated risk of invasive listeriosis following the ingestion of 1000 bacteria would be multiplied by a factor of 2.5 to 2.6, according to the sub-group and class of virulence.

The model is additionally scaled to the number of clinical cases estimated for the 42 sub-groups. Some inferences were made based on a very small, estimated number of cases. For example, inferences for females aged 1–4 years old exposed to the "less virulent" strains were made from only four reported listeriosis cases. More uncertainty could be found surrounding these lower risk groups.

Although some CCs are present globally and are very similar at the core-genome level, their accessory genome can vary from region to region [24]. Within these variations in the accessory, some genetic elements may affect virulence. Regional differences in the accessory genome could then lead to variations in virulence within a given CC, which may not be captured in this study based on data from a single geographic region.

Moreover, limited data were available to classify some CCs. For instance, the proportion of CC87 strains in the EU is quite small. The classification of CC87 in the "more virulent" category was based only on 10 strains in foods and 4 in sporadic cases. Using data from China could provide more information as CC87 strains are predominant in Chinese food isolates and in sporadic clinical infections [25]. This points to the challenge of classifying virulence based on limited data and suggests that incorporating more comprehensive datasets, especially from regions where certain strains are more prevalent, could lead to more accurate classifications.

Collectively this model suggests higher r values than those derived previously by the FAO and WHO [5], but lower than the more conservative model from Pouillot et al. [8]. The effect of higher r values is that they result in a left shift of the DR curve, i.e., to lower doses. Considering classes of virulence leads to RRs between "less virulent" and "more virulent" strains ranging from 21.6 to 24.1. These relatively low RRs should be compared to the extremely high RRs obtained by Goulet et al. for various comorbidities [13] (e.g., 373.6

for hematological cancer when compared to the less than 65-year-old individual with no known underlying conditions). This observation means that the influence of comorbidity seems much more important than the influence of the strains, as estimated in this study.

5. Conclusions

Parameters for lognormal Poisson DR models for *L. monocytogenes* were estimated for three classes of strain virulence and 14 age–sex sub-populations, as a surrogate for underlying health conditions. The updated model parameters allow better estimation of the risk of invasive listeriosis across a population of interest, provided the necessary data on population demographics and the proportional contribution of strain virulence classes in food products of interest are available. In turn, these improved estimates will facilitate better decision making in food safety risk assessments, though future focus on comorbidity data seems much more important than data on strain virulence variability.

Author Contributions: Conceptualization, R.P., L.G. and M.S.; methodology, R.P., L.G. and M.S.; software, R.P., A.K., V.C. and M.S.; validation, R.P. and A.K.; formal analysis, R.P., L.G. and A.K.; data curation, L.G.; writing—original draft preparation, R.P. and A.K.; writing—review and editing, L.G., V.C. and M.S.; visualization, R.P.; supervision, M.S.; project administration, M.S. All authors have read and agreed to the published version of the manuscript.

Funding: This research received no external funding.

Institutional Review Board Statement: Not applicable.

Informed Consent Statement: Not applicable.

Data Availability Statement: Data, code and an R package are available at www.github.com/rpouillot/FoodsDR (accessed on 1 February 2024).

Acknowledgments: The FAO/WHO Joint Expert Meeting on Microbiological Risk Assessment (JEMRA) on *Listeria monocytogenes* in foods, held in Rome, Italy from October 24 to 28, 2022, recommended the development of a new dose-response model for *Listeria monocytogenes* that considers consumer susceptibility and virulence characterization of *L. monocytogenes*. The model was finalized during JEMRA on *Listeria monocytogenes* in foods Part 2: Risk Assessment Models, which was held in Geneva, Switzerland from 29 May to 2 June 2023. The authors thank the JEMRA experts for Part 1 (https://www.who.int/publications/m/item/jemra-of-listeria-monocytogenes-in-foods) and Part 2, (https://www.who.int/publications/m/item/jemra-of-listeria-monocytogenes-in-foods-part-2-risk-assessment-models) as well as FAO/WHO secretariat for their suggestions and feedback. The authors would like to thank Ursula Gonzales-Barron.

Conflicts of Interest: Author A.K. is the director of Statistical Process Improvement Consulting and Training Pty Ltd. Author R.P. is self-employed. They participated in this work in their own personal capacity. The remaining authors declare that the research was conducted in the absence of any commercial or financial relationships that could be construed as a potential conflict of interest.

References

1. European Food Safety Authority (EFSA); European Centre for Disease Prevention and Control (ECDC). The European Union One Health 2022 Zoonoses Report. *EFSA J.* **2023**, *21*, e8442. [CrossRef]
2. FAO; WHO. *Microbiological Risk Assessment Guidance for Food*; Microbiological Risk Assessment Series, No 36; Food and Agriculture Organization of the United Nations: Roma, Italy; World Health Organization: Geneva, Switzerland, 2021; p. 288.
3. Buchanan, R.L.; Smith, J.L.; Long, W. Microbial Risk Assessment: Dose-Response Relations and Risk Characterization. *Int. J. Food Microbiol.* **2000**, *58*, 159–172. [CrossRef] [PubMed]
4. Hoelzer, K.; Chen, Y.; Dennis, S.; Evans, P.; Pouillot, R.; Silk, B.J.; Walls, I. New Data, Strategies, and Insights for *Listeria monocytogenes* Dose-Response Models: Summary of an Interagency Workshop, 2011. *Risk Anal.* **2013**, *33*, 1568–1581. [CrossRef] [PubMed]
5. FAO; WHO. *Risk Assessment of Listeria monocytogenes in Ready to Eat Foods*; Technical Report; Microbiological Risk Assessment Series, No 5; Food and Agriculture Organization of the United Nations: Roma, Italy; World Health Organization: Geneva, Switzerland, 2004; p. 269.
6. Haas, C.N.; Rose, J.B.; Gerba, C.P. *Quantitative Microbial Risk Assessment*; Wiley: New York, NY, USA, 1999.

7. Buchanan, R.L.; Damert, W.G.; Whiting, R.C.; van Schothorst, M. Use of Epidemiologic and Food Survey Data to Estimate a Purposefully Conservative Dose-Response Relationship for *Listeria monocytogenes* Levels and Incidence of Listeriosis. *J. Food Prot.* **1997**, *60*, 918–922. [CrossRef] [PubMed]
8. Pouillot, R.; Hoelzer, K.; Chen, Y.; Dennis, S.B. *Listeria monocytogenes* Dose Response Revisited—Incorporating Adjustments for Variability in Strain Virulence and Host Susceptibility. *Risk Anal.* **2015**, *35*, 90–108. [CrossRef] [PubMed]
9. EFSA Panel on Biological Hazards (BIOHAZ); Ricci, A.; Allende, A.; Bolton, D.; Chemaly, M.; Davies, R.; Fernández Escámez, P.S.; Girones, R.; Herman, L.; Koutsoumanis, K.; et al. *Listeria monocytogenes* Contamination of Ready-to-Eat Foods and the Risk for Human Health in the EU. *EFSA J.* **2018**, *16*, e05134. [CrossRef]
10. Fritsch, L.; Guillier, L.; Augustin, J.-C. Next Generation Quantitative Microbiological Risk Assessment: Refinement of the Cold Smoked Salmon-Related Listeriosis Risk Model by Integrating Genomic Data. *Microb. Risk Anal.* **2018**, *10*, 20–27. [CrossRef]
11. FAO; WHO. *Listeria monocytogenes in Ready-to-Eat (RTE) Foods: Attribution, Characterization and Monitoring—Meeting Report*; Microbiological Risk Assessment Series, No 38; Food and Agriculture Organization of the United Nations: Roma, Italy; World Health Organization: Geneva, Switzerland, 2022.
12. Haas, C.N. Conditional Dose-Response Relationships for Microorganisms: Development and Application. *Risk Anal.* **2002**, *22*, 455–463. [CrossRef] [PubMed]
13. Goulet, V.; Hebert, M.; Hedberg, C.; Laurent, E.; Vaillant, V.; De Valk, H.; Desenclos, J.C. Incidence of Listeriosis and Related Mortality among Groups at Risk of Acquiring Listeriosis. *Clin. Infect. Dis.* **2012**, *54*, 652–660. [CrossRef] [PubMed]
14. Roche, S.M.; Velge, P.; Bottreau, E.; Durier, C.; Marquet-van der Mee, N.; Pardon, P. Assessment of the Virulence of *Listeria monocytogenes*: Agreement between a Plaque-Forming Assay with HT-29 Cells and Infection of Immunocompetent Mice. *Int. J. Food Microbiol.* **2001**, *68*, 33–44. [CrossRef] [PubMed]
15. FDA; FSIS. *Quantitative Assessment of Relative Risk to Public Health from Foodborne Listeria monocytogenes among Selected Categories of Ready-to-Eat Foods*; Food and Drug Administration, United States: College Park, MD, USA; Department of Agriculture: Washington, DC, USA; Centers for Disease Control and Prevention: Atlanta, GA, USA, 2003; p. 541.
16. Møller Nielsen, E.; Björkman, J.T.; Kiil, K.; Grant, K.; Dallman, T.; Painset, A.; Amar, C.; Roussel, S.; Guillier, L.; Félix, B.; et al. Closing Gaps for Performing a Risk Assessment on *Listeria monocytogenes* in Ready-to-eat (RTE) Foods: Activity 3, the Comparison of Isolates from Different Compartments along the Food Chain, and from Humans Using Whole Genome Sequencing (WGS) Analysis. *EFSA Support. Publ.* **2017**, *14*, 1151E. [CrossRef]
17. European Food Safety Authority. Analysis of the Baseline Survey on the Prevalence of *Listeria monocytogenes* in Certain Ready-to-Eat Foods in the EU, 2010–2011 Part A: Listeria Monocytogenes Prevalence Estimates. *EFSA J.* **2013**, *11*, 3241. [CrossRef]
18. Pérez-Rodríguez, F.; Carrasco, E.; Bover-Cid, S.; Jofré, A.; Valero, A. Closing Gaps for Performing a Risk Assessment on *Listeria monocytogenes* in Ready-to-Eat (RTE) Foods: Activity 2, a Quantitative Risk Characterization on *L. monocytogenes* in RTE Foods; Starting from the Retail Stage. *EFSA Support. Publ.* **2017**, *14*, 1252E. [CrossRef]
19. Gombas, D.E.; Chen, Y.; Clavero, R.S.; Scott, V.N. Survey of *Listeria monocytogenes* in Ready-to-Eat Foods. *J. Food Prot.* **2003**, *66*, 559–569. [CrossRef]
20. R Development Core Team, R. A Language and Environment for Statistical Computing. Available online: http://www.R-project.org (accessed on 1 February 2024).
21. Pouillot, R.; Lubran, M.B. Predictive Microbiology Models vs. Modeling Microbial Growth within *Listeria monocytogenes* Risk Assessment: What Parameters Matter and Why. *Food Microbiol.* **2011**, *28*, 720–726. [CrossRef]
22. European Food Safety Authority (EFSA). Guidance of the Scientific Committee on Transparency in the Scientific Aspects of Risk Assessments Carried out by EFSA. Part 2: General Principles. *EFSA J.* **2009**, *7*, 1051. [CrossRef]
23. Pouillot, R.; Goulet, V.; Delignette-Muller, M.L.; Mahe, A.; Cornu, M. Quantitative Risk Assessment of *Listeria monocytogenes* in French Cold-Smoked Salmon: II. Risk Characterization. *Risk Anal.* **2009**, *29*, 806–819. [CrossRef]
24. Liao, J.; Guo, X.; Li, S.; Anupoju, S.M.B.; Cheng, R.A.; Weller, D.L.; Sullivan, G.; Zhang, H.; Deng, X.; Wiedmann, M. Comparative Genomics Unveils Extensive Genomic Variation between Populations of *Listeria* Species in Natural and Food-Associated Environments. *ISME Commun.* **2023**, *3*, 85. [CrossRef] [PubMed]
25. Zhang, H.; Chen, W.; Wang, J.; Xu, B.; Liu, H.; Dong, Q.; Zhang, X. 10-Year Molecular Surveillance of *Listeria Monocytogenes* Using Whole-Genome Sequencing in Shanghai, China, 2009–2019. *Front. Microbiol.* **2020**, *11*, 551020. [CrossRef] [PubMed]

foods

MDPI

Review

A Critical Review of Risk Assessment Models for *Listeria monocytogenes* in Produce

Ursula Gonzales-Barron [1,2,*], Vasco Cadavez [1,2], Juliana De Oliveira Mota [3], Laurent Guillier [4] and Moez Sanaa [3,*]

1 Centro de Investigação de Montanha (CIMO), Instituto Politécnico de Bragança, Campus de Santa Apolónia, 5300-253 Bragança, Portugal; vcadavez@ipb.pt
2 Laboratório para a Sustentabilidade e Tecnologia em Regiões de Montanha, Instituto Politécnico de Bragança, Campus de Santa Apolónia, 5300-253 Bragança, Portugal
3 Nutrition and Food Safety Department, World Health Organization, 1202 Geneva, Switzerland
4 Risk Assessment Department, French Agency for Food, Environmental and Occupational Health & Safety (Anses), 14 rue Pierre et Marie Curie Maisons-Alfort, 94700 Maisons-Alfort, France; laurent.guillier@anses.fr
* Correspondence: ubarron@ipb.pt (U.G.-B.); sanaam@who.int (M.S.); Tel.: +351-273-303-325 (U.G.-B.)

Abstract: A review of quantitative risk assessment (QRA) models of *Listeria monocytogenes* in produce was carried out, with the objective of appraising and contrasting the effectiveness of the control strategies placed along the food chains. Despite nine of the thirteen QRA models recovered being focused on fresh or RTE leafy greens, none of them represented important factors or sources of contamination in the primary production, such as the type of cultivation, water, fertilisers or irrigation method/practices. Cross-contamination at processing and during consumer's handling was modelled using transfer rates, which were shown to moderately drive the final risk of listeriosis, therefore highlighting the importance of accurately representing the transfer coefficient parameters. Many QRA models coincided in the fact that temperature fluctuations at retail or temperature abuse at home were key factors contributing to increasing the risk of listeriosis. In addition to a primary module that could help assess current on-farm practices and potential control measures, future QRA models for minimally processed produce should also contain a refined sanitisation module able to estimate the effectiveness of various sanitisers as a function of type, concentration and exposure time. Finally, *L. monocytogenes* growth in the products down the supply chain should be estimated by using realistic time–temperature trajectories, and validated microbial kinetic parameters, both of them currently available in the literature.

Keywords: systematic review; exposure assessment; listeriosis; leafy greens; vegetables; fruits

1. Introduction

Invasive listeriosis is a rare but severe foodborne disease that affects certain population groups such as the elderly, pregnant women and neonates, and immunocompromised individuals (i.e., patients with diabetes, AIDS, cancer and inflammatory diseases) [1]. According to EFSA BIOHAZ [2], demographic changes and health status factors, as well as a preference switch of the consumers over high-risk ready-to-eat (RTE) foodstuffs would produce an increase in listeriosis incidence. Surveillance data have shown that the occurrence of *Listeria monocytogenes* in RTE products has been responsible for outbreaks worldwide. According to EU data [3], in the period between 2010 and 2020, a share of 10% of the total strong-evidence outbreaks of listeriosis was linked to produce, having raw vegetables, juices and black olives as sources of the disease. Unlike the EU scenario, in the USA, produce presented a higher share (30%) in the ten-year span, being the second topmost food category (following dairy) linked to strong-evidence listeriosis outbreaks [4]. In the USA, pre-cut celery, cantaloupe, frozen vegetables, peaches/nectarine, mung bean sprouts, pre-packed leafy greens, sprouts, pre-packed lettuce, enoki mushrooms, avocado, and stone fruit were the food vehicles associated with listeriosis outbreaks.

Many case–control studies of sporadic listeriosis in susceptible populations pointed out that vegetables could be significantly linked to listeriosis. Carrots, retail vegetables and RTE mixed salads presented odds ratios (OR) of 2.00 (95% CI: 0.90–4.54) in the perinatal and non-perinatal US population [5], 1.92 (95% CI: 1.32–2.78) in the perinatal and non-perinatal UK population [6], and 1.72 (95% CI: 1.20–2.47) in the UK non-perinatal UK population [7], respectively. In the case of fruits, a meta-analysis investigation [8] obtained a significant pooled OR of 1.54 (95% CI: 1.14–2.07) for the association between sporadic listeriosis and the consumption of watermelons, melons and cantaloupes by susceptible population in the USA [5] and Australia [9].

Various listeriosis quantitative risk assessment (QRA) models have been produced for RTE foods in general [10] in an attempt to assess the efficacy of current practices or potential strategies that retard or prevent the development of this pathogen. The objectives of this study are: (i) to accomplish a critical review of the published QRA models of listeriosis linked to RTE and non-RTE produce; (ii) to compare the relative effectiveness of the control measures or strategies evaluated in the various QRA models as simulated scenarios; and (iii) to elicit, thereof, recommendations for future QRA models in produce.

2. Materials and Methods

QRA models were retrieved through a literature search on Scopus and PubMed® considering 1998 as the starting year of publication. The searches in title, keywords and abstract were performed on 18 May 2022, using logically connected terms ((“risk assessment” OR exposure OR quantitative microbial OR risk modelling OR modeling OR simulation* OR second-order OR “second order” OR “risk management”) AND (“*L. monocytogenes*” OR “*Listeria monocytogenes*” OR listeriosis)) properly structured according to the syntaxes of the literature search engines. The full systematic review process and extraction of information have been described in Gonzales-Barron et al. [7]. The present review focuses only on produce, which is the subject of 13 QRA models described in 12 publications [11–22].

3. Results

A total of 13 QRA models on produce as a source of listeriosis were recovered in the literature search of models published between January 1998 and May 2022. Table 1 compiles the main features of the 13 QRA models, whereas Table 2 summarises the predictive microbiology models used in their construction, and the key outcomes from what-if scenarios and sensitivity analysis.

Eight models (61.5%) evaluated RTE produce, including lettuce salad [12,13], fresh-cut romaine lettuce [14], leafy greens from salad bars [16,18], leafy vegetables [17], fruits and vegetables [19] and fresh-cut cantaloupe [14]; whereas, five models (38.5%) focused on non-RTE produce, namely, fresh lettuce [11,20], fresh baby spinach [15] and frozen vegetables [21,22].

Most of the available models (11/13) characterised the conditions of the USA [14,15,19,21] and European countries, namely Spain [12,20], France [13], and the Netherlands [16,18], in addition to a model using data from the EU [22]. The other two QRA models pertained to the risk of listeriosis evaluated in Korea [11] and Brazil [17] (Table 1).

Table 1. Features of quantitative risk assessment models of *L. monocytogenes* (LM) from consumption of produce by scope.

Scope	Food	RTE	Cross-Contamination	DR–End-Point	Type of DR Model	DR Sub-Populations	Strain Variability	Temp Profiles/ Lagtime	Country	Source
Farm-to-table	Lettuce	No	Yes: Transport, market, restaurants and at home	Exp–I	FAO/WHO [23]	High-risk/Low-risk	LM strain diversity implicit in r	No/Yes	Korea	Ding et al. [11]
Processing-to-table	RTE lettuce salad	Yes	No	WG–I	Farber et al. [24]	High-risk/Low-risk	-	No/No	Spain	Carrasco et al. [12]
	RTE lettuce salad	Yes	No	Exp–I	FAO/WHO [23]	High-risk/Low-risk	LM strain diversity implicit in r	Yes/No	France	Crèpet [13]
	Fresh-cut romaine lettuce	Yes	Yes: Processing—during packaging	WG–I	Farber et al. [24]	High-risk/Low-risk	-	No/Yes	USA	Guzel [14]
	Fresh-cut cantaloupe	Yes	Yes: Processing—after cutting	WG–I	Farber et al. [24]	High-risk/Low-risk	-	No/Yes	USA	Guzel [14]
	Fresh baby spinach	No	Yes: Processing—before packaging	Exp–I	Chen et al. [25]	General	-	No/Yes	USA	Omac et al. [15]
End processing-to-table	Leafy green salads from salad bars	Yes	No	Exp–I	Chen et al. [25]	General	-	Yes/No	Netherlands	Franz et al. [16]
	RTE leafy vegetables	Yes	No	Exp–I	Buchanan et al. [26]	General	-	No/No	Brazil	Sant'Ana et al. [17]
	Leafy green salads from salad bars	Yes	No	Exp–I Log–D	Chen et al. [25]; Williams et al. [27]	General Perinatal	-	Yes/No	Netherlands	Tromp et al. [18]
Retail-to-table	Fruits and vegetables	Yes	No	Mouse Epi–I	FDA-FSIS [19]	Multiple	Variability in the virulence of different LM strains represented in DR	No/No	USA	FDA-FSIS [19]
	Lettuce	No	Yes: Handling at home	Exp–I	FAO/WHO [23]	High-risk	Strain diversity implicit in r	No/No	Spain	Domenech et al. [20]
Consumption	Non-RTE frozen vegetables	No	No	Exp/WG–I	Buchanan et al. [26] Bemrah et al. [28]	High-risk/Low-risk	-	No/Yes	USA	Zoellner et al. [21]
	Blanched frozen vegetables	No	No	Exp–I	EFSA BIOHAZ [2] based on Pouillot et al. [29]	Elderly population (male, female)	Distribution for EGR 5 °C modelled from LM growth data in corn, green peas, carrots, broccoli, beans and asparagus. LM strain virulence and host susceptibility explicit in r distribution	No/No	EU	EFSA BIOHAZ [22]

DR: dose–response; Exp: exponential model; WG: Weibull-gamma model; Log: logistic model; Mouse-Epi: mouse-epidemiological model; I: illness endpoint; D: death endpoint.

Table 2. Predictive microbiology models and main outcomes related to what-if scenarios and sensitivity analysis from quantitative risk assessment models of *L. monocytogenes* (LM) from consumption of produce.

Scope	Food	Predictive Microbiology Models	What-If Scenarios	Sensitivity Analysis [1]	Model Complexity [2]	Source
Farm-to-table	Lettuce	Growth (Gompertz model, polynomial model for lag phase, growth square root model)	NA	NA	Low	Ding et al. [11]
Processing-to-table	RTE lettuce salad	Growth (linear model, growth square root model)	(1) Use of MAP (5.5%, CO_2, 3% O_2; 92.5% N_2) as opposed to no packaging (baseline) reduce mean number of listeriosis cases by 95%; (2) Reducing the shelf-life from a maximum of 12 days to 4 days reduces number of cases by 84%; (3) Preventing high-risk consumers from consuming RTE salads reduces number of cases by 75%; (4) Applying microbiological criterion at primary production (n = 20; c = 0; absence in 25 g) reduces cases by 43%.	Outcome—number of listeriosis cases, ranked in this order: serving size, storage temperature at home, storage time at home, LM concentration at consumption (no r provided)	Medium: An approximation is given to solve growth for dynamic temperatures	Carrasco et al. [12]
	RTE lettuce salad	Growth (logistic model with delay and rupture, cardinal parameter model for temperature). Three models were proposed for the maximum levels of contamination (one based on observed challenge tests, the two others based on the initial contamination in the pack). A model was established to consider lag phase. The models are fully described in [30].	The model enables the assessment of the effect of water chlorination during the washing of lettuce.	The impact of different hypothesis for the risk characterisation was carried. It includes the way of modelling maximum level (y_{max}), the consideration of lag, and the clustering of contamination in packed salad (b parameter). The non-treatment or water with chlorine multiplied the risk of listeriosis by 2.	High: Second-order Monte Carlo simulation is used to assess uncertainty of risk of listeriosis. The model starts from the lettuce and takes into account the effect of washing. The model reproduces the cold chain itinerary of the lettuce.	Crèpet [13]
	Fresh-cut romaine lettuce	Growth (Baranyi model, growth square root model)	(1) LM counts at consumption is reduced by >99% after exposure to ionising radiation (1 kGy at room temperature) and reduces log risk of illness by 1.66 log in the susceptible population; (2) Cold atmospheric plasma reduces LM population by 92% and log risk in 1.34 log; (3) Peroxyacetic acid reduces LM counts by 28%, and log risk by 0.35 log; (4) Cross-contamination during processing increases LM counts in 18% and log risk by 0.06 (because the transfer coefficient was very low at 0.002); (5) Home temperature abuse (20 °C × 24 h) increases mean LM counts by 56% and log risk in 1.1; (6) Consumption time up to a maximum of 14 days increases LM counts by 2100% and log risk in 2.6.		Low	Guzel [14]

Table 2. *Cont.*

Scope	Food	Predictive Microbiology Models	What-If Scenarios	Sensitivity Analysis [1]	Model Complexity [2]	Source
	Fresh-cut cantaloupe	Growth (Baranyi model, growth square root model)	(1) Implementation of irradiation reduces LM at consumption by 99.9%, and mean cases of listeriosis by 99%; (2) Cross-contamination during processing increases cases by 300%; (3) Home temperature abuse at home (20 °C × 24 h) increases LM at consumption by 300% and cases by 220%; (4) Extending consumption time up to a maximum of 10 days increases LM at consumption by 2300% and cases in 39,000%.	NA	Low	Guzel [14]
	Fresh baby spinach	Growth (Baranyi model, square root models for growth and lag phase, polynomial model for y_{max},)	Baseline scenario represents neither interventions during processing nor cross-contamination. (1) Washing with water decreases mean cases of listeriosis by 7.5%; (2) Water with PAA or ClO2 reduces mean cases by 22%; (3) Washing and cross-contamination still reduces mean cases by 12%; (4) Washing plus temperature abuse (at home, ambient temperature for 1.2 h) increases mean cases by 55%; (5) Washing plus irradiation reduce cases by 56%; (6) Washing plus irradiation plus MAP reduce cases by 65%; (7) Washing plus cross-contamination plus irradiation plus MAP plus temperature abuse reduce mean cases by 35%.	NA	Medium: Various scenarios tested by sanitiser and combinations	Omac et al. [15]

Table 2. *Cont.*

Scope	Food	Predictive Microbiology Models	What-If Scenarios	Sensitivity Analysis [1]	Model Complexity [2]	Source
End process-to-table	Leafy green salads from salad bars	Growth (Baranyi, growth square root model)	(1) A breakdown in the salad bar's cooling unit (temperature of 18 °C from the moment that the salad bar is filled) increases the mean number of cases in 23% (In the baseline scenario temperature of salad bar is assumed to fluctuate normally between 0 and 13 °C).	Outcome—probability of infection from any serving: Initial prevalence (r = 0.75) and portion size (r = 0.62).	Medium: temperature profiles	Franz et al. [16]
	RTE leafy vegetables	Growth (linear model, growth square root model)	(1) Reducing mean initial prevalence of LM from 1.7% to 0.17% decreases mean cases of listeriosis by 84%; (2) Reducing initial mean prevalence and keeping temperature strictly between 1 and 5 °C along processing and storage reduces cases by 85%; (3) Reducing maximum initial counts of LM from 2.74 to −1.04 log CFU/g reduces mean cases by 91%; (4) Reducing maximum initial counts and keeping temperature strictly between 1 and 5 °C reduce cases by 92%; (5) Reducing both prevalence and counts decreases mean cases by 98.7%.	NA	Low	Sant'Ana et al. [17]
	Leafy green salads from salad bars	Growth (linear model, growth square root model)	(1) The delivery frequency towards the restaurant was increased from 2 days a week to 5 days a week. In this scenario, the catering outlet is allowed to order leafy green–based salad products every weekday. This scenario halved the mean number of cases.	Outcome—the desired service level with regard to "out-of-stock" (z parameter; the greater z is, the smaller the	Medium: temperature profiles	Tromp et al. [18]

Table 2. *Cont.*

Scope	Food	Predictive Microbiology Models	What-If Scenarios	Sensitivity Analysis [1]	Model Complexity [2]	Source
Retail-to-table	Fruits and vegetables	Growth (linear model, square root model for EGR)	NA	NA	Medium: Fruits and vegs considered in separate; dose–response models developed for three subpopulations	FDA-FSIS [19]
	Lettuce	Growth (linear at 6 °C and 23 °C), Survival (empirical equation for water treatment)	NA	Outcome—LM counts at consumption: probability of washing (r = −0.46–−0.43), surface contamination (r = 0.23–0.29), time under running tap water (r = −0.09–−0.14), board/knife transfer rate (r = 0.07–0.13), contamination at retail (r = 0.02–0.04)	Low	Domenech et al. [20]
Consump-tion	Non-RTE frozen vegetables	Growth (linear, EGR square root model, empirical model for lag phase)	The median log risk of listeriosis from consumption of frozen vegetables contaminated with LM is −12.7. (1) Within-package clustering parameter between 0.01 and 0.1 in the baseline—as opposed to 1 in the baseline—reduces median log risk to −15/−14.1; (2) Number of packages tested per lot of 20 or 10—as opposed to 5 in the baseline—reduces median log risk to −14.4/−13.7; (3) Thawing at ambient temperature or in the fridge has negligible effect on the risk; (4) Changing the number of servings per meal (s = 0.5, 2) also resulted in no difference from the baseline (s = 1) risk of listeriosis.	Outcome—dose of LM consumed: cooking the serving (r = −0.87), log reduction due to proper cooking (r = −0.48), LM counts in a serving from contaminated package (r = 0.46), time stored at room temperature (r = 0.02), time/temperature in the refrigerator (r = 0.01)	Medium: Bacterial clustering in a package is represented; partitioning of the package into portions is modelled; handling prior to consumption such as cooking and thawing is included	Zoellner et al. [21]

Table 2. *Cont.*

Scope	Food	Predictive Microbiology Models	What-If Scenarios	Sensitivity Analysis [1]	Model Complexity [2]	Source
	Blanched frozen vegetables	Growth (linear, square-root model using EGR 5 °C)	(1) In elderly females, cooking the vegetables reduces the risk of listeriosis per serving by 3.2 log (from −9.4 to −12.6 log), and the number of cases per 10^{12} servings from 400 to 0.23; (2) In elderly males, cooking the vegetables reduces the risk of listeriosis per serving by 3.6 log (from −8.7 to −12.3 log), and the number of cases per 10^{12} servings from 1900 to 0.53; (3) Reducing the proportion of uncooked servings from 23% to 4% reduces the predicted listeriosis cases per year from 1.62 to 0.041 in elderly females.	Outcome—Probability of illness per serving from uncooked frozen vegetables: MPD from 7.8 to 9.8 log CFU/g increases risk by 2.5; serving size from 49 g to 106 g increases risk by 2.2; initial prevalence from 9.8% to 13.3% increases risk by 1.2.	Low: Generic model; only demands some knowledge in R software to utilise it	EFSA BIOHAZ [22]

[1] The outcome variable on which sensitivity analysis is carried out is indicated; "r" refers to the Pearson's correlation coefficient. [2] Model complexity was assessed by the authors in terms of easiness of reproducibility. When considered as of medium or high complexity, explanations are provided. aw: water activity; LPD: lag phase duration; RLT: relative lag time; GR: maximum growth rate; EGR_x: exponential growth rate at x °C; LAB: lactic acid bacteria; LAC: lactic acid concentration; RR: risk reduction; NA: not addressed in the study.

The QRA models were variable in terms of the scope of the food chain they represented. Four models simulated the growth of *L. monocytogenes* from processing to table [12–15], three models from end of processing to table [16–18], two models from retail to table [19,20], and two models represented only the consumption module [21,22]. Only one QRA model evaluated the contamination from production to consumption [11], although the actual utility of the model for the assessment of control measures is not clear since they did not assess risk factors or intervention strategies (Table 1).

The construction of models varied in complexity. Five out of the 13 QRA models mimicked cross-contamination taking place at processing (during cutting [14], before packaging [15], and during packaging [14]), during transport and retail [11], and during consumer's handling [11,20]. All the QRA models relied on predictive microbiology models to estimate the concentrations of *L. monocytogenes* as the raw materials (i.e., vegetables or fruits) went through processing, and as end products went down the supply chain. The most widely used primary models for growth were the loglinear [12,17–21] and the Baranyi and Robert's model [14–16], whereas the square-root model was the preferred choice of secondary model in all cases [11,12,14–21], except that of Crèpet [13] where the cardinal parameter model for temperature was employed (Table 2). The lag phase duration of *L. monocytogenes* in produce was taken into account in six models [11,13–15,21]; while only three QRA models employed time–temperature profiles to estimate the growth of *L. monocytogenes* during cold storage, including display at deli salads establishments [13,16,18] (Table 1). Crépet [13] explored the impact of maximum bacterial levels. The comparison of predicted and observed counts of naturally contaminating *L. monocytogenes* in minimally processed fresh lettuce showed inconsistency. The maximum levels observed in challenge tests were higher than those observed in naturally contaminated products. The maximum level during the stationary phase was better predicted by considering the initial contamination [30]. All the 13 QRA models regarded illness as the endpoint for risk estimation. The most commonly chosen dose–response function for risk characterisation was the exponential function, although in different approaches, namely the exponential model of FAO-WHO [23] (used in [11,13,20]), the one of Chen et al. [25] (used in [15,16,18]), the one of FDA-FSIS [19] (used in their own QRA model), the one of Buchanan et al. [26] (used in [17,21]), and the one of EFSA-BIOHAZ [2], based on the model of Pouillot et al. [29] (used in [22]).

The early Weibull-gamma model proposed by Farber et al. [24] was used in the three QRA models [12,14] and the dose–response of Bemrah et al. [28] used in the QRA model of Zoellner et al. [21]. The only QRA model that, in addition to illness, considered death as endpoint for risk estimation was that of Tromp et al. [18], which used a dose–response animal model in pregnant guinea pigs [27] (Table 1).

In total, 10 out of 13 QRA models assessed the effect of what-if scenarios on the final risk [12–18,21,22], whereas sensitivity analysis on response variables such as *L. monocytogenes* concentration at consumption, risk per serving or number of listeriosis cases was undertaken in 7 QRA models [12,13,16,18,20–22] (Table 2).

Although the outcomes of the QRA models are not directly comparable due to differences in model architecture, data and assumptions, taking together the results from the what-if scenarios and sensitivity analysis, the factors pointed out as having the biggest influence on reducing the likelihood of illness in the consumers are (Table 2): lower initial contamination and prevalence of *L. monocytogenes* in fresh vegetables, processing interventions such as ionising radiation or cold-atmospheric plasma, modified atmosphere packaging, reducing cross-contamination during processing, washing vegetables with chemical sanitisers, the maintenance of low storage temperatures, within-lot microbiological testing of the product at the end of processing, and a reduction in shelf-life. These factors are discussed in depth in the next section.

4. Discussion

The fact that most of the listeriosis QRA models were undertaken for leafy greens (9/13: as fresh, RTE and in salad bars) is evidence of the researchers' concern from a microbiological safety perspective. Most leafy greens have a short growing season; therefore, they may be more vulnerable to microbial contamination. They are commonly grown outdoors, where they may be exposed to contaminated soil, fertilisers and irrigation water. In addition, produce such as leafy greens are mostly consumed raw. When minimally processed, the processing steps of washing, sanitising and modified atmosphere packaging do not guarantee the elimination of *L. monocytogenes*. Leafy greens and other produce have therefore far fewer barriers against *L. monocytogenes* than the more traditional vehicles of foodborne illness (i.e., cheese, RTE meat products).

4.1. Risk Factors on Farm

L. monocytogenes is ubiquitous in natural environments; it has been isolated from soil, water ways, vegetation, and from the faeces of domestic and wild animals [31]. Such environments are associated with the production of produce, and hence, the pathogen can be transferred from multiple transmission routes such as air, soil, water, insects, animals and human activity [32]. Furthermore, production activities including irrigation, fertilisation, and other on-farm management practices can affect *L. monocytogenes* prevalence on farm.

The type of cultivation system is the first on-farm risk factor that should be taken into account, because from a microbiological perspective, protected cultivation (i.e., greenhouse) is considered safer than open field, which is related to the minimisation of the risk factors linked with the sources of pre-harvest contamination, and a greater control of water disinfection with water being recirculated and cleaned periodically [33]. Another major risk factor, particularly in the contamination of leafy crops eaten raw as salads, is the agricultural water [34]. The main irrigation water sources are municipal water, rainwater, groundwater and surface water. The prevalence of *L. monocytogenes* in irrigation water appears to be variable; yet, surface water represents one of the riskiest water sources [35]. Irrigation methods also influence the microbiological safety of fresh produce. Song et al. [36] found out that furrow irrigation resulted in greater microbial contamination of lettuce and soil surface than subsurface drip irrigation. Settanni et al. [37] demonstrated that watering pots containing soil artificially contaminated with *L. monocytogenes* from sub-irrigation did not determine the contamination of the aerial parts of leafy vegetables, and, for this reason, the transmission of pathogens by drip irrigation is very low in comparison to the irrigation performed by overhead sprinklers. Even if direct contact between irrigation water and the edible part of the leafy crop is avoided, irrigation water may contaminate the soil, where the bacteria can survive for some time, and irrigation (or rainfall) splash may contaminate the crop [35]. Furthermore, fields that contain animal manure are more likely to be contaminated with *L. monocytogenes*, as shown by Szymczak et al. [38], because of their ability to survive in soils for months. Inadequate composting is also responsible for the transmission of undesired microorganisms from manures to the soil [33]. Strawn et al. [39] and Weller et al. [40] concluded that good manure management practices, such as aging, treating, and handling before application, are essential, since they can significantly influence the risk of *L. monocytogenes* contamination.

Despite the broad understanding that the microbial contamination of produce begins at the pre-harvest stage, none of the QRA models retrieved has looked into pre-harvest factors as sources of *L. monocytogenes* on farm. The only on-farm strategy tested among all available QRA models was the application of a microbiological criterion at primary production, consisting of $n = 20$, $c = 0$ and absence in 25 g. In the model of Carrasco et al. [12], such intervention would reduce the mean cases of listeriosis due to consumption of RTE lettuce salads by 43%. Carrasco et al. [12] argued that their QRA model covered the scope from processing to table, since on-farm factors influencing the status of *L. monocytogenes* in vegetables are challenging to control. Nonetheless, from a full supply chain model, a deeper understanding can be gained on the contributions to risk from important on-farm sources such as type of cultivation, water, fertilisers and irrigation. According to Olaimat and Holley [41], the pre-harvest microbial contamination is of utmost importance in keeping up the microbiological quality of fresh and minimally processed vegetables, because the post-harvest sanitisation applied during processing does not pursue the complete elimination of pathogens on plant surfaces. In fact, the sanitising washing procedure achieves a limited reduction in spoilage and pathogenic microorganisms. Instead, the application of on-farm safety procedures can reduce the burden of *L. monocytogenes* at the processing stage. Therefore, the construction of a QRA model that encompasses an on-farm module is needed, as it can help assess the effects of current on-farm management practices and preventive measures as well the effects of the type of cultivation on the final risk to the consumers associated with fresh and minimally processed produce. Now that dose–response relationships based on strain genetics have been proposed [42,43], it would be important to look at the genetic diversity of *L. monocytogenes* in agricultural soils. Clonal complexes (CC) 1, CC2, CC4, and CC6 are infection-associated clones usually causing sporadic or outbreak listeriosis, while CC9 and CC121 are strongly food-associated clones that mostly infect immunocompromised individuals [44].

4.2. Risk Factors at Processing

L. monocytogenes may enter the processing facilities through contaminated produce and personnel. Once introduced in the environment, the pathogen can grow at operational temperatures and resist cleaning and disinfection of the processing plant, given its ability to persist on abiotic surfaces such as stainless steel and polystyrene. Furthermore, *L. monocytogenes* can survive on the surface of fresh produce for extended periods of time [45], whereas contamination on an injured leaf may lead to growth and colonisation [46].

Except for irradiation after packaging, there is no step that can be highly lethal against *L. monocytogenes* in the processing of RTE produce. Therefore, the microbiological safety of leafy greens should rely on sanitising, washing and the maintenance of cold temperatures, in addition to good management practices and good hygiene practices.

As opposed to the void observed among the QRA models retrieved for an on-farm module, five QRA models simulated processing modules [12–15] and assessed the effects of a range of processing stages or interventions, namely, ionising radiation, cold atmospheric plasma (CAP), sanitisation and modified atmosphere packaging (MAP). Among the strategies tested by Guzel [14] and Omac et al. [15], irradiation appeared as the most effective intervention to reduce the exposure and the risk of listeriosis. In the QRA model for RTE fresh-cut romaine lettuce, exposure to ionising radiation of 1 kGy at room temperature decreased *L. monocytogenes* counts at the point of consumption by >99% and risk of illness in 1.66 log in the susceptible population, whereas in a QRA model for fresh cantaloupe by the same authors, the implementation of irradiation also reduced *L. monocytogenes* counts at the point of consumption by >99.9% and the mean cases of listeriosis by 99%, in comparison to baseline scenarios without an irradiation process [14]. In a QRA model for non-RTE baby spinach, Omac et al. [15] compared scenarios of water washing and washing plus irradiation with a baseline scenario representing neither interventions nor cross-contamination during processing. They found that irradiation was far more effective than water washing since washing decreased mean cases of listeriosis by 7.5% whereas adding irradiation to this scenario decreased mean cases of listeriosis by 56%, both in comparison to the baseline scenario (Table 2). Although the irradiation technology has been proven to be more effective than mild technologies [14,15], many consumers are still reluctant to accept irradiated foods, therefore making its adoption by the industry difficult. Crépet [13] assessed the impact of the chlorination of water. It was found that removing the treatment of water (baseline scenario) increased the risk by two.

Mild technologies of produce preservation such as CAP and MAP were also assessed in QRA models; where the use of MAP (5.5% CO_2, 3% O_2; 92.5% N_2) as opposed to no packaging (baseline) reduced the mean number of listeriosis cases associated with RTE lettuce salad by 95% [12], treatment of romaine lettuce with CAP eliminated on average

92% of the population of *L. monocytogenes* at the point of consumption, reducing in 1.34 log units the risk of illness in the susceptible population [14].

Mild decontamination treatments such as washing and sanitising with chemicals take place during processing. Such a decontamination step plays an important role in the reduction or prevention of pathogenic contamination in fresh produce since the processing of produce does not include any critical control point such as heat treatment, sterilisation or freezing to control the microbial load. Eco-friendly sanitisers, as opposed to the widely known chlorine which forms by-products that may cause negative effects on human health and environment (i.e., trihalomethanes, halo acetic acids and haloketones), are being applied in fresh cut industry, including chlorine dioxide (ClO_2), peroxyacetic acid (PAA) and electrolysed water (EW) [47]. In his QRA model, Guzel [14] estimated that, when no cross-contamination during processing was considered, the use of PAA for sanitising reduced *L. monocytogenes* mean concentration at consumption by 28% and reduced the risk of listeriosis by 0.35 log. According to Omac et al. [15], sanitising baby fresh spinach either with aerosolised peracetic acid (80 ppm for 20 min) or with ClO_2 gas (10 ppmv for 20 min) reduced the estimated mean cases of listeriosis by 22%. The same authors determined that combining the three processes, water washing, irradiation and MAP, reduced the mean cases of listeriosis by 65% (Table 2).

Nonetheless, none of the QRA models simulating chemical sanitising took into account that the concentration of the sanitiser, and the time and temperature of exposure of the produce, drive the level of microbial reduction attained by the sanitising treatment [48]. A realistic representation of washing and sanitising in the processing module should disaggregate the effects of sanitising by type of sanitiser, attending to their effectiveness to reduce *L. monocytogenes* depending upon treatment time and temperature, sanitiser concentration and the interaction with the product itself. A sanitisation module could be built upon the mixed-effects meta-regression models estimating *L. monocytogenes* reductions in various types of produce by specific chemical sanitisers presented in Prado-Silva et al. [48].

4.3. Cross-Contamination during Processing

In the processing environment, cross-contamination can occur from human carriers, from harbourage sites, and from surfaces in contact with the produce such as conveyor belts, shredders, centrifuges, sorting tables, containers and packaging machines. The survival of *L. monocytogenes* in the processing environments is key to its transmission to foodstuffs. According to Leong et al. [49], *L. monocytogenes* can persist in a processing facility for weeks, and then re-contaminate the product passing through that facility. From the pool of produce QRA models, cross-contamination events were evaluated in five QRA models. In all cases, cross-contamination was modelled by transfer coefficients, which constitute a simple empirical approach that depends heavily on source, recipients and number of contacts [50]. Transfer coefficients during processing stages were defined for packaging and handling steps [14], and for handling mistakes, conveyor belts, and packing equipment [15]. The QRA models did not simulate cross-contamination events during shredding.

In the simulation models, the contribution of cross-contamination during processing to the final risk was variable; a low contribution was found for fresh-cut romaine lettuce [14] and fresh baby spinach [15]. In the former, the occurrence of cross-contamination increased the risk in the susceptible population by 0.06 log (by increasing only in 18% the concentration of *L. monocytogenes* at consumption), whereas, in the latter, the reduction in *L. monocytogenes* attained by sanitising baby spinach was found to counterbalance the increase in *L. monocytogenes* due to cross-contamination, resulting in a reduction in mean listeriosis cases by 12%. On the contrary, Guzel [14], in his QRA model for fresh-cut cantaloupe, estimated that cross-contamination during packaging and handling has a greater contribution to the annual cases of listeriosis (300% increase) than a possible scenario of home temperature abuse of maintaining the product at an ambient temperature for 1 day (220% increase), both measured against a baseline scenario without interventions, cross-contamination and temperature abuse. Nonetheless, the measured effects of cross-contamination on the final risk cannot be directly comparable since they depend on the base assumptions. For instance, it is probable that the low transfer coefficient (0.002) used to model cross-contamination in the QRA model for fresh-cut romaine lettuce [14] has driven the low increase in the final risk, since the occurrence of the cross-contamination leads only to a 0.1 CFU/g increase in the mean concentration of *L. monocytogenes* at the time of consumption in comparison to the baseline scenario.

More complex cross-contamination models, such as discrete-event models [51], have not been applied in any of the QRA models retrieved for produce. Recently, agent-based models of contamination dynamics of *Listeria* spp. for produce packing facilities have been proposed with agents that represent equipment surfaces and employees, each having customised characteristics [52,53]. The agents are assumed to operate autonomously with other agents and the environment, including floors, walls and ceiling. Such models have been proven to be useful for understanding the spread of *Listeria* within the simulated facility, for assessing the efficacy of a variety of sampling plans schemes, and for helping determine corrective actions. However, since these agent-based models are built upon the facility's design, employees' shifts, location of employees and equipment, which change between processing facilities, and even in time within a facility, their outcomes are specific to an individual processing facility, allowing an industry to make better decisions on how to detect potential issues quicker and with fewer samples and resources.

The relevance of cross-contamination during the processing stage can be illustrated by a cantaloupe outbreak investigation [54], where subtyping analysis revealed that the processing environment was the main source of contamination. The fact that *L. monocytogenes* forms biofilm on several abiotic surfaces contributes to its persistence and subsequent contamination of post-harvest produce [55]. Another trait of *L. monocytogenes* aiding to persistence in the processing environment is its ability to enter a viable-but-nonculturable (VBNC) state. The cleaning and disinfection procedure leads to a loss in the culturability of *L. monocytogenes* and the appearance of VBNC populations. Subsequently, upon entry into suitable environments (i.e., harbouring site), VBNC *L. monocytogenes* can recover its culturability and begin to proliferate, even remaining infectious [56].

4.4. Risk Factors at Retail

The QRA models of a shorter scope (i.e., retail-to-table) start with an estimation of the initial prevalence and/or counts of *L. monocytogenes* in the product, modelled either from data obtained at the end of processing or at the point of retail. Although these listeriosis QRA models commonly tested what-if scenarios reducing the initial prevalence or counts, their authors did not indicate how these reductions could be achieved in the processing stage or earlier in the primary production. For instance, in their retail-to-table model, Franz et al. [16] recognised that the risk of listeriosis from the consumption of leafy greens from salad bars in the Netherlands was heavily driven by the initially assumed prevalence (r = 0.75). In the consumption model of Zoellner et al. [21] for non-RTE frozen vegetables, classification tree analysis highlighted that the initial concentration of *L. monocytogenes* in the lot was the main predictor of illness in the frozen vegetables lot, whereas in the model of EFSA BIOHAZ [22] for non-RTE blanched frozen vegetables, a change in the initial prevalence of *L. monocytogenes* in the product from 9.8% to 13.3% increased the risk by a factor of 1.2. Similarly, for RTE leafy vegetables, Sant'Ana et al. [17] estimated that by reducing the mean initial prevalence of *L. monocytogenes* (at end of processing) from 1.7% to 0.17%, the mean cases of listeriosis in Brazil would decrease by 84%, whereas reducing the maximum initial counts from 2.74 to −1.04 log CFU/g would decrease the mean cases by 91% (Table 2). A discussion on which processes or interventions would be more efficient in helping attain such rather large reductions in prevalence and concentrations was not provided in Sant'Ana et al. [17]. Nevertheless, bringing together the results from the models with processing-to-table scope (Table 2), it could be suggested that highly effective intervention strategies are ionising radiation and cold atmospheric plasma. Milder technologies to control *L. monocytogenes* are sanitisation and MAP; nonetheless, they are more practicable in the produce industry. However, none of the QRA models evaluated the combined effectiveness of sanitisation and MAP (a very likely scenario to occur in industry) to reduce exposure and risk. Moreover, none of the QRA simulations separately modelled the effect of the various sanitisers used in processing. For leafy vegetables processing, chlorine is still widely used for being low cost, and PAA for its activity over a wide pH range and limited reaction with organic matter. For fresh-cut products, there are a growing number of sanitising compounds, including chlorine dioxide, hydrogen peroxide, organic acids, ozone, and electrolysed water, among others [57].

Two QRA models investigated the risk of listeriosis associated with leafy green salads from salad bars in the Netherlands under an end processing-to-table scope, therefore covering inputs that characterise the processing facility for processing fresh-cut vegetables, the distributor and a catering outlet [16,18]. Both studies coincided in that the supply chain of leafy green vegetables for salad bars can be considered as a risk amplifier with respect to *L. monocytogenes*. In Franz et al. [16], to model exposure, actual time–temperature data from the processing plant to restaurant were acquired as well as time–temperature trajectories of the vegetables kept in the catering establishment. This model showed that, despite the fact that maintaining the temperature controlled at 2–5 °C from the processing plant to the restaurant, *L. monocytogenes* in leafy greens could grow from a mean of 250 CFU/g at the end of processing to 735 CFU/g (95% CI: 690–780 CFU/g) at consumption because, in the salad bar, temperatures can fluctuate up to 13 °C. This resulted in a mean dose per contaminated portion of 85,000 CFU, which increased to 105,000 CFU (95% CI: 28,500 to 251,000 CFU) in a hypothetical scenario of a breakdown in the salad bar's cooling unit that increased the temperature up to 18 °C. Despite the temperature fluctuations having a direct impact on the contaminated servings, the risk of infection from any serving of leafy green salads from salad bars was driven mainly by the initial prevalence of *L. monocytogenes* at retail ($r = 0.75$) and the portion size ($r = 0.62$). Tromp et al. [18] refined the model of the leafy greens salads from salad bars, (1) by introducing the notion that in reality storage times are interdependent, meaning that if a product remains for a long time in one step, it will remain a shorter time in a following step due to less remaining shelf-life; and (2) by modelling logistics in order to incorporate logistics performance indicators such as shrinkage (product loss) and out-of-stock. Tromp et al. [18] found that modelling logistics—as opposed to Franz et al. [16]—increased the estimated risk of listeriosis 4-fold. In addition, from a logistics point of view, increasing the delivery frequency in the catering establishment from 2 to 5 times per week halved the mean number of cases of listeriosis in the general population from 1.43 to 0.70 cases per annum. Through the results of this model, it was clear that having shorter storage times with greater delivery frequency resulted in reduced growth of *L. monocytogenes*.

Time–temperature measurements indicate that the cold storage in display cabinets at retail is generally not the most efficient step in the cold chain, and that temperatures frequently rises above the recommended limit [58,59]. In any case, at retail or in salad bars, temperature abuse has been often pointed out as a key factor contributing to increasing the risk of listeriosis. Therefore, in a QRA model, it is very important to be able to capture the growth of *L. monocytogenes* during distribution and retail as realistically as possible by representative time–temperature profiles as well as accurate kinetic data and validated growth models of *L. monocytogenes* in the specific product.

4.5. Risk Factors at Home

At the consumer level, temperature abuse has been tested by some of the QRA models, whereas in Carrasco et al. [12], the home storage temperature had a stronger effect on the number of listeriosis cases associated with RTE lettuce salad than home storage time and *L. monocytogenes* concentration at consumption; Guzel [14] predicted that keeping the fresh-cut romaine lettuce or the fresh-cut cantaloupe for 24 h at 20 °C (ambient temperature) would increase *L. monocytogenes* populations at consumption by 56% or 300%, respectively, with the consequent increase in risk of listeriosis in the susceptible population by 1.1 log or 2.95 log, respectively. Nonetheless, keeping the fresh produce for one day on the countertop is an unusual consumer practice. In a more credible scenario, where fresh spinach was left at ambient temperature for 1.2 h, Omac et al. [15] estimated that this practice would increase the mean cases of listeriosis by 55% (no temperature abuse in the baseline scenario). The only QRA model where domestic temperature did not determine the exposure to *L. monocytogenes* was that of Zoellner et al. [21] for non-RTE frozen vegetables, where in sensitivity analysis neither the time stored at room temperature (Pearson's coefficient of correlation r = 0.02) nor the time or temperature in the refrigerator (r = 0.01) correlated with the dose of *L. monocytogenes* per serving. Furthermore, they found that the consumer practice of thawing frozen vegetables did not affect the risk of illness. Domenech et al. [20] assessed the effects of other habitual consumer practices, and found that the probability of washing (r = −0.46–−0.43) followed by the time under running tap water (r = −0.09–−0.14) were parameters that could reduce the prevalence of *L. monocytogenes* the most in fresh lettuce at the domestic level (Table 2).

Fresh and RTE minimally processed produce are products that have a very short shelf-life because of their high perishability caused by their high moisture content; in particular, their wounding caused by processing leads to many physical and physiological changes that noticeably affect their quality [60]. Notwithstanding these facts, some QRA models have assessed unrealistic what-if scenarios of extending consumption times, probably to make the point that home storage time is determinant of the risk of listeriosis. For instance, the models of Guzel [14] estimated that increasing the consumption time of fresh-cut romaine lettuce and fresh-cut cantaloupe to a maximum of 14 and 10 days, increased the mean *L. monocytogenes* concentration (CFU/g) at consumption by 2100% and 2300%, respectively. On the contrary, according to Carrasco et al. [12], reducing the shelf-life of RTE salad from a maximum of 12 days to a maximum of 4 days decreased by 84% the annual cases of listeriosis (Table 2).

The two QRA models consisting of a sole consumption module were conducted for non-RTE frozen vegetables [21,22]. The QRA model of Zoellner et al. [21] was prompted by the multi-country outbreak of listeriosis linked to frozen corn that caused 53 cases and 10 deaths over the period of 2015–2018. Using whole-genome-sequencing (WGS), the outbreak investigation concluded that the environmental contamination of a frozen vegetable plant in Hungary was the source of the strain, which persisted despite cleaning and disinfection [61]. It was highlighted that some frozen fruits and vegetables can be added uncooked to salads or used in smoothies or other products without being subjected to any process to eliminate or reduce the level of pathogens [62]. For that reason, the QRA model of Zoellner et al. [21] focused only on the consumption module, aiming to understand to what extent consumer preparation methods different from the packaging instructions impact on the risk of listeriosis. A similar objective was pursued by the subsequent QRA model of EFSA BIOHAZ [22] for blanched frozen vegetables. Zoellner et al. [21] found that cooking the vegetables before serving (r = −0.87) and log reduction due to proper cooking (r = −0.48) were the most important drivers for reducing the dose of *L. monocytogenes* consumed; when at least half the servings are properly cooked, the number of illnesses per consumed lot is 0 (95% CI: 0–1) in the susceptible population. Likewise, the QRA model of EFSA BIOHAZ [22] estimated that reducing the proportion of uncooked servings from 23% (baseline) to 4% (best case scenario) decreases the predicted listeriosis cases per year from 1.62 to 0.041 in elderly females. Furthermore, they estimated that cooking the frozen vegetables reduces the listeriosis cases per 10^{12} servings from 400 to 0.23 in the female elderly population, and from 1900 to 0.53 listeriosis cases in the male elderly population (Table 2).

Serving size was evaluated in Carrasco et al. [12] and Franz et al. [16] for RTE lettuce and salad bar leafy greens, respectively, and in both cases, it had a moderate effect on the risk of infection. In the model of EFSA BIOHAZ [22], a change in serving size from 49 to 106 g was found to increase the probability of illness per serving of uncooked frozen vegetables by a factor of 2.2. In contrary, in the model for frozen vegetables by Zoellner et al. [21], serving size did not drive the final risk of listeriosis. This difference arises from the fact that frozen vegetables are mostly eaten cooked, which attenuates the effect of increasing the serving size.

4.6. Cross-Contamination during Consumer's Handling

The cross-contamination during the consumer's handling was modelled using transfer rates for contaminated boards, hands and knives [20], and for unwashed cutting boards or countertops, kitchen tools, and unwashed hands [11]. From the QRA models retrieved, the relative importance of cross-contamination during the consumer's handling was assessed only by Domenech et al. [20]. They found that the parameters related to cross-contamination such as the contamination of surfaces ($r = 0.23$–0.29) and chopping board and knife transfer rates ($r = 0.07$–0.13) were the parameters that increased the consumers' exposure to *L. monocytogenes* the most, in comparison to contamination at retail ($r = 0.02$–0.04), and refrigeration temperature and storage time ($r = 0$). These assessments highlighted the importance of accurately representing the transfer coefficients in the cross-contamination modules, since they have been shown to moderately drive the final risk estimate of listeriosis.

4.7. Availability of QRA Scripts

Supplementary Material Table S1 provides information about model availability. Most of the models were developed using @risk software. Only the EFSA model (R model) is directly accessible [63]. Zoellner et al. [21] indicate that the models will be made available on request to the authors. This lack of availability of the models raises questions about the ability of other risk assessors to construct new developments. Each time, it seems necessary to start from scratch without being able to rely easily on existing models. The more systematic sharing of spreadsheets and scripts would benefit the entire community.

5. Conclusions

It would be advantageous that a QRA model for produce includes a primary production module so that the contributions to risk from important on-farm factors such as type of cultivation, water, fertilisers and irrigation method could be introduced. This would enable the evaluation of current on-farm practices, control measures and the potential application of more stringent standards of production to reduce the exposure to *L. monocytogenes*. Despite the availability of data on the effectiveness of the use of sanitisers to reduce the load of *L. monocytogenes* in produce, none of the QRA modules contained a module for washing that could estimate microbial reduction as a function of the type of sanitiser, its concentration and the time of exposure. A realistic QRA model for minimally processed produce should take into account such an informative module. Furthermore, since secondary contamination can occur in the processing plants from equipment and environmental elements, the most relevant stages and occasions for cross-contamination during processing should be pinpointed and modelled. This would enable the assessment of control measures that can limit the occurrence of cross-contamination events, such as the implementation of more stringent controls for raw materials, environmental monitoring programs and/or sanitation procedures. Since the safety of produce, in particular for RTE products, heavily relies on the maintenance of low temperatures in the supply chain, *L. monocytogenes* growth should be simulated using representative data of time and temperature alongside distribution, retail and home storage, as well as accurate microbial kinetic parameters for the specific foodstuff.

Supplementary Materials: The following supporting information can be downloaded at: https://www.mdpi.com/article/10.3390/foods13071111/s1, Table S1: Accessibility to listeriosis QRA published models for produce.

Author Contributions: Conceptualisation, M.S. and U.G.-B.; methodology, M.S. and U.G.-B.; software, U.G.-B.; validation, V.C., J.D.O.M. L.G., M.S. and U.G.-B.; formal analysis, M.S. and U.G.-B.;

investigation, V.C., L.G., M.S. and U.G.-B.; resources, J.D.O.M., M.S.; data curation, L.G., V.C. and U.G.-B.; writing—original draft preparation, M.S. and U.G.-B.; writing—review and editing, U.G.-B.; visualisation, V.C., J.D.O.M., L.G., M.S. and U.G.-B.; supervision, M.S. and U.G.-B.; project administration, M.S.; funding acquisition, M.S. All authors have read and agreed to the published version of the manuscript.

Funding: This research received no external funding.

Data Availability Statement: Data is contained within the article. The BibTeX file containing records used in this systematic review is available.

Acknowledgments: WHO commissioned this critical review to serve as background document for the FAO/WHO Joint Expert Meeting on Microbiological Risk Assessment (JEMRA), which was held in Rome, Italy, from 24 to 28 October 2022. The authors are grateful to the JEMRA experts * and the FAO/WHO secretariat ** for their suggestions and feedback. * Hamzah Al-Qadiri, Sukhadeo B Barbuddhe, Ursula A. Gonzales-Barron, Heidy M. W. den Besten, Qingli Dong, Laurent Guillier, Claudia Guldimann, Ludwig Maximilian, Jovana Kovacevic, Alejandra Latorre, Yumiko Okada, Taran Skjerdal, Matthew J. Stasiewicz, and Paula Teixeira. ** Juliana De Oliveira Mota, Akio Hasegawa, Christine Kopko, Jeffrey LeJeune, Moez Sanaa, and Kang Zhou.

Conflicts of Interest: The authors declare no conflicts of interest.

References

1. Sampedro, F.; Pérez-Rodríguez, F.; Servadio, J.L.; Gummalla, S.; Hedberg, C.W. Quantitative risk assessment model to investigate the public health impact of varying *Listeria monocytogenes* allowable levels in different food commodities: A retrospective analysis. *Int. J. Food Micro.* **2022**, *383*, 109932. [CrossRef] [PubMed]
2. EFSA Panel on Biological Hazards (BIOHAZ); Ricci, A.; Allende, A.; Bolton, D.; Chemaly, M.; Davies, R.; Lindqvist, R. Scientific Opinion on the *Listeria monocytogenes* contamination of ready-to-eat foods and the risk for human health in the EU. *EFSA J.* **2018**, *16*, 5134. [CrossRef]
3. ECDC. Surveillance and Disease Data for Listeriosis. European Centre for Disease Prevention and Control. Available online: https://www.ecdc.europa.eu/en/all-topics-z/listeriosis/surveillance-and-disease-data/eu-summary-reports (accessed on 15 June 2022).
4. NORS. National Outbreak Reporting System Dashboard. Centers for Disease Control and Prevention, National Center for Emerging and Zoonotic Infectious Diseases (NCEZID). Available online: https://wwwn.cdc.gov/norsdashboard/ (accessed on 15 June 2022).
5. Varma, J.K.; Samuel, M.C.; Marcus, R.; Hoekstra, R.M.; Medus, C.; Segler, S.; Anderson, B.J.; Jones, T.F.; Shiferaw, B.; Haubert, N.; et al. *Listeria monocytogenes* infection from foods prepared in a commercial establishment: A case-control study of potential sources of sporadic illness in the United States. *Clin. Infect. Dis.* **2007**, *44*, 521–528. [CrossRef] [PubMed]
6. Gillespie, I.A.; Mook, P.; Little, C.L.; Grant, K.A.; McLauchlin, J. Human listeriosis in England, 2001-2007: Association with neighbourhood deprivation. *Euro Surveill.* **2010**, *15*, 7–16. [CrossRef] [PubMed]
7. Gillespie, I.A.; Mook, P.; Little, C.L.; Grant, K.; Adak, G.K. *Listeria monocytogenes* infection in the over-60s in England between 2005 and 2008: A retrospective case-control study utilizing market research panel data. *Foodborne Pathog. Dis.* **2010**, *7*, 1373–1379. [CrossRef] [PubMed]
8. Leclercq, A.; Kooh, P.; Augustin, J.-C.; Guillier, L.; Thebault, A.; Cadavez, V.; Gonzales-Barron, U.; Sanaa, M. Risk factors for sporadic listeriosis: A systematic review and meta-analysis. *Microbial Risk Analysis* **2021**, *17*, 100128. [CrossRef]
9. Dalton, C.B.; Merritt, T.D.; Unicomb, L.E.; Kirk, M.D.; Stafford, R.J.; Lalor, K.; OzFoodNet Working, G. A national case-control study of risk factors for listeriosis in Australia. *Epidemiol. Infect.* **2011**, *139*, 437–445. [CrossRef] [PubMed]
10. Gonzales-Barron, U.; Cadavez, V.; Guillier, L.; Sanaa, M. A critical review of risk assessment models for *Listeria monocytogenes* in dairy products. *Foods* **2023**, *12*, 4436. [CrossRef] [PubMed]
11. Ding, T.; Iwahori, J.; Kasuga, F.; Wang, J.; Forghani, F.; Park, M.-S.; Oh, D.-H. Risk assessment for *Listeria monocytogenes* on lettuce from farm to table in Korea. *Food Control* **2013**, *30*, 190–199. [CrossRef]
12. Carrasco, E.; Pérez-Rodríguez, F.; Valero, A.; García-Gimeno, R.M.; Zurera, G. Risk assessment and management of *Listeria monocytogenes* in ready-to-eat lettuce salads. *Compr. Rev. Food Sci. Food Saf.* **2010**, *9*, 498–512. [CrossRef]
13. Crèpet. Statistique Bayesienne et Monte-Carlo de Second Ordre pour l'Evaluation des Risques Microbiologiques: Le Cas de *Listeria monocytogenes* dans les Produits de IVeme Gamme. Ph.D. Thesis, L'Institut des Sciences et Industries du Vivant et de l'Environnement (Agro Paris Tech), Paris, France, 2007; p. 204.
14. Guzel, M. Quantitative Risk Assessment for *Listeria monocytogenes* on Fresh-Cut Lettuce and Fresh-Cut Cantaloupe. Master's Thesis, Texas A&M University, College Station, TX, USA, 2015.
15. Omac, B.; Moreira, R.; Puerta-Gomez, A.; Castell-Perez, E. Effect of intervention strategies on the risk of infection from *Listeria monocytogenes* due to consumption of fresh baby spinach leaves: A quantitative approach. *LWT Food Sci. Technol.* **2017**, *80*, 208–220. [CrossRef]

16. Franz, E.; Tromp, S.O.; Rijgersberg, H.; van der Fels-Klerx, H.J. Quantitative microbial risk assessment for *Escherichia coli* O157:H7, *Salmonella*, and *Listeria monocytogenes* in leafy green vegetables consumed at salad bars. *J. Food Prot.* **2010**, *73*, 274–285. [CrossRef] [PubMed]
17. Sant'Ana, A.; Franco, B.; Schaffner, D. Risk of infection with *Salmonella* and *Listeria monocytogenes* due to consumption of ready-to-eat leafy vegetables in Brazil. *Food Control* **2014**, *42*, 1–8. [CrossRef]
18. Tromp, S.O.; Rijgersberg, H.; Franz, E. Quantitative microbial risk assessment for *Escherichia coli* O157:H7, *Salmonella enterica*, and *Listeria monocytogenes* in leafy green vegetables consumed at salad bars, based on modeling supply chain logistics. *J. Food Prot.* **2010**, *73*, 1830–1840. [CrossRef] [PubMed]
19. FDA-FSIS. *Quantitative Assessment of Relative Risk to Public Health from Foodborne Listeria monocytogenes among Selected Categories of Ready-to-Eat Foods;* Center for Food and Safety and Applied Nutrition, Food and Drug Administration, U.S. Department of Health and Human Services, and Food Safety and Inspection Service, U.S. Department of Agriculture: Washington, DC, USA, 2003; pp. 1–541.
20. Doménech, E.; Conchado, A.; Escriche, I. Evaluation of risk impact of consumers' behaviour in terms of exposure to *Listeria monocytogenes* in lettuce. *Int. J. Food Scie. Technol.* **2014**, *49*, 2176–2183. [CrossRef]
21. Zoellner, C.; Wiedmann, M.; Ivanek, R. An assessment of listeriosis risk associated with a contaminated production lot of frozen vegetables consumed under alternative consumer handling scenarios. *J. Food Prot.* **2019**, *82*, 2174–2193. [CrossRef] [PubMed]
22. EFSA Panel on Biological Hazards (BIOHAZ); Koutsoumanis, K.; Alvarez-Ordóñez, A.; Bolton, D.; Bover-Cid, S.; Chemaly, M.; Allende, A. The public health risk posed by *Listeria monocytogenes* in frozen fruit and vegetables including herbs, blanched during processing. EFSA Panel of Biological Hazards (BIOHAZ). *EFSA J.* **2020**, *8*, 6092.
23. FAO/WHO. *Risk Assessment of Listeria Monocytogenes in Ready-to-Eat Foods: Technical Report;* World Health Organization and Food and Agriculture Organization of the United Nations: Geneve, Switzerland, 2004; pp. 1–269.
24. Farber, J.M.; Ross, W.H.; Harwig, J. Health risk assessment of *Listeria monocytogenes* in Canada. *Int. J. Food. Microbiol.* **1996**, *30*, 145–156. [CrossRef]
25. Chen, Y.; Ross, W.H.; Gray, M.J.; Wiedmann, M.; Whiting, R.C.; Scott, V.N. Attributing risk to *Listeria monocytogenes* subgroups: Dose response in relation to genetic lineages. *J. Food Prot.* **2006**, *69*, 335–344. [CrossRef] [PubMed]
26. Buchanan, R.L.; Damert, W.G.; Whiting, R.C.; van Schothorst, M. An approach for using epidemiologic and microbial food survey data to develop a "purposefully conservative" estimate of the dose-response relationship between *Listeria monocytogenes* levels and the incidence of foodborne listeriosis. *J. Food Prot.* **1997**, *60*, 918–922. [CrossRef] [PubMed]
27. Williams, D.; Castleman, J.; Lee, C.-C.; Mote, B.; Smith, M.A. Risk of fetal mortality after exposure to *Listeria monocytogenes* based on dose-response data from pregnant guinea pigs and primates. *Risk Anal.* **2009**, *29*, 1495–1505. [CrossRef] [PubMed]
28. Bemrah, N.; Sanaa, M.; Cassin, M.H.; Griffiths, M.W.; Cerf, O. Quantitative risk assessment of human listeriosis from consumption of soft cheese made from raw milk. *Prev. Vet. Med.* **1998**, *37*, 129–145. [CrossRef] [PubMed]
29. Pouillot, R.; Hoelzer, K.; Chen, Y.; Dennis, S.B. *Listeria monocytogenes* dose response revisited—Incorporating adjustments for variability in strain virulence and host susceptibility. *Risk Anal.* **2015**, *35*, 90–108. [CrossRef] [PubMed]
30. Crépet, A.; Stahl, V.; Carlin, F. Development of a hierarchical Bayesian model to estimate the growth parameters of Listeria monocytogenes in minimally processed fresh leafy salads. *Int. J. Food Microbiol.* **2009**, *131*, 112–119. [CrossRef] [PubMed]
31. Smith, A.; Moorhouse, E.; Monaghan, J.; Taylor, C.; Singleton, I. Sources and survival of *Listeria monocytogenes* on fresh, leafy produce. *J. Appl. Microbiol.* **2018**, *125*, 930–942. [CrossRef] [PubMed]
32. Szabo, E.A.; Simons, L.; Coventry, M.J.; Cole, M.B. Assessment of control measures to achieve a food safety objective of less than 100 CFU of *Listeria monocytogenes* per gram at the point of consumption for fresh precut iceberg lettuce. *J. Food Prot.* **2003**, *66*, 256–264. [CrossRef] [PubMed]
33. Miceli, A.; Settanni, L. Influence of agronomic practices and pre-harvest conditions on the attachment and development of *Listeria monocytogenes* in vegetables. *Annals Microbiol.* **2019**, *69*, 185–199. [CrossRef]
34. Andreoletti, O.; Baggesen, D.L.; Bolton, D.; Butaye, P.; Cook, P.; Davies, R.; Threlfall, J. Scientific opinion on the risk posed by some pathogens in food of non-animal origin. Part 2 (*Salmonella* and Norovirus in leafy greens eaten raw as salads. *EFSA J.* **2014**, *12*, 3600. [CrossRef]
35. Allende, A.; Monaghan, J. Irrigation water quality for leafy crops: A perspective of risks and potential solutions. *Int. J. Environ. Res. Public Health* **2015**, *12*, 7457–7477. [CrossRef] [PubMed]
36. Song, I.; Stine, S.; Choi, C.; Gerba, C. Comparison of crop contamination by microorganisms during subsurface drip and furrow irrigation. *J. Environ. Eng.* **2006**, *132*, 1243–1248. [CrossRef]
37. Settanni, L.; Miceli, A.; Francesca, N.; Moschetti, G. Investigation of the hygienic safety of aromatic plants cultivated in soil contaminated with *Listeria monocytogenes*. *Food Control* **2012**, *26*, 213–219. [CrossRef]
38. Szymczak, B.; Szymczak, M.; Sawicki, W.; Dabrowski, W. Anthropogenic impact on the presence of *L. monocytogenes* in soil, fruits and vegetables. *Folia Microbiol.* **2014**, *59*, 23–29. [CrossRef] [PubMed]
39. Strawn, L.K.; Gröhn, Y.T.; Warchocki, S.; Worobo, R.W.; Bihn, E.A.; Wiedmann, M. Risk factors associated with *Salmonella* and *Listeria monocytogenes* contamination of produce fields. *Appl. Environm. Microbiol.* **2013**, *79*, 7618–7627. [CrossRef] [PubMed]
40. Weller, D.; Wiedmann, M.; Strawn, L.K. Spatial and temporal factors associated with an increased prevalence of *Listeria monocytogenes* in spinach fields in New York State. *Appl. Environm. Microbiol.* **2015**, *81*, 6059–6069. [CrossRef] [PubMed]

41. Olaimat, A.N.; Holley, R.A. Factors influencing the microbial safety of fresh produce: A review. *Food Microbiol.* **2012**, *32*, 1–19. [CrossRef] [PubMed]
42. Fritsch, L.; Guillier, L.; Augustin, J.C. Next generation quantitative microbiological risk assessment: Refinement of the cold smoked salmon-related listeriosis risk model by integrating genomic data. *Microb. Risk Anal.* **2018**, *10*, 20–27. [CrossRef]
43. Pouillot, R.; Kiermeier, A.; Guillier, L.; Cadavez, V.; Sanaa, M. Updated Parameters for *Listeria monocytogenes* Dose–Response Model Considering Pathogen Virulence and Age and Sex of Consumer. *Foods* **2024**, *13*, 751. [CrossRef] [PubMed]
44. Song, Z.; Ji, S.; Wang, Y.; Luo, L.; Wang, Y.; Mao, P.; Li, L.; Jiang, H.; Ye, C. The population structure and genetic diversity of Listeria monocytogenes ST9 strains based on genomic analysis. *Front. Microbiol.* **2020**, *13*, 982220. [CrossRef] [PubMed]
45. Sheng, L.; Edwards, K.; Tsai, H.C.; Hanrahan, I.; Zhu, M.J. Fate of Listeria monocytogenes on fresh apples under different storage temperatures. *Front. Microbiol.* **2017**, *8*, 1396. [CrossRef] [PubMed]
46. Koukkidis, G.; Haigh, R.; Allcock, N.; Freestone, P.P.E. Salad leaf juices enhance Salmonella growth, fresh produce colonisation and virulence. *Appl. Environ. Microbiol.* **2016**, *83*, e02416-16. [PubMed]
47. Nguyen, T.V.; Ross, T.; Van Chuyen, H. Evaluating the efficacy of three sanitizing agents for extending the shelf life of fresh-cut baby spinach: Food safety and quality aspects. *AIMS Agric. Food* **2019**, *4*, 320–339. [CrossRef]
48. Prado-Silva, L.; Cadavez, V.; Gonzales-Barron, U.; Rezende, A.C.B.; Sant'Ana, A.S. Meta-analysis of the effects of sanitizing treatments on *Salmonella, Escherichia coli* O157:H7 and *Listeria monocytogenes* inactivation in fresh produce. *Appl. Environ. Microbiol.* **2015**, *81*, 8008–8021. [CrossRef] [PubMed]
49. Leong, D.; NicAogáin, K.; Luque-Sastre, L.; McManamon, O.; Hunt, K.; Alvarez-Ordóñez, A.; Scollard, J.D.; Schmalenberger, A.; Fanning, S.; O'Byrne, C.; et al. A 3-year multi-food study of the presence and persistence of Listeria monocytogenes in 54 small food businesses in Ireland. *Int. J. Food Microbiol.* **2017**, *249*, 18–26. [CrossRef] [PubMed]
50. Hoelzer, K.; Pouillot, R.; Gallagher, D.; Silverman, M.B.; Kause, J.; Dennis, S. Estimation of *L. monocytogenes* transfer coefficients and efficacy of bacterial removal through cleaning and sanitation. *Int. J. Food Microbiol.* **2012**, *157*, 267–277. [CrossRef] [PubMed]
51. Tenenhaus-Aziza, F.; Daudin, J.-J.; Maffre, A.; Sanaa, M. Risk-based approach for microbiological food safety in the dairy industry: The case of *Listeria monocytogenes* in soft cheese made from pasteurized milk. *Risk Anal.* **2014**, *34*, 56–74. [CrossRef] [PubMed]
52. Barnett-Neefs, C.; Sullivan, G.; Zoellner, C.; Wiedmann, M.; Ivanek, R. Using agent-based modeling to compare corrective actions for *Listeria* contamination in produce packinghouses. *PLoS ONE* **2022**, *17*, e0265251. [CrossRef] [PubMed]
53. Sullivan, G.; Zoellner, C.; Wiedmann, M.; Ivanek, R. In silico models for design and optimization of science-based *Listeria* environmental monitoring programs in freshcut produce facilities. *Appl. Environ. Microbiol.* **2021**, *87*, e00799-21. [CrossRef] [PubMed]
54. McCollum, J.T.; Cronquist, A.B.; Silk, B.J.; Jackson, K.A.; O'Connor, K.A.; Cosgrove, S.; Mahon, B.E. Multistate outbreak of listeriosis associated with cantaloupe. *N. Engl. J. Med.* **2013**, *369*, 944–953. [CrossRef] [PubMed]
55. Colagiorgi, A.; Bruini, I.; Di Ciccio, P.A.; Zanardi, E.; Ghidini, S.; Ianeri, A. *Listeria monocytogenes* biofilms in the wonderland of food industry. *Pathogens* **2017**, *6*, 41. [CrossRef]
56. Highmore, C.J.; Warner, J.C.; Rothwell, S.D.; Wilks, S.A.; Keevil, C.W. Viable-but-nonculturable *Listeria monocytogenes* and Salmonella enterica serovar Thompson induced by chlorine stress remain infectious. *MBio* **2018**, *9*, 10–1128. [CrossRef] [PubMed]
57. Sethi, S.; Nayak, S.L.; Joshi, A.; Sharma, R.R. 5—Sanitizers for fresh-cut fruits and vegetables. In *Fresh-Cut Fruits and Vegetables. Technologies and Mechanisms for Safety Control*; Academic Press: Cambridge, MA, USA, 2020; pp. 99–119. [CrossRef]
58. Mercier, S.; Villeneuve, S.; Mondor, M.; Uysal, I. Time—Temperature management along the food cold chain: A review of recent developments. *Compr. Rev. Food Sci. Food Saf.* **2017**, *16*, 647–667. [CrossRef] [PubMed]
59. Ndraha, N.; Hsiao, H.I.; Vlajic, J.; Yang, M.-F.; Lin, H.-T.V. Time-temperature abuse in the food cold chain: Review of issues, challenges, and recommendations. *Food Control* **2018**, *89*, 12–21. [CrossRef]
60. Miceli, A.; Gaglio, R.; Francesca, N.; Ciminata, A.; Moschetti, G.; Settanini, L. Evolution of shelf life parameters of ready-to-eat escarole (*Cichorium endivia* var. latifolium) subjected to different cutting operations. *Sci. Hortic.* **2019**, *247*, 175–183. [CrossRef]
61. EFSA (European Food Safety Authority); ECDC (European Centre for Disease Prevention and Control). Multi-country outbreak of *Listeria monocytogenes* serogroup IVb, multi-locus sequence type 6, infections linked to frozen corn and possibly to other frozen vegetables—First update. *EFSA Support. Publ.* **2018**, *15*, 1448E. [CrossRef]
62. Truchado, P.; Gil, M.I.; Querido-Ferreira, A.P.; López Capón, C.; Álvarez-Ordoñez, A.; Allende, A. Frozen vegetable processing plants can harbour diverse *Listeria monocytogenes* populations: Identification of critical operations by WGS. *Foods* **2022**, *11*, 1546. [CrossRef] [PubMed]
63. Lindqvist, R.; Messens, W.; Bover-Cid, S. *Modified Listeria Monocytogenes Generic QMRA Model (mgQMRA)*; Zenodo: Meyrin, Switzerland, 2020. [CrossRef]

foods

MDPI

Review

A Critical Review of Risk Assessment Models for *Listeria monocytogenes* in Seafood

Ursula Gonzales-Barron [1,2,*], Vasco Cadavez [1,2], Juliana De Oliveira Mota [3], Laurent Guillier [4] and Moez Sanaa [3,*]

1 Centro de Investigação de Montanha (CIMO), Instituto Politécnico de Bragança, Campus de Santa Apolónia, 5300-253 Bragança, Portugal; vcadavez@ipb.pt
2 Laboratório para a Sustentabilidade e Tecnologia em Regiões de Montanha, Instituto Politécnico de Bragança, Campus de Santa Apolónia, 5300-253 Bragança, Portugal
3 Department of Nutrition and Food Safety, World Health Organization (WHO), CH-1211 Geneva, Switzerland
4 Risk Assessment Department, French Agency for Food, Environmental and Occupational Health & Safety (Anses), 14 Rue Pierre et Marie Curie, 94701 Maisons-Alfort, France; laurent.guillier@anses.fr
* Correspondence: ubarron@ipb.pt (U.G.-B.); sanaam@who.int (M.S.); Tel.: +351-273-303-325 (U.G.-B.)

Abstract: Invasive listeriosis, due to its severe nature in susceptible populations, has been the focus of many quantitative risk assessment (QRA) models aiming to provide a valuable guide in future risk management efforts. A review of the published QRA models of *Listeria monocytogenes* in seafood was performed, with the objective of appraising the effectiveness of the control strategies at different points along the food chain. It is worth noting, however, that the outcomes of a QRA model are context-specific, and influenced by the country and target population, the assumptions that are employed, and the model architecture itself. Studies containing QRA models were retrieved through a literature search using properly connected keywords on Scopus and PubMed®. All 13 QRA models that were recovered were of short scope, covering, at most, the period from the end of processing to consumption; the majority (85%) focused on smoked or gravad fish. Since the modelled pathways commenced with the packaged product, none of the QRA models addressed cross-contamination events. Many models agreed that keeping the product's temperature at 4.0–4.5 °C leads to greater reductions in the final risk of listeriosis than reducing the shelf life by one week and that the effectiveness of both measures can be surpassed by reducing the initial occurrence of *L. monocytogenes* in the product (at the end of processing). It is, therefore, necessary that future QRA models for RTE seafood contain a processing module that can provide insight into intervention strategies that can retard *L. monocytogenes'* growth, such as the use of bacteriocins, ad hoc starter cultures and/or organic acids, and other strategies seeking to reduce cross-contamination at the facilities, such as stringent controls for sanitation procedures. Since risk estimates were shown to be moderately driven by growth kinetic parameters, namely, the exponential growth rate, the minimum temperature for growth, and the maximum population density, further work is needed to reduce uncertainties.

Keywords: systematic review; exposure assessment; simulation; fish; smoked salmon; listeriosis

1. Introduction

Listeria monocytogenes is a ubiquitous microorganism that is widely distributed in the environment. Whereas soil and water are considered to be the primary sources of *L. monocytogenes* for transmission to plant material, feed, animals and the food chain, this pathogen has also demonstrated the ability to resist adverse environmental conditions and persist in the processing environment [1]. Ready-to-eat (RTE) foods that are not heat-treated or do not undergo any listericidal treatment before consumption are of significance in the transmission of foodborne listeriosis [1]. Recently, using a generic quantitative risk assessment (QRA) model, EFSA [2] compared the probability of listeriosis in the elderly population in the European Union and its link to products such as RTE fish, pâté, cooked meats, sausages, soft and semi-soft cheeses, and blanched frozen vegetables. They found out that gravad fish in normal-atmosphere packaging, and hot-/cold-smoked fish in reduced-oxygen packaging (ROP) ranked as the most high-risk products. The occurrence of *L. monocytogenes* in RTE products has been responsible for outbreaks and product recalls in the EU and the USA. For instance, as reported by the Rapid Alert System for Food and Feed (RASFF), over 40% of the seafood notifications between 2008 and 2016 were related to *L. monocytogenes* [1]. According to the latest EFSA and ECDC report [3] for the year 2022, the highest occurrences of *L. monocytogenes* were found for the categories of fish (2.6%; n = 971) and fishery products (2.5%; n = 842) sampled at the manufacturing and the distribution stages in the EU MS, with the proportion of samples exceeding the limit of the food safety criteria at distribution (100 CFU/g) generally being low (0.05% and 0.08%, respectively).

In terms of reported outbreaks, according to EU surveillance data [4], in the period between 2010 and 2020, fish and fish products (namely, crab meat, crustaceans, shellfish and molluscs, smoked fish and non-specified seafood) caused 23% of the 53 cases with strong evidence of outbreaks in the EU. Unlike the EU scenario, in the USA during the ten-year span, fish and fish products (namely, smoked fish and raw sushi) had a lower than 6% share in the 50 strong-evidence outbreaks [5]. Considering that most of the listeriosis cases occur sporadically [1,6], the results of a recent meta-analysis on case-control studies of sporadic listeriosis [7] cannot be overlooked. Combining the odds ratio (OR) outcomes from 12 primary studies, Leclercq et al. [7] found that RTE seafood presented the highest association with sporadic listeriosis, with a pooled OR of 10.95 ($p < 0.001$) for non-perinatal population, and pooled OR of 6.273 ($p < 0.001$) for the entire susceptible population (in comparison to processed meats, cheese, vegetables, fruits, and composite foods).

Various listeriosis QRA models have been produced for RTE seafood [8] in an attempt to provide guidance to reduce the occurrence of *L. monocytogenes* via practices or strategies that retard or prevent the growth of this pathogen. The objectives of this study are as follows: (i) to undertake a critical review of the published QRA models of listeriosis acquired from the consumption of RTE seafood; (ii) to contrast the control measures or strategies evaluated in the various QRA models as what-if scenarios; and (iii) to derive important lessons and recommendations for future QRA models in RTE seafood.

2. Materials and Methods

QRA models were retrieved through a literature search on Scopus and PubMed®, using 1998 as the starting year of publication. The searches of the title, keywords, and abstract were carried out on 18 May 2022, using logically connected terms (("risk assessment" OR exposure OR quantitative microbial OR risk modelling OR modeling OR simulation* OR second-order OR "second order" OR "risk management") AND ("*L. monocytogenes*" OR "*Listeria monocytogenes*" OR listeriosis)) properly arranged in the syntaxes of the literature search engines. Studies were considered eligible if (1) they presented a quantitative risk or exposure assessment for listeriosis linked to seafood, with explicitly indicated formulae and assumptions, and (2) they were written in English or Spanish language. The full systematic review process and extraction of information are described in the work of Gonzales-Barron et al. [8]. The present review focuses only on seafood products, which were the subject of 13 QRA models described in 14 publications [1,6,9–20].

3. Results

A total of 13 QRA models on seafood as a source of listeriosis were recovered in the literature search of models published between January 1998 and May 2022. Table 1 compiles the main features of the 13 QRA models, whereas Table 2 summarises the predictive microbiology models and main outcomes related to what-if scenarios and sensitivity analyses. None of the 13 QRA models comprised simulations of cross-contamination, nor did a single QRA model include a processing stage module. All the models represented short supply chains, from the end of processing or retail to table, or focused on consumption only. Three QRA models assessed the growth of *L. monocytogenes* from the end of processing until consumption [9–12], and the other three began the analysis at retail [1,6,13]. Seven out of the thirteen models solely represented the consumption module [14–20].

Table 1. Features of quantitative risk assessment models of *L. monocytogenes* regarding the consumption of seafood products by scope.

Scope	Food	RTE	Cross-Contamination	DR—End-Point	Type of DR Model	DR Sub-Populations	Strain Variability	Temp Profiles/ Lagtime	Country	Source
End-processing-to-table	Cold-smoked salmon	Yes	No	Exp—I	Pouillot et al. [10]	Multiple	NA	Yes/No	France	Pouillot et al. [9,10]
	Cold-smoked salmon	Yes	No	Exp—I	Fritsch et al. [11]: r values from Pouillot et al. [21] were re-scaled to three diff-erent groups of virulence (according to CCs)	General	Specific prevalence for each LM genotypic sub-group (CC) in Europe; two different distributions for Tmin to represent "low-growing" and "high-growing" strains; three virulence levels in the DR r values	Yes/No	France	Fritsch et al. [11] (model based on Pouillot et al. [9,10] integra-ting genomic data)
	Cold-smoked salmon	Yes	No	None	NA	NA	Variable proportion of contaminated packages and growth kinetics parameters according to LM serotypes 1/2a, 1/2b, and 4b	No/Yes	USA	Chen et al. [12]
Retail-to-table	Various: smoked seafood, raw seafood, preserved fish, cooked RTE crustaceans	Yes	No	Mouse Epi—I	FDA-FSIS [6]	Multiple	Variability in the virulence of different strains represented in DR	No/No	USA	FDA-FSIS [6]
	Packaged cold-/hot-smoked fish and gravad fish	Yes	No	Exp—I	Pouillot et al. [21]	Multiple	Challenge test data from a mixture of strains; h0 distribution of variability in physiological state of cells; variability in strain virulence and in susceptibility across population subgroups	Yes/Yes	EU	Pérez-Rodríguez et al. [13]
	Cold-, hot-smoked fish, gravad fish	Yes	No	Exp—I	EFSA BIOHAZ [1] based on Pouillot et al. [21]	Multiple (sex/age group)	Challenge test data from a mixture of strains; strain virulence and host susceptibility explicit in r distribution	No/No	EU	EFSA BIOHAZ [1]

Table 1. *Cont.*

Scope	Food	RTE	Cross-Contamination	DR—End-Point	Type of DR Model	DR Sub-Populations	Strain Variability	Temp Profiles/Lagtime	Country	Source
Consumption	Smoked/gravad salmon/rainbow trout	Yes	No	Exp—I	Buchanan et al. [22]	General	All strains are virulent vs. a proportion of virulent strains	No/No	Sweden	Lindqvist and Westöö [14]
	Cold-smoked fish	Yes	No	Exp—I	FAO-WHO [15]	High-risk/low-risk	Strain diversity implicit in r	No/Yes	Non-specific	FAO-WHO [15]
	Smoked fish and sliced cooked ham	Yes	No	Exp—I	FAO-WHO [15]	High-risk/low-risk	Strain diversity implicit in r	No/No	Spain	Garrido et al. [16]
	Cold-smoked salmon	Yes	No	BP—I	Haas et al. [23]	General	NA	No/Yes	Non-specific	Gospavic et al. [17]
	VP cold-smoked salmon	Yes	No	WG—I	Farber et al. [24]	High-risk/low-risk	Challenge test data from a mixture of strains	No/Yes	Ireland	Dass [18]
	Traditional processed fish	No	No	WG—I	Farber et al. [24]	High-risk/low-risk	NA	No/No	Ghana	Bomfeh [19]
	Cold-smoked and salt-cured fishery products	Yes	No	Exp—I	Pasonen et al. [20]	High-risk/low-risk	NA	No/No	Finland	Pasonen et al. [20]

DR: dose–response; Exp: exponential model; Mouse-Epi: mouse epidemiological model; I: illness endpoint; D: death endpoint; NA: not addressed in the study.

Table 2. Predictive microbiology models and main outcomes related to what-if scenarios and sensitivity analysis from quantitative risk assessment models of *L. monocytogenes* (LM) from consumption of seafood products.

Scope	Food	Predictive Microbiology Models	What-If Scenarios	Sensitivity Analysis	Model Complexity	Source
End processing-to-table	Cold-smoked salmon	Growth (Jameson effect LM and background microflora, growth square root models for LM and background microflora)	**EXPOSURE ASSESSMENT**: (1) Reducing theoretical shelf-life from 28 days to 15 days reduced mean LM/g in contaminated servings by 10%; (2) the baseline scenario of 21.4% of shelf lives at home being longer than 7 days was compared to a scenario of consumption within 7 days maximum, which reduced the mean LM/g by 10%; (3) better refrigeration at retail, reducing the mean temperature from 5.6 to 4 °C, reduces the mean LM counts by 19%; (4) better refrigeration at home, reducing the mean temperature from 7 to 4 °C, reduces the mean LM counts by 36%; (5) a lower initial concentration, from 0.46% of values above 1 CFU/g to a distribution truncated at 1 CFU/g, reduces the mean LM counts by 8%. **RISK ASSESSMENT**: Output—Listeriosis cases compared to a base 100 for the baseline model: (1) shelf-life 15 days = 23; (2) prevalence of LM to a quarter= 25; (3) mean home refrigerator temperature 4 °C = 34; (4) consumed 7 days after purchase = 37; (5) prevalence of LM to a half = 50; (6) mean retail temperature at 4 °C = 67.	**EXPOSURE ASSESSMENT: Output—concentration of LM in contaminated servings**: (1) total duration at the consumer phase ($p = 10^{-30}$); (2) mean temperature at the consumer phase ($p = 10^{-20}$); (3) initial LM counts ($p = 10^{-20}$); (4) mean temperature at retail phase ($p = 10^{-14}$); (5) total duration of retail phase ($p = 10^{-8}$); (6) Tmin for growth ($p = 10^{-8}$); (7) Tmin microflora ($p = 10^{-6}$); (8) initial background flora counts ($p = 0.002$); (9) serving size ($p = 0.003$); (10) MPD ($p = 0.008$); (11) ref. GR at 25 °C ($p = 0.015$); (12) ref. GR of flora at 25 °C ($p = 0.025$). **RISK ASSESSMENT: Output—listeriosis cases in the reference population**: (1) r value of DR model ($p = 10^{-300}$); (2) SD (MPD) ($p = 10^{-137}$); (3) ref. of GR 25 °C for LM ($p = 10^{-101}$); (4) MPD of LM ($p = 10^{-76}$); (5) Tmin of LM ($p = 10^{-12}$); (6) GR of flora 25 °C ($p = 10^{-8}$); (7) prevalence of LM ($p = 10^{-6}$); (8) servings/year ($p = 10^{-2}$).	medium: complex predictive microbiology model, a new method for solving growth under dynamic temperature profiles, was proposed.	Pouillot et al. [9, 10]
	Cold smoked salmon	Growth (Jameson effect LM and background microflora, growth square root models for LM and background microflora)	Baseline predicted 978 listeriosis cases after consumption of 50 g cold-smoked salmon with an initial LM prevalence of 10.4%, considering a single prevalence distribution. (1) Taking into account specific prevalences for each LM genotypic sub-group lowered the listeriosis cases to 574. (2) A total of 97% of listeriosis cases were caused by the hypervirulent group, despite their low prevalence (12.6%) in contaminated salmon. Inversely, the most prevalent (hypovirulent) group (51.7%) was responsible for only 0.02% of the listeriosis cases. (3) The effect of the low/high-growth strains (two distributions for Tmin) was lower than the effect of the virulence: mean exposure from the high-growth LM group was 25 CFU/g compared to the low-growth groups (13 CFU/g).	NA	Medium: Same as Pouillot et al. [9,10] but with the further complexity of adding phenotypic characteristics of LM by subgroup and the virulence properties of LM.	Fritsch et al. [11] (model based on Pouillot et al. [9, 10], integrating genomic data)

Table 2. *Cont.*

Scope	Food	Predictive Microbiology Models	What-If Scenarios	Sensitivity Analysis	Model Complexity	Source
	Cold-smoked salmon	Growth models (Buchanan, Gompertz and Baranyi as primary models, and secondary square root model); and Die-off and re-growth models (Weibull-Buchanan, Weibull-Gompertz and Weibull-Baranyi)	End point of the model is the *regulatory and recall risk* (RRR), defined as the overall risk of a lot sampled found positive for LM. (1) Treatment of salmon with 5 or 20 ppm nisin reduced RRR to 0.109 or 0.017 (in comparison to baseline RRR of 0.333); (2) reducing prevalence to half decreased RRR to 0.182; (3) the use of inhibitors (2% potassium lactate + 0.14% sodium diacetate) slightly reduced RRR to 0.313; (4) keeping cold storage below 5 °C did not reduce RRR.	**Output—regulatory and recall risk**: (1) initial contamination level (r = 0.404); (2) GR at 25 °C (r = 0.275); (3) storage temperature (r = 0.177); (4) Tmin (r = −0.169); (5) MPD (r = 0.053)	Medium: Uncertainty and variability are separated; the die-off and/or growth kinetics are too compartment-alised.	Chen et al. [12]
Retail-to-table	Various: smoked seafood, raw seafood, preserved fish, cooked RTE crus-taceans	Growth (linear model, EGR5 square root models)	(1) For cold-smoked salmon, reducing the max. home storage time from 45 to 30 days reduces the mean cases by 38% in the elderly population.	NA	Medium: Various foods	FDA-FSIS [6]
	Packaged cold-/hot-smoked fish and graved fish	Growth (Baranyi model with Jameson effect LM and LAB, EGR5 square root model and effect of lactate)	(1) Decreasing the maximum initial LM concentration by 2 log decreases listeriosis cases per million servings in >99%; (2) decreasing time to consumption by 25% decreases listeriosis by 80%; (3) decreasing 1–2 °C in the dynamic temperature profiles reduces cases by 75%; (4) including lag time in the model has no effect on listeriosis cases.	NA	Medium: Dynamic time–temperature profiles from retail to consumption, and microbial competition models used were solved with the RK4 algorithm.	Pérez-Rodríguez et al. [13]
	Cold-, hot-smoked fish, gravad fish	Growth (Rosso model, EGR 5 °C)	(1) Across the 3 RTE fish products, there is no strong difference in the probability of a product exceeding 100 CFU/g at the time of consumption between normal packaging (0.066–0.112) and reduced-oxygen packaging (0.040–0.115); (2) in both reduced-oxygen and normal packaging, hot-smoked fish presented with a higher probability of exceeding 100 CFU/g at the point of consumption (0.115, 0.112) than cold-smoked fish (0.080, 0.074) and gravad fish (0.047, 0.066).	Risk is very sensitive to MPD. A shift in 0.5 log CFU/g can double the estimated risk. However, a sensitivity analysis was conducted, taking various RTE food classes into account.	Low: Generic model; only demands some knowledge of R software to utilise it	EFSA BIOHAZ [1]

Table 2. *Cont.*

Scope	Food	Predictive Microbiology Models	What-If Scenarios	Sensitivity Analysis	Model Complexity	Source
Consumption	Smoked/gravad salmon/rainbow trout	NA	(1) The minimum level of LM resulting in a risk of illness greater than 10^{-7} or 10^{-8} was 25 or 2 CFU/g; (2) if the assumption that all strains are virulent is reduced to 1–10%, the annual listeriosis cases are reduced by 84% in the both high-risk and the low-risk populations.	**Output—annual risk of illness**: ranked as initial LM counts, prevalence, serving size, and proportion of virulent strains	Low	Lindqvist and Westöö [14]
	Cold-smoked fish	Growth (LM growth model affected by LAB growth, square root model for GR as a function of temperature, pH, aw, un-dissociated lactic acid)	(1) If the LM growth rate inhibition due to LAB growth is between 80 and 100%, the increase in listeriosis per 100,000 people is 684-fold in the overall population, in comparison to the baseline scenario of no growth in LM between purchase and consumption; (2) if the LM growth rate inhibition due to LAB growth is 95%, the increase in listeriosis per 100,000 people is 67-fold in the overall population in comparison to the no growth in the LM baseline scenario; (3) reducing the mean shelf-life of smoked fish from 14 to 7 days results in an 80% reduction in listeriosis.	NA	Medium: Relative lag time concept for LM and LAB	FAO-WHO [15]
	Smoked fish (salmon and trout)	Growth (logistic model without delay, growth cardinal model)	(1) Reducing home storage time from a max of 30–7 days reduces the annual cases by 15% for salmon and 45% for trout; (2) if all domestic temperatures had a mean temperature of 4.5 °C, the mean annual cases are reduced by 65% for salmon and by 70% for trout; (3) Combining the two measures above reduces the mean annual cases by 75% for salmon and 87% for trout; (4) If, at purchase, LM counts do not exceed 100 CFU/g (truncating the baseline N~(1.01, 0.71) for smoked salmon and N~(1.35, 1.40) for smoked trout, the mean annual cases would decrease by 22% in salmon and 99% in trout.	NA	Low	Garrido et al. [16]

Table 2. *Cont.*

Scope	Food	Predictive Microbiology Models	What-If Scenarios	Sensitivity Analysis	Model Complexity	Source
	Cold-smoked salmon	Growth (Baranyi model with Jameson effect LM and background microflora, extended GR models for LM and LAB as a function of temperature, pH, aw, un-dissociated lactic acid, undissociated diacetate, phenols, dissolved CO_2 and nitrite)	(1) At a mean initial LM count of 4 CFU/g, reducing the time of consumption from 28 to 14 days reduces the risk of illness by 64%; (2) if the mean time of consumption is 14 days, reducing the mean initial counts from 25 CFU/g to 4 CFU/g reduces the risk of illness by 67%.	NA	Medium: stochastic fluctuations in the GR of LM are taken into account by using white noise and the Winner process.	Gospavic et al. [17]
	Vacuum-packed cold-smoked salmon	Growth (Baranyi model, growth square root model)	(1) If initial LM counts at retail (1–1000) are truncated at >100 CFU/g, the risk of illness would reduce by 0.3/0.9 log (high-risk and low-risk populations); (2) reducing the maximum consumer shopping time from 3 h to 30 min reduces risk of illness by 0.8/0.8 log; (3) reducing consumer storage days from 21–30 to 7–15 days reduces risk of illness by 0.5/0.6 log; (4) fixing storage temperature from 3–10 °C to 4 °C reduces risk of illness by 1.0/1.1 log; (5) if LM counts are not higher than 2 log CFU/g and the maximum shopping time is reduced to 30 min, reducing consumer storage days to 7–15 days and storage temperature to 4 °C reduces risk of illness by 1.32/1.39 log.	**Output—annual risk of illness in the high-risk population**: (1) LM counts at retail (r = 0.97); (2) temperature in consumer fridge (r = 0.13); (3) time in consumer fridge (r = 0.06).	Low: Lag: Baranyi model with bacterial adaptation	Dass [18]
	Traditional processed fish	NA	NA	NA	Low	Bomfeh [19]

Table 2. *Cont.*

Scope	Food	Predictive Microbiology Models	What-If Scenarios	Sensitivity Analysis	Model Complexity	Source
	Cold-smoked and salt-cured fishery products	Growth (logistic growth model, growth cardinal parameter model as a function of temperature, salt content, pH and phenolic compounds)	(1) If home storage temperature decreased from 7 °C to 3 °C, the median cases of listeriosis per 100,000 elderly population would decrease by 70%; (2) if home storage temperature decreased from 7 °C to 3 °C, the median cases of listeriosis per 100,000 working-age population would decrease by 40%.	NA	High: Parameters, including r, were estimated from a Bayesian model consisting of three linked modules: a model for the occurrence data, a model for the consumption data and a predictive model for the total number of cases in the population. The current model takes into account the possibility of continuing consumption of the same (contamina-ted) package of CSS/SCS, rather than assuming independent consumption days.	Pasonen et al. [20]

aw: water activity; LPD: lag-phase duration; RLT: relative lag time; MPD: maximum population density; GR: maximum growth rate; EGR_x: exponential growth rate at x °C; LAB: lactic acid bacteria; LAC: lactic acid concentration; RR: risk reduction; r: Pearson's correlation coefficient; NA: not addressed in the study.

Most of the models focused on RTE seafood, except for one [19], which dealt with a traditional processed fish from Ghana. Within the RTE seafood, the majority of models (75%) estimated the risk of listeriosis from smoked fish (or salmon), while the other 25% estimated the risk associated with gravad fish. In addition, other products were investigated, such as raw seafood, preserved fish, cooked RTE crustaceans by FDA-FSIS [6], and salt-cured fishery products, as studied by Pasonen et al. [20].

More than half of the available models (9/13) represented the conditions of European countries, namely France [9–11], Finland [20], Ireland [18], Sweden [14], and Spain [16]—where these products are highly consumed—in addition to models that used data from the EU [1,13]. Two QRA models pertained to the risk of listeriosis in the USA's population [6,12], and two models were not linked to any specific geographical location [15,17] (Table 1).

The QRA models that were retrieved varied in the degree of complexity in their construction. All QRA models, except two [14,19], employed predictive microbiology models, including microbial growth, survival, and competition models, and die-off and re-growth models (Table 2). The lag-phase duration of *L. monocytogenes* was considered in five models [12,13,15,17,18], whereas only three QRA models [9–11,13] employed time–temperature trajectories to more realistically estimate the kinetics of *L. monocytogenes* during cold storage and their shelf life (Table 1).

One model did not perform any risk estimation, as the authors targeted the estimation of the regulatory and recall risk [12], while the other twelve models considered illness as the end-point for risk estimation. The exponential dose–response function was the equation of choice for risk characterisation in nine QRA models, although they followed different approaches, namely the dose–response models of Pouillot et al. [21] (used by Pérez-Rodriguez et al. [13], and modified versions by Fritsch et al. [11] and EFSA-BIOHAZ [1]), FAO-WHO [15] (used by themselves [15] and Garrido et al. [16]), FDA-FSIS [6] (used by themselves), Pasonen et al. [20] (used by themselves), and Buchanan et al. [22] (used by Lindqvist and Westöö [14]). The early Weibull–gamma model proposed by Farber et al. [24] was used in the QRA models of Dass [18] and Bomfeh [19], whereas the beta-Poisson model of Haas et al. [23] was employed in the QRA model of Gospavic et al. [17] (Table 1).

Except for Bomfeh [19], all seafood QRA models assessed the effect of what-if scenarios on the final risk. A sensitivity analysis of response variables, such as *L. monocytogenes'* concentration at consumption, regulatory and recall risk, or final risk measures, was undertaken in 38.0% of the models [1,9,10,12,14,18] (Table 2).

4. Discussion

Listeriosis QRA models in seafood focused mainly on RTE smoked and gravad fish (i.e., generic fish, salmon, or trout), because these are products that have considerable public health implications regarding listeriosis, for the following reasons: (1) they are not heat-treated; (2) they are given a relatively long shelf-life; (3) they are mostly vacuum-packed—which does not preclude *L. monocytogenes* growth; and (4) they are generally eaten with no prior cooking. Many reports and surveys have indicated that seafood products are frequently contaminated with *L. monocytogenes* [25]. A recent genomic-based epidemiological study [26] determined that, from 2018 to 2020, 27% of all listeriosis cases in Germany with suspected food vehicles were caused by smoked or gravad salmon products. These authors demonstrated that, despite the considerable efforts that have been made to improve the safety of smoked fish, outbreaks linked to the presence of *L. monocytogenes* at infective levels in these seafood products continue to occur. Controlling *L. monocytogenes* in smoked fish is challenging because this pathogen is widely distributed in a variety of environments, including natural ones [27,28], and processing facilities [29]. This underscores the need to assess new technological interventions, post-lethality treatments, and intensified sanitation programmes to reduce the risk of listeriosis in RTE seafood.

4.1. Risk Factors at Processing

Apart from *L. monocytogenes* contamination in raw materials (fish) as a primary source, smoked fish can acquire the pathogen from contact with surfaces in the processing environment. Moreover, the fate of *L. monocytogenes* is variable along the processing steps of evisceration and filleting [30,31], brining [32], smoking [33], and slicing [34]. Nevertheless, despite the extensive data that are available, none of these effects has been simulated in the seafood QRA models that were retrieved. The first QRA model of the broadest scope was that of Pouillot et al. [9,10], which followed the cold-smoked salmon supply chain from the end of processing until consumption. Such a model was innovative in the following ways: (1) it used a Jameson effect to account for the inhibitory effect of the background microflora on *L. monocytogenes* in the product within the vacuum-packaged atmosphere; (2) it used dynamic time–temperature profiles to represent realistic temperature oscillations between cold storage at the end of processing and home refrigeration. The other two QRA models represented the supply chain from the end of processing: Fritsch et al.'s [11] model, which has the same structure as Pouillot et al.'s [9,10] model, but was refined by the introduction of phenotypic growth characteristics of *L. monocytogenes* according to subgroup and virulence properties; and Chen et al.'s [12] model, which did not proceed to the risk characterisation stage.

In relation to the assessment of post-lethality treatments, such as product reformulation involving the application of lactate or diacetate, nisin, or specific starter cultures, these were not assessed in any of the QRA models that were retrieved, except in Chen et al.'s model [12], where they pointed out that the addition of nisin (5–20 ppm) is far more effective than the addition of 2% potassium lactate plus 0.14% sodium diacetate in decreasing the overall risk of a sampled lot that was found positive for *L. monocytogenes* (67–95% reduction versus 6% reduction, respectively). Although the QRA models of Pouillot et al. [9,10], Fritsch et al. [11], Pérez-Rodríguez et al. [13], and Gospavic et al. [17] were equipped with Jameson-effect models that were able to characterise the inhibitory effect of lactic acid bacteria on *L. monocytogenes*, none of these QRA models evaluated scenarios related to the addition of ad hoc cultures of lactic acid bacteria (Table 2). Although the cold-smoked fish model of FAO-WHO [15] did not employ any Jameson-effect competition model, it considered the effect of indigenous lactic acid bacteria, assuming that, at high concentrations, they can suppress the growth of *L. monocytogenes*. This model demonstrated the ability of cold-smoked fish to support the growth of *L. monocytogenes*. Even under the optimistic assumption that the growth rate inhibition of *L. monocytogenes* due to the growth of lactic acid bacteria is 95%, the listeriosis cases per thousand people would be ~70-fold greater than if cold-smoked fish did not support their growth (Table 2).

4.2. Cross-Contamination in Processing Plants

Abundant literature has demonstrated that cross-contamination can occur during the different processing stages of RTE seafood. During head-cutting, evisceration, and filleting, there are many opportunities for *L. monocytogenes* to be transferred from the exterior of fish to the cut surfaces of fillets or sides [30,35]. At this stage of processing, the flesh areas of fish can be inoculated by contact with the contaminated skin sides of fillets, filleting tables, and knives and gloves. For instance, Dass [18] detected *L. monocytogenes* types c and b on filleting boards, deboning pins, conveyor belts, and knives. In another study, Chen et al. [31] monitored the contamination of *L. monocytogenes* in catfish fillets and in environmental samples collected from various areas of the processing plant. They isolated serotype 1/2b (47%) from trimming boards, conveyor belts, and holding tables, and found that conveyors were contaminated with *L. monocytogenes* in a total of 16.6% of the samplings (6/36). In a processing facility of gravad salmon in Brazil, Cruz et al. [36] 80% of swabs of *L. monocytogenes* from handlers were found to be positive, and 25% from knives and tables. Lundén et al. [37] explained that the contamination on tables and cutting surfaces can adhere strongly within a short period of time. This suggests that filleted fish become contaminated during the first stages of processing.

Slicing machines can be regarded as a source of *L. monocytogenes'* contamination [34]. Di Ciccio et al. [38] repeatedly isolated *L. monocytogenes* serotypes 1/2a and 1/2b from slicer belts, distribution trays, slicing machines, and slicing covers for three years in a smoked-salmon production facility. Out of the 95 tested environmental samples, slicing machines (37%) and working tables (43%) had the highest frequencies of detection. In the USA, in a processing plant of catfish fillets, Chen et al. [31] determined that 15% (7/45) of the samples, monitored at skinning, slicing, and blending equipment, were contaminated with *L. monocytogenes*.

The review published by Jami et al. [25] showed that drains and floors, as non-direct food-contact surfaces, are the most frequently contaminated sources, with prevalences of 2–80% and 1–53%, respectively. Drains and floors may, therefore, represent niches of contamination. Many authors have also found that cross-contamination can occur during packaging from surfaces that are in direct contact with the food being packaged. In cold-smoked salmon plants, Autio et al. [39], Vogel et al. [40], Klaeboe et al. [41], Nakamura et al. [42], Thimothe et al. [43], and Hu et al. [44] recovered *L. monocytogenes* from direct food-contact surfaces of packaging equipment at frequencies of 20/84, 140/818, 23/155, 9/101, 6/125, and 5/344, respectively.

Nonetheless, despite the actual environmental contamination that occurs in processing plants, none of the 13 QRA models assembled cross-contamination modules that could help assess the effects of implementing more stringent controls for environmental monitoring programmes, good manufacturing practices, and standard operation procedures for sanitation.

4.3. Shelf-Life and Risk Factors at Retail and Home

As the seafood QRA models had a short scope, the typical scenarios that were assessed were those related to reductions in *L. monocytogenes'* initial prevalence/concentration, storage temperature, time of consumer storage, and shelf-life (Table 2). Vacuum-packaging is widely used in the smoked/gravad fish industry as it delays the proliferation of aerobic spoilage bacteria and minimises oxidative reactions. However, although vacuum-packaging is used to extend the shelf-life of these products, microaerophilic or facultative anaerobic microorganisms, such as *L. monocytogenes*, may thrive under such condition, and an extended shelf-life may provide sufficient time for the pathogen to increase to infective levels. The EFSA BIOHAZ generic QRA model [1] clearly made this point by showing that, across three RTE fish products (cold-smoked, hot-smoked and gravad fish), there is no strong difference in the probability of a product exceeding 100 CFU/g at the time of consumption between normal packaging (0.066–0.112) and reduced-oxygen packaging (0.040–0.115; Table 2). Within this context, many of the QRA models tested what-if scenarios with a shorter time of storage before consumption. Researchers obtained different estimates of the degree of reduction in the number of listeriosis cases for smoked fish.

Decreasing the consumption time of unopened packages to a maximum of 7 days at cold storage led to reductions in the cases of listeriosis of 80% for cold-smoked fish (FAO-WHO [15]; baseline 14 days), 80% for packaged cold-/hot-smoked fish and gravad fish (Pérez-Rodríguez et al. [13]; baseline not clear), 63% for cold-smoked salmon (Pouillot et al. [10]; baseline 32 days maximum), 45% for smoked trout (Garrido et al. [16]; baseline 30 days), and 15% for smoked salmon (Garrido et al. [16]; baseline 30 days). Reducing the shelf life to 14–15 days for cold-smoked salmon was estimated to reduce the listeriosis cases by 77% (Pouillot et al. [9,10]; baseline 32 days) and the risk of illness by 64% (Gospavic et al. [17]; baseline 28 days). FDA-FSIS [6] estimated that by reducing the maximum home storage time from 45 to 30 days, the mean cases of listeriosis would be reduced by 38% in the elderly population (Table 2).

Shelf-life reduction therefore appears to be a good strategy to decrease the risk of listeriosis, although its effectiveness can be surpassed by effectively maintaining the product at 4.0–4.5 °C during home storage, as attested in the what-if scenarios presented by Dass [18], Garrido et al. [16], Pérez-Rodríguez et al. [13], and Pouillot et al. [9,10]. Furthermore, in a sensitivity analysis carried out on the annual risk of illness in the high-risk population linked to smoked salmon, the temperature in the consumer's fridge was more determinant of risk (r = 0.13) than the time spent in the consumer's fridge (r = 0.06) [18]. A comparable rank correlation coefficient for storage temperature (r = 0.177) was estimated by Chen et al. [12]. In a Bayesian QRA model [20], although no comparison was made between a temperature-lowering scenario with a shelf-life reduction scenario, the authors underscored the importance of maintaining a cold temperature at the domestic level by predicting that if the home mean storage temperature decreased from 7 °C to 3 °C, the median cases of listeriosis per 100,000 people would decrease by 70% for the elderly population and 40% for the working-age population (Table 2).

Pasonen et al.'s [20] model was different from the other models in that it allowed for the possibility of continuing consumption of the same (contaminated) package of cold-smoked and salt-cured salmon, rather than assuming independent consumption days, and in that the model was not solved as a forward problem according to Monte Carlo simulation. This QRA model was built on a Bayesian, two-state, Markov chain approach consisting of three modules: a module for occurrence data, a module for consumption data, and a predictive model for the total number of listeriosis in the population. As a Markov-chain Monte Carlo (MCMC) simulation, bottom-up and top-down approaches are combined, and thus all unknown parameters can be jointly estimated from a single compact model. As a result, Pasonen et al. [20] could estimate the uncertainty distribution of the parameters, truly reflecting the information contained in the data. The r parameter of the exponential dose–response model was estimated from Finnish data by using the reported number of listeriosis cases to calibrate the dose–response function for the target populations. Nonetheless, in this calculation, it was assumed that all these listeriosis cases resulted from the consumption of the smoked fish, because the attribution of other sources was not modelled. This fact does not invalidate the MCMC approach proposed by Pasonen et al. [20], since the parameter(s) of the dose–response function could be assumed to be known. Although it demands a more complex programming, Bayesian inference features many advantages such as utilising the whole data set jointly, handling censored values, taking uncertainty into account, and the possibility of using prior knowledge.

Decreasing the initial mean prevalence or concentration of *L. monocytogenes* was another of the frequent what-if scenarios, which, in most of the QRA models of smoked salmon, turned out to be at least as effective in reducing the risk of listeriosis as maintaining the consumer's fridge temperature at 4 °C. This was observed in the models of Pouillot et al. [9,10], Chen et al. [12], Pérez-Rodríguez et al. [13], Gospavic et al. [17], Dass [18], and Lindqvist and Westöö [14], and will be described as follows. In the hypothetical scenario where the prevalence of contaminated packages of cold-smoked salmon decreased from 1.0 to 0.25, the number of listeriosis cases would drop by 75% (Pouillot et al. [9,10]). Chen et al. [12] estimated that reducing the initial prevalence to half would cause a reduction of 45% in the regulatory and recall risk. Pérez-Rodríguez et al. [13] predicted that decreasing the maximum initial concentration of *L. monocytogenes* by 2.0 log would ensure a drop in listeriosis cases by >99.9%; Gospavic et al. [17] estimated that reducing the mean initial concentration from 25 CFU/g to 4 CFU/g would reduce the risk of listeriosis by 67%. In line with the QRA models presented above, looking at sensitivity analysis outcomes, Dass [18] also showed that the concentration of *L. monocytogenes* at retail was a stronger determinant (r = 0.97) of the annual risk of illness in the high-risk population than the temperature/time in the consumer's fridge (r = 0.06–0.13). Similarly, Chen et al. [12] estimated that the initial contamination level (r = 0.404) was a stronger determinant of the regulatory and recall risk than the storage temperature (r = 0.177). Without providing coefficients of correlation, Lindqvist and Westöö [14] indicated that the variables impacting the annual risk of listeriosis, in decreasing order, were as follows: the initial counts of *L. monocytogenes* at retail; the prevalence of *L. monocytogenes*; the serving size and proportion of virulent strains (Table 2).

The results of these short-scope QRA models have shown, as a whole, that even when storage temperatures can be kept as low as 3–4 °C by the consumers, *L. monocytogenes* can proliferate in smoked/graved fish. However, reducing the initial contamination or ensuring that prevalence is low at the beginning of retail (or end of processing) would result in a significantly lower risk of listeriosis. Therefore, a robust processing module should be built to represent the strategies, combination of strategies, and/or sanitation control schemes leading to reductions in the prevalence/concentration of *L. monocytogenes*.

4.4. Microbial Growth Kinetic Parameters as Drivers of the Final Risk

Interesting outcomes from Pouillot's QRA model included the high impact of the kinetic parameters of both *L. monocytogenes* and background microflora on both the concentration of *L. monocytogenes* in contaminated servings [9] and the listeriosis cases in the reference population [10]. Right after the initial *L. monocytogenes* prevalence ($p = 10^{-20}$) and the mean temperature at retail phase ($p = 10^{-20}$), the concentration of *L. monocytogenes* in contaminated servings was highly sensitive to the minimum temperatures regarding the growth of *L. monocytogenes* ($p = 10^{-8}$) and the background microflora ($p = 10^{-6}$), followed by the reference growth rates at 25 °C of *L. monocytogenes* ($p = 0.015$) and the background microflora ($p = 0.025$), and the maximum population density ($p = 0.008$). Likewise, the listeriosis cases in the reference population were heavily impacted by the mean and the standard deviation of the maximum population density ($p = 10^{-137}$; $p = 10^{-76}$), the reference growth rates at 25 °C of *L. monocytogenes* ($p = 10^{-101}$) and the background microflora ($p = 10^{-8}$), and the minimum temperature for the growth of *L. monocytogenes* ($p = 10^{-12}$). Likewise, Chen et al. [12] found that, after the initial contamination level (r = 0.404), important determinants of the recall risk included the microbial kinetic parameters of exponential growth rate at 25 °C (r = 0.275) and the minimum temperature for growth (r = −0.169).

The distribution of the minimum temperature for growth was also demonstrated to have an impact on the mean concentration of *L. monocytogenes* in servings in Fritsch et al.'s [11] model. The ability to multiply in the cold was correlated with the presence of a genetic marker for cold adaptation. The authors looked for this marker in a collection of strains representative of smoked salmon at the distribution stage [44]. By defining two distributions of minimum temperature for growth for slow-growing strains and fast-growing strains, Fritsch et al. [11] showed that the mean exposure of the consumer was two times more important in the high-growth groups (25 CFU/g) compared to the low-growth groups (13 CFU/g). However, the importance of the type of strain on exposure was less significant for the risk of listeriosis than the presence of virulence markers.

In addition to the QRA model of Pouillot et al. [9,10], two other models ascertained that the maximum population density of *L. monocytogenes* is not a parameter of minor importance. Whereas Chen et al. [12] found a certain association with the risk recall (r = 0.053), EFSA BIOHAZ [1] determined that the listeriosis risk was very sensitive to the maximum population density, and quantified that a shift in 0.5 log CFU/g can double the estimated risk (Table 2). In order to obtain more precise estimates of the risk, it is therefore important to reduce the uncertainties regarding the characteristics of *L. monocytogenes*, including the parameters associated with the exponential growth rate, the nominal minimum temperature for growth, and the maximum population density.

4.5. Models' Availabiltity

The most widespread use of risk assessment approaches involves the sharing and description of models [45]. As we have recently shown for other foods [8], most QMRA models for seafood are often unavailable. Details of the software used and links describing the available models are provided in Table S1 of the Supplementary Material of this article.

5. Conclusions

Eighty-five percent of the QRA models focused on cold-smoked/gravad fish because of the considerable and continuous public health implications of the seafood products.

Despite the availability of *L. monocytogenes* data on stages such as filleting, brining, smoking, and slicing, none of the QRA models contained a processing module; therefore, cross-contamination events were not represented. Since secondary contamination can occur in the processing plants from equipment and environmental elements, the most relevant opportunities for cross-contamination during processing should be identified and modelled. Thus, in addition to cold chain (distribution and retail) and consumption stages, which have been represented by most QRA models, a future smoked/gravad fish model should also comprise a processing module that is robust enough to allow for an assessment of the following: (1) intervention strategies that can retard the growth of *L. monocytogenes*, such as the application of bacteriocins (nisin), suitable starter cultures, and/or organic acids; (2) control measures that can reduce the frequency of cross-contamination events, such as the implementation of more stringent controls for raw materials, environmental monitoring programs, and/or sanitation procedures. Given the strong evidence of the inhibitory effect of background microflora on *L. monocytogenes* in vacuum-packed RTE smoked fish, predictive microbiology models that describe microbial competition should be employed after the reduced-oxygen packaging stage. Furthermore, since the growth kinetic parameters of *L. monocytogenes* and microflora have been demonstrated to have a heavy influence on the estimated risk of listeriosis, efforts should be made to accurately model such growth kinetic parameters.

Supplementary Materials: The following supporting information can be downloaded at: https://www.mdpi.com/article/10.3390/foods13050716/s1: Table S1: Accessibility to published listeriosis QRA models for seafood products.

Author Contributions: Conceptualization, M.S. and U.G.-B.; methodology, M.S. and U.G.-B.; software, U.G.-B.; validation, V.C., J.D.O.M., L.G., M.S. and U.G.-B.; formal analysis, M.S. and U.G.-B.; investigation, V.C., L.G., M.S. and U.G.-B.; resources, M.S.; data curation, L.G., V.C. and U.G.-B.; writing—original draft preparation, U.G.-B., L.G. and M.S.; writing—review and editing, J.D.O.M., V.C. and U.G.-B.; visualization, V.C., L.G., M.S. and U.G.-B.; supervision, M.S. and U.G.-B.; project administration, J.D.O.M. and M.S.; funding acquisition, M.S. All authors have read and agreed to the published version of the manuscript.

Funding: This research received no external funding.

Data Availability Statement: Data are contained within the article. The BibTeX file containing records used in this systematic review is available.

Acknowledgments: WHO commissioned this critical review to serve as a background document for the FAO/WHO Joint Expert Meeting on Microbiological Risk Assessment (JEMRA), which was held in Rome, Italy, from 24–28 October 2022. The authors are grateful to the JEMRA experts* and the FAO/WHO secretariat** for their suggestions and feedback. (*) Hamzah Al-Qadiri, Sukhadeo B Barbuddhe, Ursula A. Gonzales-Barron, Heidy M. W. den Besten, Qingli Dong, Laurent Guillier, Claudia Guldimann, Ludwig Maximilian, Jovana Kovacevic, Alejandra Latorre, Yumiko Okada, Taran Skjerdal, Matthew J. Stasiewicz, and Paula Teixeira. (**) Juliana De Oliveira Mota, Akio Hasegawa, Christine Kopko, Jeffrey LeJeune, Moez Sanaa, and Kang Zhou.

Conflicts of Interest: The authors declare no conflicts of interest.

References

1. EFSA BIOHAZ Panel. Scientific Opinion on the *Listeria monocytogenes* contamination of ready-to-eat foods and the risk for human health in the EU. *EFSA J.* **2018**, *16*, 5134. [CrossRef]
2. EFSA. The public health risk posed by *Listeria monocytogenes* in frozen fruit and vegetables including herbs, blanched during processing. EFSA Panel of Biological Hazards (BIOHAZ). *EFSA J.* **2020**, *8*, 6092. [CrossRef]
3. EFSA and ECDC (European Food Safety Authority and European Centre for Disease Prevention and Control). The European Union One Health 2022 Zoonoses Report. *EFSA J.* **2023**, *21*, e8442. [CrossRef]
4. ECDC. Surveillance and Disease Data for Listeriosis. European Centre for Disease Prevention and Control. Available online: https://www.ecdc.europa.eu/en/all-topics-z/listeriosis/surveillance-and-disease-data/eu-summary-reports (accessed on 15 June 2022).

5. NORS. National Outbreak Reporting System Dashboard. Centers for Disease Control and Prevention, National Center for Emerging and Zoonotic Infectious Diseases (NCEZID). Available online: https://wwwn.cdc.gov/norsdashboard/ (accessed on 15 June 2022).
6. FDA-FSIS. *Quantitative Assessment of Relative Risk to Public Health from Foodborne Listeria monocytogenes among Selected Categories of Ready-to-Eat Foods*; Center for Food and Safety and Applied Nutrition, Food and Drug Administration, U.S. Department of Health and Human Services, and Food Safety and Inspection Service, U.S. Department of Agriculture: Washington, DC, USA, 2003; pp. 1–541.
7. Leclercq, A.; Kooh, P.; Augustin, J.-C.; Guillier, L.; Thebault, A.; Cadavez, V.; Gonzales-Barron, U.; Sanaa, M. Risk factors for sporadic listeriosis: A systematic review and meta-analysis. *Microb. Risk Anal.* **2021**, *17*, 100128. [CrossRef]
8. Gonzales-Barron, U.; Cadavez, V.; Guillier, L.; Sanaa, M. A critical review of risk assessment models for *Listeria monocytogenes* in dairy products. *Foods* **2023**, *12*, 4436. [CrossRef] [PubMed]
9. Pouillot, R.; Miconnet, N.; Afchain, A.-L.; Delignette-Muller, M.L.; Beaufort, A.; Rosso, L.; Denis, J.-B.; Cornu, M. Quantitative risk assessment of *Listeria monocytogenes* in French cold-smoked salmon: I. Quantitative exposure assessment. *Risk Anal.* **2007**, *27*, 683–700. [CrossRef] [PubMed]
10. Pouillot, R.; Goulet, V.; Delignette-Muller, M.L.; Mahé, A.; Cornu, M. Quantitative risk assessment of *Listeria monocytogenes* in French cold-smoked salmon: II. Risk characterization. *Risk Anal.* **2009**, *29*, 806–819. [CrossRef]
11. Fritsch, L.; Guillier, L.; Augustin, J.C. Next generation quantitative microbiological risk assessment: Refinement of the cold smoked salmon-related listeriosis risk model by integrating genomic data. *Microb. Risk Anal.* **2018**, *10*, 20–27. [CrossRef]
12. Chen, R.; Orsis, R.H.; Guariglia-Oropeza, V.; Wiedmann, M. Development of a modeling tool to assess ad reduce regulatory ad recall risks for cold-smoked salmon due to *Listeria monocytogenes* contamination. *J. Food Prot.* **2022**, *85*, 1335–1354. [CrossRef]
13. Pérez-Rodríguez, F.; Carrasco, E.; Bover-Cid, S.; Joffré, A.; Valero, A. Closing gaps for performing a risk assessment on *Listeria monocytogenes* in ready-to-eat (RTE) foods: Activity 2, a quantitative risk characterization on *L. monocytogenes* in RTE foods; starting from the retail stage. *EFSA Support. Publ.* **2017**, *2017*, EN-1252. [CrossRef]
14. Lindqvist, R.; Westöö, A. Quantitative risk assessment for *Listeria monocytogenes* in smoked or gravad salmon and rainbow trout in Sweden. *Int. J. Food Microbiol.* **2000**, *58*, 181–196. [CrossRef] [PubMed]
15. FAO/WHO. *Risk Assessment of Listeria monocytogenes in Ready-to-Eat Foods*; Technical Report; World Health Organization and Food and Agriculture Organization of the United Nations: Geneve, Switzerland, 2004; pp. 1–269.
16. Garrido, V.; García-Jalón, I.; Vitas, A.; Sanaa, M. Listeriosis risk assessment: Simulation modelling and "what if" scenarios applied to consumption of ready-to-eat products in a Spanish population. *Food Control* **2010**, *21*, 231–239. [CrossRef]
17. Gospavic, R.; Haque, M.; Leroi, F.; Popov, V.; Lauzon, H. Quantitative microbial risk assessment for *Listeria monocytogenes* in cold smoked salmon. *WIT Trans. Inf. Commun. Technol.* **2010**, *43*, PI563–PI572.
18. Dass, S. Exposure Assessment of *Listeria monocytogenes* in Vacuum Packed Cold-Smoked Salmon in the Republic of Ireland. Ph.D. Thesis, Technological University Dublin, Dublin, Ireland, 2011.
19. Bomfeh, K. Risk Assessment for *Listeria monocytogenes* in Traditionally Processed Fish from Informal Markets in Accra and Tema. Master's Thesis, University of Ghana, Accra, Ghana, 2011.
20. Pasonen, P.; Ranta, J.; Tapanainen, H.; Valsta, L.; Tuominen, P. *Listeria monocytogenes* risk assessment on cold smoked and salt-cured fishery products in Finland—A repeated exposure model. *Int. J. Food Microbiol.* **2019**, *304*, 97–105. [CrossRef] [PubMed]
21. Pouillot, R.; Hoelzer, K.; Chen, Y.; Dennis, S.B. *Listeria monocytogenes* dose response revisited—incorporating adjustments for variability in strain virulence and host susceptibility. *Risk Anal.* **2015**, *35*, 90–108. [CrossRef]
22. Buchanan, R.L.; Damert, W.G.; Whiting, R.C.; van Schothorst, M. An approach for using epidemiologic and microbial food survey data to develop a "purposefully conservative" estimate of the dose-response relationship between *Listeria monocytogenes* levels and the incidence of foodborne listeriosis. *J. Food Prot.* **1997**, *60*, 918–922. [CrossRef]
23. Haas, C.N.; Thayyar-Madabusi, A.; Rose, J.B.; Gerba, C.P. Development and validation of dose-response relationship for *Listeria monocytogenes*. *Quant. Microbiol.* **1999**, *1*, 89–102. [CrossRef]
24. Farber, J.M.; Ross, W.H.; Harwig, J. Health risk assessment of *Listeria monocytogenes* in Canada. *Int. J. Food Microbiol.* **1996**, *30*, 145–156. [CrossRef]
25. Jami, M.; Ghanbari, M.; Zunabovic, M.; Domig, K.J.; Kneifel, W. *Listeria monocytogenes* in aquatic food products—A review. *Compr. Rev. Food Sci. Food Saf.* **2014**, *13*, 798–813. [CrossRef]
26. Lachman, R.; Halbedel, S.; Luth, S.; Holzer, A.; Adier, M.; Pietzka, A.; Al Dahouk, S.; Stark, K.; Flieger, A.; Kieta, S.; et al. Invasive listeriosis outbreaks and salmon products: A genomic, epidemiological study. *Emerg. Microbes Infect.* **2022**, *11*, 1308–1315. [CrossRef]
27. Terentjeva, M.; Eizenberga, I.; Valcina, O.; Novoslavskij, A.; Strazdina, V.; Bērzins, A. Prevalence of foodborne pathogens in freshwater fish in Latvia. *J. Food Prot.* **2015**, *78*, 2093–2098. [CrossRef] [PubMed]
28. Johansson, T.; Rantala, L.; Palmu, L.; Honkanen-Buzalski, T. Occurrence and typing of *Listeria monocytogenes* strains in retail vacuum-packed fish products and in a production plant. *Int. J. Food Microbiol.* **1999**, *47*, 111–119. [CrossRef] [PubMed]
29. Tocmo, R.; Krizman, K.; Khoo, W.J.; Kim, M.; Yuk, H.-G. *Listeria monocytogenes* in vacuum-packed smoked fish products: Occurrence, routes of contamination and potential intervention measures. *Compr. Rev. Food Sci. Food Saf.* **2014**, *13*, 172–188. [CrossRef] [PubMed]

30. Eklund, M.W.; Poysky, F.T.; Paranjpye, R.N.; Lashbrook, L.C.; Peterson, M.E.; Pelroy, G.A. Incidences and sources of *Listeria monocytogenes* in cold-smoked fishery products and processing plants. *J. Food Prot.* **1995**, *58*, 502–508. [CrossRef]
31. Chen, B.Y.; Pyla, R.; Kim, T.J.; Silva, J.L.; Jung, Y.S. Prevalence and contamination patterns of *Listeria monocytogenes* in catfish processing environment and fresh fillets. *Food Microbiol.* **2010**, *27*, 645–652. [CrossRef] [PubMed]
32. Porsby, C.H.; Vogel, B.F.; Mohr, M.; Gram, L. Influence of processing steps in cold-smoked salmon production on survival and growth of persistent and presumed non-persistent *Listeria monocytogenes*. *Int. J. Food Microbiol.* **2008**, *122*, 287–295. [CrossRef]
33. Neunlist, M.R.; Ralazamahulco, M.; Cappelier, J.M.; Besnard, V.; Federighi, M.; Leroi, F. Effect of salting and cold-smoking process on the culturability, viability and virulence of *Listeria monocytogenes* strain Scott A. *J. Food Prot.* **2005**, *68*, 85–91. [CrossRef]
34. Aarnisalo, K.; Sheen, S.; Raaska, L.; Tamplin, M. Modelling transfer of *Listeria monocytogenes* during slicing of 'gravad' salmon. *Int. J. Food Microbiol.* **2007**, *118*, 69–78. [CrossRef]
35. Rorvik, L.M.; Caugant, D.A.; Yndestad, M. Contamination pattern of *Listeria monocytogenes* and other *Listeria* spp. in a salmon slaughterhouse and smoked salmon processing plant. *Int. J. Food Microbiol.* **1995**, *25*, 19–27. [CrossRef]
36. Cruz, C.D.; Silvestre, F.A.; Kinoshita, E.M.; Landgraf, M.; Franco, B.D.G.M.; Destro, M.T. Epidemiological survey of *Listeria monocytogenes* in a gravlax salmon processing line. *Braz. J. Microbiol.* **2008**, *39*, 375–383. [CrossRef] [PubMed]
37. Lundén, J.; Autio, T.J.; Sjoberg, A.M.; Kokeala, H.J. Persistent and nonpersistent *Listeria monocytogenes* contamination in meat and poultry processing plants. *J. Food Prot.* **2003**, *66*, 2062–2069. [CrossRef] [PubMed]
38. Di Ciccio, P.; Meloni, D.; Festino, A.R.; Conter, M.; Zanardi, E.; Ghidini, S.; Vergara, A.; Mazzette, R.; Ianieri, A. Longitudinal study on the sources of *Listeria monocytogenes* contamination in cold-smoked salmon and its processing environment in Italy. *Int. J. Food Microbiol.* **2012**, *158*, 79–84. [CrossRef] [PubMed]
39. Autio, T.; Hielm, S.; Miettinen, M.; Sjoberg, A.-M.; Aarnisalo, K.; Bjorkroth, J.; Mattila-Sandholm, T.; Korkeala, H. Sources of *Listeria monocytogenes* contamination in a cold-smoked rainbow trout processing plant detected by pulsed-field gel electrophoresis typing. *Appl. Environ. Microbiol.* **1999**, *65*, 150–155. [CrossRef] [PubMed]
40. Vogel, B.; Huss, H.; Ojeniyi, B.; Ahrens, P.; Gram, L. Elucidation of *Listeria monocytogenes* contamination routes in cold-smoked salmon processing plants detected by DNA-based typing methods. *Appl. Environ. Microbiol* **2001**, *67*, 2586–2595. [CrossRef] [PubMed]
41. Klaeboe, H.; Rosef, O.; Saebo, M. Longitudinal studies on *Listeria monocytogenes* and other *Listeria* species in two salmon processing plants. *Int. J. Environ. Health Res.* **2005**, *15*, 71–77. [CrossRef]
42. Nakamura, H.; Tokuda, Y.; Sono, A.; Koyama, T.; Ogasawara, J.; Hase, A.; Haruki, K.; Nishikawa, Y. Molecular typing to trace *Listeria monocytogenes* isolated from cold-smoked fish to a contamination source in a processing plant. *J. Food Prot.* **2006**, *69*, 835–841. [CrossRef]
43. Thimothe, J.; Nightingale, K.K.; Gall, K.; Scott, V.N.; Wiedmann, M. Tracking of *Listeria monocytogenes* in smoked fish processing plants. *J. Food Prot.* **2002**, *67*, 328–341. [CrossRef]
44. Hu, Y.; Gall, K.; Ho, A.; Ivanek, R.; Grohn, Y.T.; Wiedman, M. Daily variability of *Listeria* contamination patterns in a cold-smoked salmon processing operation. *J. Food Prot.* **2006**, *69*, 2123–2133. [CrossRef]
45. Filter, M.; Nauta, M.; Pires, S.M.; Guillier, L.; Buschhardt, T. Towards efficient use of data, models and tools in food microbiology. *Curr. Opin. Food Sci.* **2022**, *46*, 100834. [CrossRef]

foods

MDPI

Review

A Critical Review of Risk Assessment Models for *Listeria monocytogenes* in Meat and Meat Products

Ursula Gonzales-Barron [1,2,*], Vasco Cadavez [1,2], Juliana De Oliveira Mota [3], Laurent Guillier [4] and Moez Sanaa [3,*]

[1] Centro de Investigação de Montanha (CIMO), Instituto Politécnico de Bragança, Campus de Santa Apolónia, 5300-253 Bragança, Portugal; vcadavez@ipb.pt

[2] Laboratório para a Sustentabilidade e Tecnologia em Regiões de Montanha, Instituto Politécnico de Bragança, Campus de Santa Apolónia, 5300-253 Bragança, Portugal

[3] Department of Nutrition and Food Safety, World Health Organization (WHO), CH-1211 Geneva, Switzerland; deju@who.int

[4] Risk Assessment Department, French Agency for Food, Environmental and Occupational Health & Safety (Anses), 14 Rue Pierre et Marie Curie, 94701 Maisons-Alfort, France; laurent.guillier@anses.fr

* Correspondence: ubarron@ipb.pt (U.G.-B.); sanaam@who.int (M.S.); Tel.: +351-273-303-325 (U.G.-B.)

Abstract: A review of the published quantitative risk assessment (QRA) models of *L. monocytogenes* in meat and meat products was performed, with the objective of appraising the intervention strategies deemed suitable for implementation along the food chain as well as their relative effectiveness. A systematic review retrieved 23 QRA models; most of them (87%) focused on ready-to-eat meat products and the majority (78%) also covered short supply chains (end processing/retail to consumption, or consumption only). The processing-to-table scope was the choice of models for processed meats such as chorizo, bulk-cooked meat, fermented sausage and dry-cured pork, in which the effects of processing were simulated. Sensitivity analysis demonstrated the importance of obtaining accurate estimates for lag time, growth rate and maximum microbial density, in particular when affected by growth inhibitors and lactic acid bacteria. In the case of deli meats, QRA models showed that delicatessen meats sliced at retail were associated with a higher risk of listeriosis than manufacture pre-packed deli meats. Many models converged on the fact that (1) controlling cold storage temperature led to greater reductions in the final risk than decreasing the time to consumption and, furthermore, that (2) lower numbers and less prevalence of *L. monocytogenes* at the end of processing were far more effective than keeping low temperatures and/or short times during retail and/or home storage. Therefore, future listeriosis QRA models for meat products should encompass a processing module in order to assess the intervention strategies that lead to lower numbers and prevalence, such as the use of bio-preservation and novel technologies. Future models should be built upon accurate microbial kinetic parameters, and should realistically represent cross-contamination events along the food chain.

Keywords: systematic review; exposure assessment; simulation; deli meats; sausages; listeriosis

1. Introduction

Although invasive listeriosis is a rare disease, it is its high rate of mortality which makes it of significant public health concern worldwide. In 2021, listeriosis occupied the fifth place among the most frequent zoonoses in the European Union (EU), with 2183 confirmed cases in 27 EU Member States and with a high fatality rate of 13.7% [1]. The overall EU trend of listeriosis rates were fairly constant in the period 2017–2020; however, in 2021, the EU notification rate increased by 14% (this is, from 0.43 per 100,000 population in the year 2020, up to 0.49 per 100,000 population in the year 2021). According to the US Centers for Disease Control and Prevention [2], the reported incidence rates of listeriosis in the USA are significantly lower than those of the EU; nonetheless the USA data showed a growing trend between the years 2012/2013 and 2020. During this time span, the crude incidence rates increased from 0.23 to 0.27 cases per 100,000 population.

Reasons for the increase in the incidence of listeriosis could be as follows: (i) the increase in the share of the elderly population in the demographics of industrialised countries [3]; (ii) the greater purchase of convenience/RTE foods due to the limited time available for preparing meals at home [4]; and (iii) the increased consumption of high-risk RTE foods [5]. Consumers have increased preference for "trendy foods" or foods perceived as healthy, whose safety heavily relies on mild treatments, such as plant-based foods (e.g., sprouts, raw quinoa grains), seafood preparations containing raw/macerated ingredients (e.g., gravad fish), etc.

According to the EFSA report [1], the food vehicles implicated in the strong-evidence listeriosis outbreaks in 2021 belonged to the categories "Fish and fish products" (four outbreaks), "Broiler meat and products thereof" (one outbreak), and "Other mixed red meat and products thereof' (one outbreak). Thus, meat products and seafood continue to be important food vehicles, as they were in the decade 2010–2020, with a pooled share of 30.2% and 22.6%, respectively, of the total strong-evidence listeriosis outbreaks in the EU.

In this context, many quantitative risk assessment (QRA) models have focused on meat products—as important vehicles of transmission—to estimate the risk and incidence of invasive listeriosis linked to their consumption. The present study aims (i) to perform a critical review of the QRA models currently published for listeriosis linked to the consumption of meat and meat products; (ii) to discuss the relative effectiveness of the risk mitigation measures evaluated in the various QRA models as what-if scenarios; and (iii) to extract key messages and suggestions for future QRA models.

2. Materials and Methods

QRA models were retrieved through a literature search on Scopus and PubMed® published between 1 January 1998 and 18 May 2022 (date of the searches). The searches in title, keywords and abstract were carried out using logically connected terms (("risk assessment" OR exposure OR quantitative microbial OR risk modelling OR modeling OR simulation* OR second-order OR "second order" OR "risk management") AND ("*L. monocytogenes*" OR "*Listeria monocytogenes*" OR listeriosis)) properly arranged in the syntaxes of the literature search engines. The full systematic review and information extraction processes are described in Gonzales-Barron et al. [6]. QRA models conducted in any region of the world were included. The present review focuses only on meat products, which were the subject of 25 studies [7–31].

3. Results

Description of Collected QRA Models

Table 1 summarises the main features of the 24 QRA models of *Listeria monocytogenes* retrieved for meat or meat products, while Table 2 compiles the predictive microbiology models and summarised results from what-if scenarios and sensitivity analysis, excerpted from the models. Most of the models developed in the American continent represented the food production conditions of the USA (9). Other 4 models were developed in Argentina (2), Canada (1) and Chile (1). Five QRA models investigated the risk of listeriosis in Europe (Italy, Greece, Spain, France and EU), whereas the other three investigated in Asia (one from China and two from Malaysia). One QRA model pertained to the risk of listeriosis in the Australian population, whereas two models were not linked to any specific geographical location (Table 1).

Table 1. Characteristics of quantitative risk assessment models of *L. monocytogenes* from consumption of meat or meat products by scope.

Scope	Food	RTE	Cross-Conta-Mination	DR—End-Point	Type of DR Model	DR Sub-Populations	Strain Variability	Temp Profiles/ Lag Time	Country	Source
Processing-to-table	Deli meats: turkey, ham and beef—pre-packed and retail-sliced	Yes	Yes: Processing: a transfer coefficient modelled; and for retail-sliced meats, cross-contamination was modelled at retail	Exp—I	FAO/WHO [18]	High-risk/Low-risk	Variability of strains implicit in r	No/Yes	USA	Tang [7]
	HPP-treated chorizo	Yes	Yes: Processing—slicing—transfer coefficient	NA	NA	NA	No strain variability considered	No/No	Spain	Possas et al. [8]
	Bulk-cooked meat	Yes	Yes: Retail: from other products and from the retail environment	Exp—I	FAO/WHO [18]	High-risk/Low-risk	Variability of strains implicit in r	No/No	China	Sun et al. [9]
	Fermented sausage	Yes	No	Exp—I	Pouillot et al. [32]	Multiple	Virulence and host susceptibility explicit in r distribution	No/Yes	Argentina	Brusa et al. [10]
	Dry-cured pork shoulder	Yes	No	Exp—I	Pouillot et al. [32]	Multiple	Virulence and host susceptibility explicit in r distribution	No/Yes	Argentina	Brusa et al. [10]
End Process-to-table	Deli meats: ham, turkey and roasted beef	Yes	No	Mouse-Epi—I, D	FDA-FSIS [17]	Multiple	Virulence of different strains represented in DR	No/Yes	USA	Pradhan et al. [11]
	Luncheon meats, cooked sausages, pâtés	Yes	No	Exp—I	FAO/WHO [18]	High-risk/Low-risk	Variability of strains implicit in r; Some strain variability in LM growth rate	No/Yes	Australia	Ross et al. [12,13]
	Manufacture-(pre-packaged) and retai—sliced ham and turkey	Yes	No	Mouse-Epi—D	FDA-FSIS [17]	High-risk	Virulence of different strains represented in DR	No/Yes	USA	Pradhan et al. [14]
	Deli meats: RTE turkey, ham and roast beef	Yes	Yes: Retail	Exp—I	FAO/WHO [18]	High-risk/Low-risk	Variability of strains implicit in r	No/Yes	USA	Gallagher et al. [15,16]

Table 1. *Cont.*

Scope	Food	RTE	Cross-Conta-Mination	DR—End-Point	Type of DR Model	DR Sub-Populations	Strain Variability	Temp Profiles/ Lag Time	Country	Source
Retail-to-table	Processed meats: Frankfurters, fermented sausages, deli meats, pâté	Yes	No	Mouse-Epi—I	FDA-FSIS [17]	Multiple	Virulence of different strains represented in DR	No/No	USA	FDA-FSIS [17]
	Fermented meat	Yes	No	Exp—I	FAO/WHO [18]	High-risk/Low-risk	Variability of strains implicit in r	No/No	Non-specific	FAO-WHO [18]
	Hams	Yes	No	Exp—I	FAO/WHO [33]	Multiple	NA	No/No	Italy	Giovaninni et al. [19]
	Pre-packed deli meats/retail-sliced deli meats	Yes	No	Mouse-Epi—I, D	FDA-FSIS [17]	Multiple	Virulence of different strains represented in DR	No/No	USA	Endrikat et al. [20]
	RTE meat and poultry deli meat	Yes	No	Mouse-Epi—I, D	FDA-FSIS [17]	Multiple	Virulence of different strains represented in DR	No/No	USA	FSIS [21]
	Poultry and beef	No	Yes: Handling at home	Exp—I	FAO/WHO [18]	General	Variability of strains implicit in r	No/No	Chile	Foerster et al. [22]
	Retail delicatessens	Yes	Yes: Retail: series of events when serving deli meat	Exp—I	FAO/WHO [18]	High-risk/Low-risk	Variability of strains implicit in r	Yes/No	USA	Pouillot et al. [23]
	Delicatessen meats/hotdogs	Yes	Yes: Handling at home	Exp—I	Pouillot et al. [32]	Multiple	Virulence and host susceptibility explicit in r distribution	No/Yes	Canada	Falk et al. [24]
	Packaged heat-treated meat products (cooked meat, sausage, pâté)	Yes	No	Exp—I	Pouillot et al. [32]	Multiple	Challenge test data from a mixture of strains; Virulence and host susceptibility explicit in r distribution; h0 distribution of variability in physiological state of cells; LM DR modelling explicitly considers variability in strain virulence and in susceptibility across populations	Yes/Yes	Non-specific	Pérez-Rodríguez et al. [25]

Table 1. *Cont.*

Scope	Food	RTE	Cross-Conta-Mination	DR—End-Point	Type of DR Model	DR Sub-Populations	Strain Variability	Temp Profiles/Lag Time	Country	Source
Retail-to-table	Retail-sliced cooked meats	Yes	Yes: Retail: from slicing machine	Exp—I	Pouillot et al. [32]	High-risk/Low-risk	NA	No/No	Greece	Tsaloumi et al. [26]
	RTE cooked meat, RTE sausage, patê	Yes	No	Exp—I	EFSA BIOHAZ [11] based on Pouillot et al. [32]	Multiple (sex/age group)	Mixture of strains used in challenge test; Virulence and host susceptibility explicit in r distribution	No/No	Non-specific	EFSA BIOHAZ [27]
	RTE cooked meat	Yes	No	Exp—I	Pouillot et al. [34]	Multiple	NA	Yes/Yes	France	Duret et al. [28]
Consumption	Vacuum-packed and freshly sliced deli meats	Yes	Yes: Handling at home: transfer rates from refrigerator and from hands	Mouse-Epi—D	FDA-FSIS [17]	Intermediate-age population	Virulence of different strains represented in DR	No/No	USA	Yang et al. [29]
	Cooked chicken offal	No	No	Exp—I	FAO/WHO [18]; Lindqvist and Westöö [35]; FDA-FSIS [17]	Multiple	Strain diversity implicit in r	No/No	Malaysia	Kuan et al. [30]
	Cooked chicken offal	No	Yes: Handling at home: transfer rate to cooked samples	Exp—I	FAO/WHO [18]; Lindqvist and Westöö [35]; FDA-FSIS [17]	Multiple	Strain diversity implicit in r	No/No	Malaysia	Wai et al. [31]

DR: dose–response; Exp: exponential model; Mouse-Epi: Mouse-Epidemiological model; I: illness endpoint; D: death endpoint; NA: not available.

Table 2. Microbial kinetic models and main results related to scenarios and sensitivity analysis from quantitative risk assessment models of listeriosis acquired from consumption of meat or meat products.

Scope	Food	Predictive Microbiology Models	What-If Scenarios	Sensitivity Analysis	Model Complexity	Source
Processing-to-table	Deli meats: turkey, ham and beef—pre-packed and retail-sliced	Growth (linear, square root for EGR5)	(1) Maximum sampling frequency of 60 samples per month in small, medium and large facilities that produce two lots per day (60–60–60), reduces listeriosis cases by 15% in comparison to the no-testing baseline; (2) Sampling every lot in small, medium and large facilities (60–60–60L) reduces listeriosis cases by 12%; (3) Implementing 100% post-processing lethality (pasteurisation, UV) reduces 38% of the listeriosis cases; (4) Implementing 100% growth inhibiting packaging (1.5–3% lactate alone or in combination with 0.125–0.25% diacetate) reduces cases by 80%; (5) Post-processing lethality and growth inhibiting packaging reduces cases by 92.5%.	ND	Low	Tang [7]
	HPP-treated chorizo	For HPP processing (inactivation of LM as a function of a_w, pressure intensities and pressure holding time); Storage, retail, transport (survival: biphasic model, secondary model, survival rate as function of temperature)	(1) When no high-pressure processing (HPP) is applied, if LM in pork meat batter is below 1 log CFU/g, the prevalence of contaminated sausage packs at consumption is 3%; (2) When no HPP is applied, 2 log CFU/g LM in pork meat batter increases prevalence in sausage packs at consumption to 10%; (3) HPP at 600 MPa for 3 min reduces the prevalence of contaminated sausage packs at consumption by 90%; (4) HPP at 600 MPa for 6 min reduces prevalence of contaminated sausage packs by 97%; (5) HPP at 600 MPa for 9 min reduces prevalence of contaminated sausage packs by >99.9%; (6) If nitrites 150 ppm is eliminated (0 ppm) and HPP is applied at 600 MPa for 6 min, the prevalence of contaminated sausage packs increases by 66% (from 0.09% to 0.15%); (7) If nitrites 150 ppm is eliminated (0 ppm) and HPP is applied at 600 MPa for 9 min, the prevalence in sausage packs increases by 100% (from 0.01% to 0.02%).	ND	Low	Possas et al. [8]

Table 2. *Cont.*

Scope	Food	Predictive Microbiology Models	What-If Scenarios	Sensitivity Analysis	Model Complexity	Source
Processing-to-table	Bulk-cooked meat	Growth (linear, square root for EGR5)	(1) If LM at retail had a mean of −2.7 log CFU/g (instead of −1.4 log in the baseline), the final concentration at consumption would be 0 log CFU/g (instead of 0.20 log CFU/g; (2) If cooked meat was stored under unfavourable conditions that causes LM to go >−0.52 log CFU/g at retail, the final mean concentration at consumption would be 1.5 log CFU/g.	ND	Low	Sun et al. [9]
	Fermented sausage	Growth (both growth probability and GR as functions of temp, a_w, pH, LAC and LAB_cpd)	(1) When LAB culture was added, the final LM concentration was <100 CFU/g in 98.2% of the portions (as opposed to 73.7% when LAB was not added); (2) Higher values of final pH during fermentation led to higher risk of listeriosis. At pH values of 5.3, 5.5, 5.7 or 5.9, the calculated odds ratios were 1.03, 1.61, 1.97 or 2.52 times higher compared with pH < 5.1; (3) When a_w was $\leq$0.92 during ripening, the risk of listeriosis was 1.73 times lower than with $a_w \geq 0.93$. For increasing a_w (0.94, 0.95 and 0.96), the risk of listeriosis was 39, 56 and 60 times greater compared with $a_w \leq 0.92$.	**Output—risk of listeriosis**: (1) Use of LAB (r = −0.51), (2) Prevalence of LM in raw meat (r = 0.28), (3) Fermentation temperature (r = 0.24), (4) pH reached during fermentation (r = 0.24), (5) LAB counts in the fermented sausage (r = −0.16), (6) a_w of sausage (r = 0.13)	Low	Brusa et al. [10]
	Dry-cured pork shoulder	Growth (both growth probability and GR as functions of temp, a_w, pH, LAC and LAB_cpd)	(1) At $a_w \leq 0.93$, the listeriosis risk was 27 times lower compared with the product with a_w not reduced in the process.	**Output—risk of listeriosis**: (1) a_w reached during salting (r = 0.18), (2) Prevalence of LM in raw meat (r = 0.13), (3) Counts of LM in raw meat (r = 0.10), (4) Temperature at which salting was carried out (r = 0.02), (5) pH during salting (r = −0.01)	Low	Brusa et al. [10]
End Process-to-table	Deli meats: ham, turkey and roasted beef	Growth (linear model for growth; secondary models for EGR_5 °C, LPD_5 °C as square root models)	(1) The mean deaths and illnesses for ham and roast beef elaborated without GIs were 2.4- and 1.9-fold lower when lag phase was considered than those obtained without lag phase (mean lag phases of 5.9 and 5.1 days for ham and roast beef without GI); (2) If RTE products were formulated with GIs, the mean deaths in the elderly populations would reduce by factors of 7.8, 3.7 and 2.5 for RTE turkey, roast beef and ham, respectively.	ND	Low	Pradhan et al. [11]

Table 2. *Cont.*

Scope	Food	Predictive Microbiology Models	What-If Scenarios	Sensitivity Analysis	Model Complexity	Source
End Process-to-table	Luncheon meats, cooked sausages, pâtés	Growth (competiion growth model for LM and spoilage bacteria)	(1) A 90% reduction in LM prevalence leads to an 80% risk reduction. Similarly, a 67% reduction in prevalence leads to a 50% reduction in the risk; (2) Any treatment that reduces LM growth rate by 50% reduces the risk by 80–90%; (3) A 3- to 4-log reduction in initial LM concentration (i.e., heat treatment or HPP in the packaged product) results in a ~600-fold reduction in the annual listeriosis cases; (4) A milder post-processing listericidal treatment, assumed to reach a 1–2-log reduction, results in a 150-fold decrease in the annual listeriosis cases.	ND	Low	Ross et al. [12,13]
	Manufacture (pre-packaged) and retail-sliced ham and turkey	NS	(1) Home storage at temperatures >10 °C causes 17 and 32% of the estimated deaths linked to pre-packed ham without and with GIs, respectively, and 20 and 41% of the deaths associated with retail-sliced ham without and with GIs, respectively; (2) If the maximum temperature was limited to 7 °C, the median numbers of deaths would be reduced by 64% and 80% for pre-packed ham elaborated without and with GIs, respectively. The median numbers of deaths would be reduced by 62% and 79% for retail-sliced ham elaborated without and with GIs, respectively; (3) When the mean storage time was reduced from 28 to 16 days, the median numbers of deaths were reduced by 24%, 51%, 32% and 57% for pre-packed ham elaborated without and with GIs, and retail-sliced ham elaborated without and with GIs, respectively; (4) Limiting storage temperature and time to 10 °C and 16 days reduced the annual number of deaths by ~50% for pre-packaged and retail-sliced ham elaborated without GIs; (5) For products elaborated with GIs, the same combination reduced the annual number of deaths by ~75 and 90% for pre-packed and retail-sliced ham, respectively.	**PRODUCTION TO RETAIL PHASE:** **Output—LM at the end of retail**: Storage temperature had the strongest influence on LM growth ($r = 0.65$), followed by lag time at reference temperature ($r = -0.49$), storage time ($r = 0.33$), and growth rate at reference temperature ($r = 0.24$) **RETAIL PHASE TO CONSUMPTION** **Output—annual deaths in the elderly population**: (1) If maximum storage temperature between retail and consumption is reduced by 2 °C (baseline 21 °C), mean deaths are reduced from 13.5 to 6.5, (2) If maximum storage time between retail and consumption is reduced by 4 days (baseline 45 days), the mean deaths are reduced from 13.5 to 10, (3) If EGR 5 °C is reduced by 0.03 log/day; mean deaths are reduced from 13.5 to 11.5.	Low: Linked to model of Pradhan et al. (2009)	Pradhan et al. [14]

Table 2. *Cont.*

Scope		Food	Predictive Microbiology Models	What-If Scenarios	Sensitivity Analysis	Model Complexity	Source
End table	**Process-to-**	Deli meats: RTE turkey, ham and roast beef	Growth (linear model, EGR_5 °C for temperature)	(1) Reformulating all RTE products sold in delis with GIs reduces the mean listeriosis risk in the susceptible population by 95.2%; (2) Setting retail deli temperatures no higher than 5 °C reduces mean risk of listeriosis by 16.3%; (3) Shortening time in retail delis from 7 to 4 days has no effect on the mean risk of listeriosis; (4) A decrease in LM concentration on incoming RTE products by a factor of 2 would decrease the listeriosis risk of RTE foods prepared at retail by 10 to 24%; (5) Keeping all home refrigerators at temperatures <5 °C reduces mean risk by 99.8%; (6) Consuming all products within up to 3–4 days reduces mean risk by 99.0%.	ND	Medium: Different products considered with and without GI.	Gallagher et al. [15,16]
Retail-to-table		Processed meats: Frankfurters, fermented sausages, deli meats, pâté	Growth (linear model, square root model for EGR)	(1) Reducing the home storage time for deli meats from 28 days (baseline) to 14 days decreases the median listeriosis cases in the elderly population from 228 to 197 (13.6%); and reducing home storage time to 10 days further decreases the cases to 154 (32.5%); (2) Eliminating storage above 8 °C or all storage times longer than 8 days, or combination of maximum 10 °C and maximum 11 days, led to a reduction in listeriosis cases of 50%; (3) Inclusion of a pre-retail lethal intervention in deli meats that produced a 1 log reduction in contamination at the start of retail would reduce the predicted deaths in the elderly population by nearly 50% (from 227 to 120 in the elderly population). Reducing contamination 2 logs would result in a 74% reduction.	ND	Medium: Different products considered	FDA-FSIS [17]
		Fermented meat	Growth (probability of growth, linear growth, GR at Tref), Survival (inactivation model as a function of lactic acid, salt and nitrate)	ND	ND	Medium: more complex predictive microbiology models	FAO-WHO [18]

Table 2. *Cont.*

Scope	Food	Predictive Microbiology Models	What-If Scenarios	Sensitivity Analysis	Model Complexity	Source
Retail-to-table	Hams	Growth (linear model, secondary model for GR as an empirical function of a_w, storage temperature)	(1) During the first 6 days of storage at 10 °C, the expected increase in the listeriosis risk is 1 log; afterwards, the expected increase is around 1 log every 2 days; (2) If storage is at 4 °C, the risk increases in 1 log after 16 days; subsequent increases in 1 log would happen every week	ND	Low	Giovaninni et al. [19]
	Pre-packed deli meats/retail-sliced deli meats	Growth (linear model, sqrt model for GRs with and without inhibitors)	(1) The relative risk, on a per annum basis, of deli meats sliced at retail versus sliced in plants is 4.89; (2) In the elderly, use of GI in pre-packed deli meats reduces the mean annual deaths by 55%; (3) In the elderly, use of GI in retail-sliced deli meats reduces the mean annual deaths by 82%.	Regression tree analysis showed the most important determinants of risk are age of consumers, slicing location (i.e., retail or pre-packed) and presence of growth inhibitor.	Medium: The sensitivity analysis was carried out through the regression trees methodology; this model revised the FSIS-FDA (2003) model	Endrikat et al. [20]
	RTE meat and poultry deli meat	Growth (linear model, sqrt model for GRs with and without inhibitors)	(1) The formulation with GI in the elderly reduces annual death cases by 78%; (2) Retail-sliced deli meats present annual death cases 80% higher than those of pre-packaged deli meats; (3) An increase in the shelf-life from 10 to 40 days decreases the annual deaths by 13% (due to proper use of effective GIs that reduce deaths); (4) None of the simulated deaths are linked to the GI product.	Regression tree analysis showed the most important determinants of risk are age of consumers, slicing location (i.e., retail or pre-packed) and presence of growth inhibitor	Medium: The sensitivity analysis was carried out through regression	FSIS [21]
	Poultry and beef	Growth (linear model sqrt for growth), Inactivation (Bigelow model)	(1) Increasing mean storage temperature from 8 to 10 °C of meat stored for 3 days after cooking increases risk of listeriosis 62-fold for beef and 60-fold for poultry	ND	Low	Foerster et al. [22]

Table 2. *Cont.*

Scope	Food	Predictive Microbiology Models	What-If Scenarios	Sensitivity Analysis	Model Complexity	Source
Retail-to-table	Retail delicatessens	Growth (Yule's model, secondary model for GR as a function of temperature, a_w, nitrites, LAC and diacetate)	(1) When the the highly contaminated RTE food does not support the growth of LM, the predicted absolute risk increases two-fold in the susceptible population; (2) The risk from stores that have a highly contaminated RTE food that supports growth of LM is 6 times higher than the risk from stores that have a highly contaminated RTE food that does not support the growth of LM; (3) Most of the increase in the risk of products from highly contaminated stores results from cross-contamination to RTE foods that supports growth. (4) (i) Retail delis without niches and retail delis that control temperature produce lower listeriosis risk; and (ii) retail delis with incoming highly contaminated RTE foods (in particular, if they support growth), or retail delis with niches, produce higher listeriosis risk.	(1) When more LM cells enter the retail deli environment, the risk increases, regardless of whether these LM cells come from niches in the environment or from LM cells on incoming RTE food from the manufacturer; (2) High frequency of cross-contamination events (daily versus weekly) has greater impact than higher LM counts per cross-contamination event (100 versus 1000 CFU per contamination event).	High: Models cross-contamination in a deli grocery establishment by a discrete-event framework representing transfer from object to object, from food to object and from object to food	Pouillot et al. [23]
	Delicatessen meats/hotdogs	Growth (linear model, secondary model for EGR at 5 °C), Inactivation (Bigelow for death rate)	(1) The use of GIs led to a 110-fold reduction in the median cases of listeriosis for turkey deli meat, 78-fold for ham delicatessen meat, 56-fold for beef delicatessen meat and 49-fold for hotdogs; (2) Lowering the initial mean contamination levels from 75 CFU/g to 1 CFU/g resulted in corresponding 952-, 279-, 381- and 116-fold reductions, respectively.	Output—listeriosis cases: (1) consumer refrigerator temperature, (2) Consumer storage time, (3) EGR, (4) Retail storage temperature, (5) Temperature prior to retail (all the above with r = 0.30–0.43), (6) Lag time (−0.13 to −0.27)	Low: However, lag considers time elapsed between end processing and retail	Falk et al. [24]
	Packaged heat-treated meat products (cooked meat, sausage, pâté)	Growth (Baranyi with Jameson effect due to LAB, EGR5 secondary model and effect of lactate)	(1) Decreasing by 1–2 °C in the dynamic temperature profiles reduced cases of listeriosis per million servings by up to 37%; (2) Reducing maximum initial LM concentration by 2 log CFU/g decreases cases by 89%; (3) Decreasing time to consumption by 25% decreases cases by up to 38%; (4) Including lag time in the model reduces cases by 57%	ND	Medium: Time–temperature profiles from retail to consumption, and microbial competition models solved with 4th order Runge–Kutta algorithm	Pérez-Rodríguez et al. [25]
	Retail-sliced cooked meats	Growth (competition growth model LM- LAB, secondary model for GR as a function of pH, a_w, nitrites)	(1) Setting a use-by date of 14 days from the time of slicing decreases the median annual listeriosis cases from 7 (no use-by date) to 0; (2) Reducing consumers' storage temperature from a mean of 6 °C to 5 °C reduces the median listeriosis cases from 7 to 0.	ND	Low	Tsaloumi et al. [26]

Table 2. *Cont.*

Scope	Food	Predictive Microbiology Models	What-If Scenarios	Sensitivity Analysis	Model Complexity	Source
Retail-to-table	RTE cooked meat, RTE sausage, patê	Growth (Rosso model, EGR 5 °C)	(1) Across the 3 meat products, there is no strong difference in the probability of a product exceeding 100 CFU/g at the point of consumption between normal packaging (0.0672–0.0691) and reduced-oxygen packaging (0.0654–0.0678)	Risk is very sensitive to MPD. A shift in 0.5 log CFU/g can double the estimated risk. However, sensitivity analysis was conducted grouping various RTE food classes.	Low: Generic model; demands some knowledge in R software to utilise it	EFSA BIOHAZ [27]
	RTE cooked meat	Growth (Baranyi, Jameson competition growth model LM- LAB, secondary model for GR as a function of T, pH, a_w, CO_2, nitrites)	(1) Current cold-chain operating conditions; (2) Home refrigerator thermostat set at 4 °C; (3) Home refrigerator thermostat set at 7 °C; (4) Better thermal insulation of the refrigerator; (5) Lower air curtain flow rate in the display cabinet (50%); (6) Lower air curtain flow rate in the display cabinet (75%); (7) Higher air curtain flow rate in the display cabinet (125%); (8) Thermostat set in the display cabinet at 2.5 °C	Sensitivity analysis of the exposure model was carried out in Duret et al. [36]	High: the model, together with liste-riosis risk, assessed the energy consumption and the spoilage at time of consumption. DALYs and costs are used to express the results, and ranking of scenarios is carried out using multicriteria decision analysis	Duret et al. [28]
Consumption	Vacuum-packed and freshly sliced deli meats	Growth (linear model, secondary model for EGR)	(1) For initial LM at retail, levels lower than −2 log CFU/g did not affect mean mortality (death/serving); (2) After 10 days of storage at a mean temperature of 3 °C, there was no increase in risk (maximum risk reached) because the mMPD was attained; (3) Risk increased only when deli meats were kept for more than 18 h at ambient temperature; (4) Reducing maximum storage temperature appeared to be more effective at reducing risk than reducing refrigeration time for deli meats; (5) Initial contamination levels at retail had stronger impact on listeriosis risk than cross-contamination in the home.	**Output—mortality in intermediate age population**: LM level at retail (r = 0.29), Repeated use of leftovers (r = 0.17), EGR5 (r = 0.12), Refrigeration temperature (r = 0.09), Serving size (r = 0.09), Refrigeration time (r = 0.03)	Low	Yang et al. [29]
	Cooked chicken offal	0	ND	ND	Low	Kuan et al. [30]
	Cooked chicken offal	0	ND	ND	Low	Wai et al. [31]

GI: growth inhibitors; a_w: water activity; LPD: lag phase duration; GR: growth rate; EGR_x: exponential growth rate at x °C; LAB: lactic acid bacteria; LAC: lactic acid concentration; r: Pearson's correlation coefficient; ND: not done; Bold: It is used to make a distinction for the grouping of the scopes of models.

Only three models investigated the risk of listeriosis from non-RTE foods; namely, poultry and beef [22] and chicken offal [30,31]. The other 21 models focused on RTE processed meats, of which 12 were explicitly deli meats. The remaining eight pertained to fermented sausages (×3: [8,10,18]), bulk-cooked meat [9], dry-cured pork shoulder [10], miscellaneous processed meats [17], and packaged heat-treated meats [25,27].

From the 12 deli meats models, four of them [7,14,20,28] compared the risk of listeriosis between manufacture pre-packed and retail-sliced deli meats (Table 1).

The majority of the listeriosis QRA models for meat products considered contamination pathways from end of processing or retail to table (15/23), with 10 of them investigating deli meats. QRA models of longer scope (processing-to-table) were reserved mostly for processed meats rather than deli meats (five models), namely, high-pressure-processed chorizo, bulk-cooked meat, fermented sausage and dry-cured pork shoulder, in which the effects of processing stages were characterised. Only one QRA model for deli meats covered the processing-to-table scope [7]. Three models consisted only of the consumption module, from which one pertained to deli meats [29]. No single QRA model included a primary production module (Table 1).

Processed meat products may be contaminated with *L. monocytogenes* at different stages: either raw materials are contaminated and processing stages are unable to reduce pathogen's populations; or contact with contaminated raw materials contaminates surfaces or operators, which may take place at any phase between the meat processing plant and the consumer's home. For such reasons, nearly half of the QRA models (11/23) attempted to characterise cross-contamination, with modules placed during food processing [7,8], at retail [7,9,15,16,23,26,28] and during handling at home [22,24,29,31] (Table 1). Except for Pouillot et al. [23], who developed a rather complex discrete event approach to model cross-contamination at retail, simple transfer coefficients were applied in the QRA models to depict cross-contamination during processing (i.e., from slicing machine), at retail (i.e., from slicing machines and from other environmental elements) and at home (i.e., from fridge, hands and to cooked meat).

The QRA models retrieved varied in degree of architectural complexity. For the estimation of the exposure or the risk, only 18% of the models established a separation between variability and uncertainty [9,14–17]. All QRA models, except two [30,31], employed predictive microbiology models, including microbial growth, survival and competition models (Table 2). The lag phase duration of *L. monocytogenes* was not represented in 14 models [8,9,17–23,26,27,29–31], whereas only 3 QRA models [23,25,28] employed time–temperature trajectories to more realistically estimate the growth of *L. monocytogenes* along the storage process (Table 1).

Five QRA models considered mortality, apart from illness, as endpoint for risk estimation, having in common that all of them used the dose-response model from FDA-FSIS [11,14,20,21,28]. In one model, no risk estimation was carried out, since authors were interested in modelling only exposure assessment [8]. The exponential "single-hit" dose function was the choice for all models, of which the most common approach was that of FAO/WHO [18] (10 models), followed by FDA-FSIS [17] (8 QRA models), Pouillot et al. [32] (5 QRA models), Lindqvist and Westoo [35] (2 QRA models) and Pouillot et al. [34]. EFSA BIOHAZ [27] proposed a dose–response model based on that of Pouillot et al. [32] but stratified by gender and age classes instead of by consumers' pre-conditions.

Most of the QRA models assessed the impact of what-if scenarios, which can be understood as risk factors or intervention strategies (20 models); sensitivity analysis on *L. monocytogenes* counts, on dose per serving or on listeriosis risk as response variables was performed in 9 models (Table 2).

4. Discussion

4.1. Risk Factors and Control Measures Assessed at Processing

According to the systematic review conducted by Gonzales-Barron et al. [6], out of 65 QRA listeriosis QRA models retrieved since 1998, most of them focused on meat and meat products (35.4%)—followed by dairy products (27.7%). The scope of the listeriosis QRA models linked to meats begins upstream in the supply chain, as processing or end-of-processing/retail, and not earlier, despite farms and slaughtering being important sources of contamination of meats and meat products [37–40]. This may happen because many tracking survey studies have shown that the contamination of the final meat products commonly occurs in the slaughterhouse or at a retail level, and rarely directly from food-producing animals (i.e., faecal contamination). As early as in 1996, Nesbakken et al. [41] concluded that post-slaughtering processing was a significant source of contamination, and that it was in the cutting room environment where contamination is amplified. Later, Kathariou [42] was more conclusive, stating that, in general, the primary source of contamination before handover to consumers was the processing stage. Thévenot et al. [43] detected both persistent and sporadic strains of *L. monocytogenes* in meat processing environments, with a high genotype diversity. Therefore, even if *L. monocytogenes* enters the processing plant at low levels, some strains may survive in biofilms, persist in the processing environment and contribute to both environment and RTE meat product contamination and recontamination.

For RTE meat products, the most frequently applied hurdles are the use/application of growth-inhibitors (GI) (i.e., nitrites, lactate, diacetate, etc.), biopreservatives (i.e., lactic acid bacteria (LAB) cultures), post-process lethality treatments (i.e., in-package pasteurisation, irradiation, high-pressure processing, etc.), vacuum-packaging and cold storage. The effectiveness of a post-process lethality treatment, the use of GI and sampling testing at the end of processing was evaluated in the only QRA model for deli meats covering processing to retail [7]. This research found that the combination of the two hurdles—post-process lethality treatment (pasteurisation or UV) and the application of GI packaging (1.5–3.0% lactate alone or in combination with 0.125–0.25% diacetate)—was more effective in reducing the annual cases of listeriosis (92.5% reduction) than when used alone (38% or 80% reduction, respectively), and far more effective than implementing a verification sampling of 60 samples per month in small, medium and large facilities (15% reduction) (Table 2). Such scenarios were tested conforming to FSIS [44] rules to effectively control *L. monocytogenes* in RTE foods.

Other hurdles applied to meat products different than deli meats were evaluated in processing-to-retail QRA models, such as the application of high-pressure processing (HPP) in chorizo [8], the addition of LAB culture in fermented sausage [10] and the effect of lower water activity during ripening (longer dehydration) in fermented sausage and dry-cured pork shoulder [10]. Possas et al. [8] demonstrated that HPP was an efficient intervention to reduce the prevalence of contaminated chorizo packs, although it depended on the application time. HPP at 600 MPa for 3, 6 or 9 min reduced the mean prevalence of *L. monocytogenes* in contaminated chorizo packs at consumption by 90, 97 or >99.9%, respectively, when initial contamination in the batter was below 100 CFU/g. Nonetheless, they showed that, if 150 ppm nitrites were removed from the chorizo formulation, and HPP was applied at 600 MPa for 6 min or 9 min, the resulting mean prevalence of contaminated chorizo packs would increase by 66% (from 0.09% to 0.15%) or 100% (from 0.01% to 0.02%), respectively, in comparison to those scenarios where HPP was applied and nitrites were kept. Other scenarios relative to alterations of processes were evaluated by Brusa et al. [10]. They estimated that, when a certain LAB cocktail is added to the fermented sausage formulation, the final concentration of *L. monocytogenes* was lower than 100 CFU/g in 98.2% of the sausages, as opposed to the 73.7% attained when LAB was not added. They also predicted a relationship between the final pH of the sausage and the odds ratios of the risk of listeriosis. As the final pH decreased from 5.9 to 5.7, 5.5 and 5.3, the odds of acquiring listeriosis diminished from 2.52 to 1.97, 1.61 and 1.03 times higher than the odds of acquiring listeriosis from sausages of pH = 5.1. Greater dehydration was also an effective hurdle; decreasing the a_w of fermented sausages during ripening to $\leq$0.92 also reduced the risk of listeriosis in a magnitude 1.7 times lower than sausages with $a_w \geq 0.93$, whereas decreasing the a_w of dry-cured pork shoulder by salting to values ≤ 0.93 reduced the risk of listeriosis to a greater extent (27 times lower). In the sensitivity analyses for both fermented sausage and dry-cured pork shoulder, the prevalence of *L. monocytogenes* in raw meat was associated with the risk of listeriosis (r = 0.28 and 0.13, respectively); however, whereas, in the fermented sausage, variables related to pH drop (i.e., the use of LAB, fermentation temperature, pH reached during fermentation, LAB concentration in the fermented sausage) were more important than the water activity of the final sausage, for the dry-cured pork shoulder, the water activity during salting was the most correlated with listeriosis risk.

Six of the QRA models [11,14–16,20,21,24] assessed the impact of GIs in deli meats, more specifically the combined application of lactate and diacetate, which have been long recognised as capable of suppressing pathogenic growth in foods with neutral pH. Their effectiveness in reducing listeriosis risk has been shown to be variable. The FSIS QRA model [21] estimated that the formulation of RTE meat and poultry deli meats with GI would reduce the mean annual death cases by 78% in the elderly; Endrikat et al. [20] predicted that the use of GI in pre-packed deli meats and in retail-sliced deli meats would reduce the mean annual deaths by 55% and 82%, respectively, in the elderly population. Pradhan et al. [11] stated that products formulated with GIs would decrease the mean annual deaths in the elderly 7.8-, 3.7- and 2.5-fold for RTE turkey, roast beef and ham, respectively, whereas Falk et al. [24] estimated greater reductions in the median cases of listeriosis, namely, 78-fold for ham delicatessen meat, 56-fold for beef delicatessen meat and 49-fold for hotdog, when they were formulated with GIs. Gallagher et al. [15,16] estimated that, if all RTE products sold in delis (RTE turkey, ham and roast beef) were reformulated with GI, the mean risk of listeriosis would be reduced by 95.2% in the susceptible population.

Many QRA models demonstrated, through what-if scenarios, how greater risk reductions can be attained by reducing the initial prevalence and/or counts of *L. monocytogenes* in foods at the end of processing or retail [12,13,17,24,25,29]. In FDA-FSIS [17], higher reduction levels, rather than decreasing storage time and storage temperature, were obtained by reducing the initial concentration of *L. monocytogenes* at the start of retail in processed meats, which they hypothesised could be achieved through a processing-level lethal intervention. This QRA model estimated that reducing the initial mean contamination by 1.0 log would reduce the number of deaths in the elderly population by 50%, whereas reducing contamination in 2.0 logs would result in a 74% reduction. Many years later, in the QRA model built by Pérez-Rodríguez et al. [25], a comparable level of reduction in listeriosis cases (89% reduction) was obtained when the maximum initial concentration of *L. monocytogenes* in packaged, heat-treated meat products was reduced in 2.0 logs. Such a reduction rate was greater than those of the scenarios of decreasing the storage temperature and time to consumption. Ross et al. [12,13], simulating what would be the application of a milder listericidal treatment during processing that would achieve 1–2 log mean reduction in *L. monocytogenes*, calculated a 150-fold decrease in the mean predicted annual listeriosis from luncheon meats, cooked sausages and pâtés. Simulating a greater reduction in *L. monocytogenes* initial counts of 3–4 logs, they estimated a ~600-fold decrease in the annual cases of listeriosis. Comparable high levels of risk reduction were found by Falk et al. [24]—952-, 279-, 381- and 116-fold decreases in the median cases of listeriosis for turkey deli meat, ham delicatessen meat, beef delicatessen meat and hotdog, respectively—if the initial mean concentration of *L. monocytogenes* at retail would be reduced from 75 CFU/g to 1 CFU/g. Furthermore, Yang et al. [29] also corroborated that the initial contamination level at retail (r = 0.29) was the main driver of mortality in the intermediate-age population due to the consumption of vacuum-packed and freshly sliced deli meats, a much stronger factor than storage temperature (r = 0.09) and storage time (r = 0.03) (Table 2).

4.2. Risk Factors and Control Measures at Retail and Home

Reducing storage temperature had a greater effect on reducing the risk of listeriosis than reducing storage time. In Pradhan et al. [14], the scenario of reducing the maximum storage temperature to 7 °C reduced the median number of deaths by 64% and 80% for pre-packaged ham elaborated without and with GIs, respectively, and by 62% and 79% for retail-sliced ham elaborated without and with GIs, respectively. In contrast, when the mean storage time was decreased from 28 to 16 days, the median numbers of deaths for the above products were reduced, to a lesser extent, by 24%, 51%, 32% and 57%, respectively.

Accordingly, these authors estimated that the concentration of *L. monocytogenes* at the end of retail was mainly driven by storage temperature (r = 0.65), followed by the lag time (r = −0.49) and the storage time (r = 0.33). A sensitivity analysis on listeriosis cases performed by Falk et al. [24] also showed that the listeriosis cases from deli meats was more affected by the consumer's refrigerator temperature than by storage time (r not provided), whereas, in Yang et al. [29], the consumers' refrigeration temperature (r = 0.09) and the refrigeration time (r = 0.03) were both poorly correlated with mortality in the intermediate-age population.

In a QRA model on deli meats, at the stage of retail, Gallagher et al. [15,16] showed that, during retail, setting the deli temperature no higher than 5 °C would reduce the mean risk of listeriosis in the susceptible population by 16.3%, whereas shortening the time in retail delis from 7 to 4 days would have no effect on the mean risk of listeriosis. Likewise, at the consumer level, keeping home refrigerators at temperatures lower than 5 °C would reduce the risk by 99.99%, whereas consuming the deli meats within 4 days after purchase would reduce the mean risk by 99.0%. On the other hand, interestingly, there were two QRA models [25,26] where decreasing the storage temperature produced a similar effect on the final risk as reducing the storage time. In Pérez-Rodríguez et al. [25], a decrease of 1–2 °C in the dynamic temperature profiles, and a decrease in the time to consumption of 25%, led both to a 37–38% reduction in the cases of listeriosis per million servings of packaged heat-treated meat products, whereas, in Tsaloumi et al. [26], setting a use-by date of 14 days from the time of slicing, or reducing consumers' storage temperature from a mean of 6 °C to 5 °C, would reduce the median cases of listeriosis from seven (baseline scenario of no use-by date and mean of 6 °C storage) to zero. Duret et al. [28] explored the link between the cold chain for cooked ham (including transport, supermarket cold storage, display cabinets, consumer transport and home refrigerator) and the associated listeriosis risk, together with the food wasted due to spoilage bacteria (LAB) and the cold-chain electricity consumption. A set of eight intervention strategies was tested to assess their effect on the three criteria investigated: food safety, food waste and energy consumption. The results showed that changing the thermostat of the home refrigerator has a high effect on listeriosis risk and food waste, but a limited effect on electricity consumption. Conversely, changing the airflow rate in the cabinet has a significant effect on electricity consumption but a negligible impact on listeriosis risk and food waste.

4.3. Cross-Contamination

Multiple studies have shown that delicatessen meats sliced in retail tend to have a higher level of bacterial contamination than deli meats sliced in factories [21,45,46]. The QRA model for pre-packed and retail-sliced deli meats from Endrikat et al. [20] estimated a relative risk of deli meats sliced at retail versus sliced in factories of 4.89 in the elderly population. A model published by FSIS [21] also found a similar result: retail-sliced deli meats presented annual death cases 80% higher than pre-packed deli meats. Both studies showed, by means of regression trees analysis, that the most important determinant of risk, after age of consumers, was slicing location (whether retail-sliced or pre-packed) and, according to them, this was more determinant than the presence of GI. Likewise, Pradhan et al. [14] demonstrated that the malpractice of storing in home refrigerators at temperatures higher than 10 °C accounts for ~17% and 32% of the predicted deaths associated with pre-packaged ham without and with Gis, respectively, whereas higher estimates of ~20% and 41% of the deaths are associated with retail-sliced ham without Gis and with Gis, respectively (Table 2).

These recurrent findings have suggested that the preparation of such RTE foods in retail shops augments the risk of contamination. In the retail environment, there are sources of contamination or cross-contamination, routes of spreading of *L. monocytogenes* and potential problems with hygiene monitoring. Pouillot et al. [23] developed a discrete-event modelling framework of cross-contamination from object to object, from food to object and from object to food; they assumed the presence of *L. monocytogenes* niches or harbourage sites allowing the release, at a given frequency, of a given number of cells to a site (i.e., utensils, slicers, food contact surfaces, scales, sinks, handles, floors, etc.). The model demonstrated that, (1) when more *L. monocytogenes* cells enter the retail environment, the risk is increased, regardless of the origin of the cells (from niches in the retail environment or from cells on incoming RTE food from the manufacturer) and that (2) a high frequency of cross-contamination events (daily versus weekly) has more impact than a high concentration of *L. monocytogenes* per cross-contamination event (100 versus 1000 CFU). Pouillot et al. [23] predicted that the risk of products from retail shops, with highly contaminated incoming RTE food that supports *L. monocytogenes* growth, is six times higher than the risk from shops that have an equally highly contaminated incoming RTE food, yet one that does not support growth. This means that most of the increase in the risk of products from these deli shops arises from cross-contamination to RTE foods supporting the growth of *L. monocytogenes*. Therefore, both the elimination of *L. monocytogenes* niches in retail deli establishments and efficient temperature control are crucial in attaining a lower risk of listeriosis. Yang et al. [29] indicated that cross-contamination at home has a lower contribution to listeriosis risk than the contamination levels that can be attained during retail.

4.4. *L. monocytogenes Lag Phase as a Driver of Risk Estimation*

Finally, some QRA models have assessed the impact of the lag phase duration of *L. monocytogenes* on the final risk. The lag phase assumption is of particular importance in the exposure assessments of foods, such as meat products, that are formulated with compounds or additives that act by extending the lag phase duration of microorganisms. In their QRA model for deli meats, Pradhan et al. [11] illustrated the importance of including the lag phase duration, by indicating that the mean numbers of deaths and illnesses for ham and roast beef formulated without Gis were 2.4- and 1.9-fold lower when lag phase was considered, than those obtained without lag phase (long mean lag phases of 5.9 and 5.1 days assumed for ham and roast beef without GI, respectively). Later, the same authors [14] estimated that lag time was the second most important variable ($r = -0.49$) driving *L. monocytogenes* concentration at the end of retail, just after retail storage temperature. Likewise, whereas, in the QRA model of Falk et al. [24] for delicatessen meats and hot dogs, the lag time duration was a moderate driver ($r = -0.13$ to -0.27) of the number of listeriosis cases, Pérez-Rodríguez et al. [25] quantified a reduction of 57% in the number of cases of listeriosis per million servings if the lag time was included in the QRA model for packaged heat-treated meat products. Conversely, using a sensitivity analysis, Duret et al. [34] showed that the latency modeling approach (population versus individual) was not an important parameter for exposure. The authors chose the simplest approach for estimating risk [28].

4.5. *Model Availability*

Sharing risk assessment models is crucial to ensure the transparency of the methodology and facilitate reusability. This holds particular significance in the realm of scientific research, where there is a growing emphasis on reproducibility and open science [47]. By sharing these models, researchers enable others to scrutinise their work, detect potential biases and use the models for their own datasets. Beyond transparency, the sharing of models streamlines reusability, a boon for researchers who may lack the resources to develop their own models [48].

Among the studies published, two grant access to scripts or spreadsheets [25,27], while two propose sharing the utilised models upon request [23,28] (details regarding model-sharing characteristics are available in the Supplementary Material of this article). For the other models, there is no indication of their availability. Several studies refer to a site that is no longer maintained [11,12,17], highlighting the challenge of reproducibility over time. As software evolves and resources vanish, such as the need for website maintenance, reproducing calculations becomes increasingly challenging [49].

5. Conclusions

None of the 23 QRA models retrieved simulated primary production or slaughtering; their scope mostly focused on end-of-processing or retail-to-consumption, although they indirectly evaluated the impact of growth inhibitor and lethal treatments on the final risk estimate. Most of the QRA models were carried out for RTE meat products such as deli meats, since these products support the growth of *L. monocytogenes*, have a long shelf-life and are very susceptible to cross-contamination in the processing and retail environments. The outputs of the QRA models agreed that deli meats sliced at retail sites lead to higher risks of listeriosis than manufacture pre-packed deli meats. Cross-contamination events are represented in approximately half the QRA models retrieved. However, cross-contamination modelling should not be overlooked, since *L. monocytogenes* has been frequently detected in processing and retail environments, revealing that persistent strains could be isolated from contact surfaces even after cleaning and disinfection, and even with recovery times of up to three years in the meat processing environment. QRA models widely agreed on the fact that controlling (reducing) the initial concentration of *L. monocytogenes* at the end of processing—which could be achieved through growth inhibitors or through the application of heat treatment or high-pressure processing—would be far more effective than keeping storage temperatures low or reducing storage times. If a meat product contains a growth inhibitor compound or a lactic acid bacteria culture starter is added, it is important to determine how these preservatives would affect the lag time duration and maximum population density of *L. monocytogenes*, because QRA models have demonstrated the moderate impact that those two parameters exert on the final risk estimate. Future QRA models should include cross-contamination modules along the food chain, should be based on accurate microbial kinetic parameters and should represent the effects of new technologies and/or intervention strategies, such as high pressure processing, functional starter cultures, bacteriocins and bioactive packaging. Models should also allow the assessment of the impact of effective cleaning and sanitation programmes in processing plants, as well as the impact of end-product batch microbiological testing.

Supplementary Materials: The following supporting information can be downloaded at: https://www.mdpi.com/article/10.3390/foods13030359/s1.

Author Contributions: Conceptualization, M.S. and U.G.-B.; methodology, M.S. and U.G.-B.; software, U.G.-B.; validation, V.C., J.D.O.M., L.G., M.S. and U.G.-B.; formal analysis, M.S. and U.G.-B.; investigation, V.C., L.G., M.S. and U.G.-B.; resources, M.S.; data curation, L.G., V.C. and U.G.-B.; writing—original draft preparation, U.G.-B., L.G. and M.S.; writing—review and editing, J.D.O.M., V.C. and U.G.-B.; visualization, V.C., L.G., M.S. and U.G.-B.; supervision, M.S. and U.G.-B.; project administration, J.D.O.M. and M.S.; funding acquisition, M.S. All authors have read and agreed to the published version of the manuscript.

Funding: This research received no external funding.

Institutional Review Board Statement: Not applicable.

Informed Consent Statement: Not applicable.

Data Availability Statement: A BibTeX file containing clean records from the systematic review is available.

Conflicts of Interest: The authors declare no conflicts of interest.

References

1. EFSA (European Food Safety Authority); ECDC (European Centre for Disease Prevention and Control). The European Union One Health 2021 Zoonoses Report. *EFSA J.* **2022**, *20*, e7666. [CrossRef]
2. U.S. Centers for Disease Control and Prevention. MMWR: Summary of Notifiable Infectious Diseases 2010–2020. Available online: https://www.cdc.gov/mmwr/mmwr_nd/index.html (accessed on 6 October 2022).
3. Pasonen, P.; Ranta, J.; Tapanainen, H.; Valsta, L.; Tuominen, P. *Listeria monocytogenes* risk assessment on cold smoked and salt-cured fishery products in Finland—A repeated exposure model. *Int. J. Food Microbiol.* **2019**, *304*, 97–105. [CrossRef] [PubMed]
4. Carrasco, E.; Pérez-Rodríguez, F.; Valero, A.; García-Gimeno, R.M.; Zurera, G. Risk assessment and management of *Listeria monocytogenes* in ready-to-eat lettuce salads. *Compr. Rev. Food Sci. Food Saf.* **2010**, *9*, 498–512. [CrossRef] [PubMed]
5. EFSA (European Food Safety Authority); Panel of Biological Hazards (BIOHAZ). The public health risk posed by *Listeria monocytogenes* in frozen fruit ad vegetables including herbs, blanched during processing. *EFSA J.* **2020**, *18*, 6092. [CrossRef]
6. Gonzales-Barron, U.; Cadavez, V.; Guillier, L.; Sanaa, M. A critical review of risk assessment models for *Listeria monocytogenes* in dairy products. *Foods* **2023**, *12*, 4436. [CrossRef]
7. Tang, J. Risk Assessment of *Listeria monocytogenes* in Ready-to-Eat Meat from Plants to Consumption. Ph.D. Thesis, Virginia Polytechnic Institute and State University, Blacksburg, VA, USA, 2013.
8. Possas, A.; Valdramidis, V.; García-Gimeno, R.; Pérez-Rodríguez, F. High hydrostatic pressure processing of sliced fermented sausages: A quantitative exposure assessment for *Listeria monocytogenes*. *Innov. Food Sci. Emerg. Technol.* **2019**, *52*, 406–419. [CrossRef]
9. Sun, W.; Liu, Y.; Wang, X.; Liu, Q.; Dong, Q. Quantitative risk assessment of *Listeria monocytogenes* in bulk cooked meat from production to consumption in China: A Bayesian approach. *J. Sci. Food Agric.* **2019**, *99*, 2931–2938. [CrossRef]
10. Brusa, V.; Prieto, M.; Campos, C.; Epszteyn, S.; Cuesta, A.; Renaud, V.; Schembri, G.; Vanzini, M.; Michanie, S.; Leotta, G.; et al. Quantitative risk assessment of listeriosis associated with fermented sausage and dry-cured pork shoulder consumption in Argentina. *Food Control* **2021**, *123*, 107705. [CrossRef]
11. Pradhan, A.; Ivanek, R.; Gröhn, Y.T.; Geornaras, I.; Sofos, J.N.; Wiedman, M. Quantitative risk assessment of *Listeria monocytogenes* in selected categories of deli meats: Impact of lactate and diacetate on listeriosis cases and deaths. *J. Food Prot.* **2009**, *72*, 978–989. [CrossRef]
12. Ross, T.; Rasmussen, S.; Sumner, J. Using a quantitative risk assessment to mitigate risk of *Listeria monocytogenes* in ready-to-eat meats in Australia. *Food Control* **2009**, *20*, 1058–1062. [CrossRef]
13. Ross, T.; Rasmussen, S.; Fazil, A.; Paoli, G.; Summer, J. Quantitative risk assessment of *Listeria monocytogenes* in ready-to-eat meats in Australia. *Int, J. Food Microbiol.* **2009**, *131*, 128–137. [CrossRef] [PubMed]
14. Pradhan, A.K.; Ivanek, R.; Gröhn, Y.T.; Bukowski, R.; Geornaras, I.; Sofos, J.N.; Wiedmann, M. Quantitative risk assessment of listeriosis-associated deaths due to *Listeria monocytogenes* contamination of deli meats originating from manufacture and retail. *J. Food Prot.* **2010**, *73*, 620–630. [CrossRef] [PubMed]
15. Gallagher, D.; Ebel, E.D.; Gallagher, O.; Labarre, D.; Williams, M.S.; Golden, N.J.; Pouillot, R.; Dearfield, K.L.; Kause, J. Characterizing uncertainty when evaluating risk management metrics: Risk assessment modeling of *Listeria monocytogenes* contamination in ready-to-eat deli meats. *Int. J. Food Microbiol.* **2013**, *162*, 266–275. [CrossRef] [PubMed]
16. Gallagher, D.; Pouillot, R.; Hoelzer, K.; Tang, J.; Dennis, S.B.; Kause, J.R. *Listeria monocytogenes* in retail delicatessens: An interagency risk assessment-risk mitigations. *J. Food Prot.* **2016**, *79*, 1076–1088. [CrossRef] [PubMed]
17. FDA-FSIS. *Quantitative Assessment of Relative Risk to Public Health from Foodborne Listeria monocytogenes among Selected Categories of Ready-to-Eat Foods*; Center for Food and Safety and Applied Nutrition, Food and Drug Administration, U.S. Department of Health and Human Services, and Food Safety and Inspection Service, U.S. Department of Agriculture: College Park, MD, USA, 2003; pp. 1–541.
18. FAO/WHO. *Risk Assessment of Listeria monocytogenes in Ready-to-Eat Foods: Technical Report*; World Health Organization and Food and Agriculture Organization of the United Nations: Geneve, Switzerland, 2004; pp. 1–269.
19. Giovannini, A.; Migliorati, G.; Prencipe, V.; Calderone, D.; Zuccolo, C.; Cozzolino, P. Risk assessment for listeriosis in consumers of Parma and San Daniele hams. *Food Control* **2007**, *18*, 789–799. [CrossRef]
20. Endrikat, S.; Gallagher, D.; Pouillot, R.; Hicks Quesenberry, H.; Labarre, D.; Schroeder, C.M.; Kause, J. A comparative risk assessment for *Listeria monocytogenes* in prepackaged versus retail-sliced deli meat. *J. Food Prot.* **2010**, *73*, 612–619. [CrossRef] [PubMed]
21. FSIS. FSIS Comparative Risk Assessment for *Listeria monocytogenes*. In *Ready-to-Eat Meat and Poultry Deli Meats: Technical Report*; Risk Assessment Division, Office of Public Health Science, Food Safety and Inspection Service, United States Department of Agriculture: College Park, MD, USA, 2010; pp. 1–60.
22. Foerster, C.; Figueroa, G.; Evers, E. Risk assessment of *Listeria monocytogenes* in poultry and beef. *British Food J.* **2015**, *117*, 779–792. [CrossRef]

23. Pouillot, R.; Gallagher, D.; Tang, J.; Hoelzer, K.; Kause, J.; Dennis, S.B. *Listeria monocytogenes* in retail delicatessens: An interagency risk assessment-model and baseline results. *J. Food Prot.* **2015**, *78*, 134–145. [CrossRef]
24. Falk, L.E.; Fader, K.A.; Cui, D.S.; Totton, S.C.; Fazil, A.M.; Lammerding, A.M.; Smith, B.A. Comparing listeriosis risks in at-risk populations using a user-friendly quantitative microbial risk assessment tool and epidemiological data. *Epidem. Inf.* **2016**, *144*, 2743–2758. [CrossRef]
25. Pérez-Rodríguez, F.; Carrasco, E.; Bover-Cid, S.; Jofré, A.; Valero, A. Closing gaps for performing a risk assessment on *Listeria monocytogenes* in ready-to-eat (RTE) foods: Activity 2, a quantitative risk characterization on *L. monocytogenes* in RTE foods; starting from the retail stage. *EFSA Support. Publ.* **2017**, *2017*, EN-1252. [CrossRef]
26. Tsaloumi, S.; Aspridou, Z.; Tsigarida, E.; Gaitis, F.; Garofalakis, G.; Barberis, K.; Tzoumanika, F.; Dandoulaki, M.; Skiadas, R.; Koutsoumanis, K. Quantitative risk assessment of *Listeria monocytogenes* in ready-to-eat (RTE) cooked meat products sliced at retail stores in Greece. *Food Microbiol.* **2021**, *99*, 103800. [CrossRef] [PubMed]
27. EFSA (European Food Safety Authority); Panel of Biological Hazards (BIOHAZ). Scientific opinion on the *Listeria monocytogenes* contamination of ready-to-eat foods and the risk for human health in the EU. *EFSA J.* **2018**, *16*, 5134. [CrossRef]
28. Duret, S.; Hoang, H.M.; Derens-Bertheau, E.; Delahaye, A.; Laguerre, O.; Guillier, L. Combining quantitative risk assessment of human health, food waste, and energy consumption: The next step in the development of the food cold chain? *Risk Anal.* **2019**, *39*, 906–925. [CrossRef] [PubMed]
29. Yang, H.; Mokhtari, A.; Jaykus, L.-A.; Morales, R.A.; Cates, S.C.; Cowen, P. Consumer phase risk assessment for *Listeria monocytogenes* in deli meats. *Risk Anal.* **2006**, *26*, 89–103. [CrossRef] [PubMed]
30. Kuan, C.; Goh, S.; Loo, Y.; Chang, W.; Lye, Y.; Puspanadan, S.; Shahril, N.; Tang, J.; Mahyudin, N.; Nishibuchi, M.; et al. Risk assessment of acquiring listeriosis from consumption of chicken offal in Selangor, Malaysia. *Int. Food Res. J.* **2015**, *22*, 1711–1718.
31. Wai, G.; Tang, J.; Alias, N.; Kuan, C.; Goh, S.; Son, R. Risk of acquiring listeriosis from consumption of chicken offal among high risk group. *Malays. J. Fundam. Appl. Sci.* **2020**, *16*, 59–63. [CrossRef]
32. Pouillot, R.; Hoelzer, K.; Chen, Y.; Dennis, S.B. *Listeria monocytogenes* dose response revisited—Incorporating adjustments for variability in strain virulence and host susceptibility. *Risk Anal.* **2015**, *35*, 90–108. [CrossRef]
33. FAO/WHO. Risk Characterization of Salmonella spp. In *Eggs and Broiler Chicken and Listeria monocytogenes in Ready-to-Eat Foods;* Food and Agriculture Organization of the United Nations: Rome, Italy, 2001; pp. 1–269.
34. Pouillot, R.; Goulet, V.; Delignette-Muller, M.L.; Mahé, A.; Cornu, M. Quantitative risk assessment of *Listeria monocytogenes* in French cold-smoked salmon: II. Risk characterization. *Risk Anal.* **2009**, *29*, 806–819. [CrossRef]
35. Lindqvist, R.; Westöö, A. Quantitative risk assessment for *Listeria monocytogenes* in smoked or gravad salmon and rainbow trout in Sweden. *Int. J. Food Microbiol.* **2000**, *58*, 181–196. [CrossRef]
36. Duret, S.; Guillier, L.; Hoang, H.-M.; Flick, D.; Laguerre, O. Identification of the significant factors in food safety using global sensitivity analysis and the accept-and-reject algorithm: Application to the cold chain of ham. *Int. J. Food Microbiol.* **2014**, *180* (Suppl. C), 39–48. [CrossRef]
37. Kanuganti, S.R.; Wesley, I.V.; Reddy, P.G.; McKean, J.; Hurd, H.S. Detection of *Listeria monocytogenes* in pigs and pork. *J. Food Prot.* **2002**, *65*, 1470–1474. [CrossRef] [PubMed]
38. Thévenot, D.; Dernburg, A.; Vernozy-Rozand, C. An updated review of *Listeria monocytogenes* in the pork meat industry and its products. *J. Appl. Microbiol.* **2006**, *101*, 7–17. [CrossRef] [PubMed]
39. Hellstrom, S.; Laukkanen, R.; Siekkinen, K.-M.; Jukka, R.; Maijala, R.; Korkeala, H. *Listeria monocytogenes* contamination in pork can originate from farms. *J. Food Prot.* **2010**, *73*, 641–648. [CrossRef] [PubMed]
40. Kurpas, M.; Wieczorek, K.; Osek, J. Ready-to-eat meat products as a source of *Listeria monocytogenes. J. Vet. Res.* **2018**, *61*, 49–55. [CrossRef]
41. Nesbakken, T.; Kapperud, G.; Caugant, D.A. Pathways of *Listeria monocytogenes* contamination in the meat processing industry. *Int. J. Food Microbiol.* **1996**, *31*, 161–171. [CrossRef]
42. Kathariou, S. *Listeria monocytogenes* virulence and pathogenicity, a food safety perspective. *J. Food Prot.* **2002**, *65*, 1811–1829. [CrossRef]
43. Thévenot, D.; Delignette-Muller, M.L.; Dernburg, A.; Christieans, S.; Vernozy-Rozand, C. Serological and molecular epidemiology of *Listeria monocytogenes* strains collected in 13 French salting plants and their products. *Int. J. Food Microbiol.* **2006**, *112*, 153–161. [CrossRef]
44. FSIS. *FSIS Rule Designed to Reduce Listeria monocytogenes in Ready-to-Eat Meat and Poultry;* Food Safety and Inspection Service, United States Department of Agriculture: College Park, MD, USA, 2003.
45. Hoelzer, K.; Pouillot, R.; Gallagher, D.; Silverman, M.B.; Kause, J.; Dennis, S. Estimation of *L. monocytogenes* transfer coefficients and efficacy of bacterial removal through cleaning and sanitation. *Int. J. Food Microbiol.* **2012**, *157*, 267–277. [CrossRef]
46. Chaitiemwong, N.; Hazeleger, W.C.; Beumer, R.R. Inactivation of *Listeria monocytogenes* by disinfectants and bacteriophages in suspension and stainless steel carrier tests. *J. Food Prot.* **2014**, *77*, 2012–2020. [CrossRef]
47. Plaza-Rodríguez, C.; Haberbeck, L.U.; Desvignes, V.; Dalgaard, P.; Sanaa, M.; Nauta, M.; Filter, M.; Guillier, L. Towards transparent and consistent exchange of knowledge for improved microbiological food safety. *Curr. Opin. Food Sci.* **2018**, *19*, 129–137. [CrossRef]

48. Filter, M.; Nauta, M.; Pires, S.M.; Guillier, L.; Buschhardt, T. Towards efficient use of data, models and tools in food microbiology. *Curr. Opin. Food Sci.* **2022**, *46*, 100834. [CrossRef]
49. Perkel, J.M. Challenge to scientists: Does your ten-year-old code still run? *Nature* **2020**, *584*, 656–659. [CrossRef] [PubMed]

foods

MDPI

Review

A Critical Review of Risk Assessment Models for *Listeria monocytogenes* in Dairy Products

Ursula Gonzales-Barron [1,2,*], Vasco Cadavez [1,2], Laurent Guillier [3] and Moez Sanaa [4,*]

[1] Centro de Investigação de Montanha (CIMO), Instituto Politécnico de Bragança, Campus de Santa Apolónia, 5300-253 Bragança, Portugal; vcadavez@ipb.pt

[2] Laboratório Para a Sustentabilidade e Tecnologia em Regiões de Montanha, Instituto Politécnico de Bragança, Campus de Santa Apolónia, 5300-253 Bragança, Portugal

[3] Risk Assessment Department, French Agency for Food, Environmental and Occupational Health & Safety (Anses), 14 Rue Pierre et Marie Curie Maisons-Alfort, 94701 Maisons-Alfort, France

[4] Nutrition and Food Safety Department, World Health Organization (WHO), CH-1211 Geneva, Switzerland

* Correspondence: ubarron@ipb.pt (U.G.-B.); sanaam@who.int (M.S.); Tel.: +351-273-303-325 (U.G.-B.)

Abstract: A review of the published quantitative risk assessment (QRA) models of *L. monocytogenes* in dairy products was undertaken in order to identify and appraise the relative effectiveness of control measures and intervention strategies implemented at primary production, processing, retail, and consumer practices. A systematic literature search retrieved 18 QRA models, most of them (9) investigated raw and pasteurized milk cheeses, with the majority covering long supply chains (4 farm-to-table and 3 processing-to-table scopes). On-farm contamination sources, either from shedding animals or from the broad environment, have been demonstrated by different QRA models to impact the risk of listeriosis, in particular for raw milk cheeses. Through scenarios and sensitivity analysis, QRA models demonstrated the importance of the modeled growth rate and lag phase duration and showed that the risk contribution of consumers' practices is greater than in retail conditions. Storage temperature was proven to be more determinant of the final risk than storage time. Despite the pathogen's known ability to reside in damp spots or niches, re-contamination and/or cross-contamination were modeled in only two QRA studies. Future QRA models in dairy products should entail the full farm-to-table scope, should represent cross-contamination and the use of novel technologies, and should estimate *L. monocytogenes* growth more accurately by means of better-informed kinetic parameters and realistic time–temperature trajectories.

Keywords: systematic review; exposure assessment; simulation; raw milk; cheese; listeriosis

1. Introduction

Listeria monocytogenes is a Gram-positive, non-spore-forming, facultatively anaerobic rod-shaped bacterium, pathogenic to both humans and animals and of great concern with regard to human foodborne illness [1]. Foodborne listeriosis is one of the severe foodborne diseases, and although it is a relatively rare disease, the high rate of mortality associated with this infection makes it a significant public health concern. In the European Union (EU), listeriosis was the fifth most commonly reported zoonosis in the year 2020, with 1876 confirmed cases in 27 EU Member States, and with the highest case fatality (13%) and hospitalization rates (97.1%) [1]. Recently, a meta-analysis investigation on case–control studies of sporadic listeriosis was conducted to summarize evidence on associations (odds ratios, OR) between risk factors and sporadic cases [2]. Meta-analysis models on the outcomes of 12 case–control studies spanning from 1988 to 2013 pointed seafood as the most likely causative agent of listeriosis (pooled OR = 6.27 for susceptible populations including perinatal/non-perinatal, immunocompromised, and the elderly), followed by cheese (pooled OR = 1.83; 95% CI: 1.27–2.64), mainly including soft cheeses.

In the past 20 years, many quantitative risk assessment (QRA) models have been developed worldwide in order to guide the decision making on food safety and risk management of *L. monocytogenes*. Models, from simple to more complex ones, attempt to assess in a structured approach the possible routes of contamination of *L. monocytogenes* at different points along the farm-to-fork chain of food products. Regardless of the scope of the QRA models (i.e., farm-to-table, processing-to-table, end-of-processing-to-table, retail-to-table, or consumption), their ultimate goals are to quantify the public health risk associated with the consumption of food products of interest and evaluate scenarios or potential risk reduction measures. The objectives of this study are as follows: (i) to carry out a critical review of the published risk assessment models of *L. monocytogenes* in milk and dairy products; (ii) to gather information on the type and relative effectiveness of the control measures and intervention strategies tested as what-if scenarios implemented at primary production, processing, retail, distribution, and consumer practices; and (iii) to identify key lessons, knowledge gaps and recommendations for improving future QRA models in dairy products. To meet the first objective, a systematic review was performed to retrieve the published listeriosis QRA models and extract information that could allow a transversal comparison of model structures, scenarios, and results. Although QRA models are science-based, they are unavoidably subjected to choices, value judgments, and assumptions, reasons as to why all of them must be documented in a systematic and transparent manner. Therefore, the QRA models presented in this critical review are those that contain a full description of assumptions and uncertainties.

2. Materials and Methods

A broad systematic review was carried out to recover published QRA models of listeriosis associated with the consumption of any foodstuff. They were sought through a literature search on Scopus and PubMed®, considering 1 January 1998 as the starting date of publication. The searches were carried out on 18 May 2022 (end date of publication), and the following search strings were applied for each of the literature engines:

SCOPUS: (TITLE-ABS-KEY ("risk assessment") OR TITLE-ABS-KEY (exposure) OR TITLE-ABS-KEY (quantitative microbial) OR TITLE-ABS-KEY (risk modeling) OR TITLE-ABS-KEY (modeling) OR TITLE-ABS-KEY (simulation*) OR TITLE-ABS-KEY (second-order) OR TITLE-ABS-KEY ("second order") OR TITLE-ABS-KEY ("risk management")) AND (TITLE-ABS-KEY ("L. monocytogenes") OR TITLE-ABS-KEY ("Listeria monocytogenes") OR TITLE-ABS-KEY (listeriosis)).

PUBMED: (("risk assessment" [Title/Abstract]) OR (exposure [Title/Abstract]) OR (quantitative microbial [Title/Abstract]) OR (risk modeling [Title/Abstract]) OR (modeling [Title/Abstract]) OR (simulation* [Title/Abstract]) OR (second-order [Title/Abstract]) OR ("second order" [Title/Abstract]) OR ("risk management" [Title/Abstract])) AND (("L. monocytogenes" [Title/Abstract]) OR ("Listeria monocytogenes" [Title/Abstract]) OR (listeriosis [Title/Abstract])).

After merging into one single database, records were deduplicated. Eligibility assessment was carried out by two senior reviewers as a two-step process: first, by evaluating the title and abstract of the records, and subsequently, by examining the full text of the remaining records. Studies were regarded as eligible if (1) they presented a quantitative risk or exposure assessment model for listeriosis linked to any foodstuff, with formulae and assumptions explicitly indicated, and (2) they were written in English or Spanish language.

From each eligible model, the following information was extracted: scope of the QRA, food, country, existence of sub-models for cross-contamination, dose–response model and endpoint, modeling of strain variability, inclusion of temperature profiles and lag time, predictive microbiology models used, outcomes of what-if scenarios, outcomes of sensitivity analysis, and observations on the overall level of complexity of the model.

3. Results

3.1. Systematic Review Process

From Scopus, 1261 articles were collected, whereas 360 articles were retrieved from PubMed®. As shown in Figure 1, at the end of the full-text assessment, forty-four records were retained. In addition, through Google search, ten records were collected, which consisted of four theses [3–6], five reports from health agencies [7–11], and one published article [12]. A total of 54 records were therefore available, in which 65 models were published. Through this literature search, listeriosis QRA models were retrieved for any foodstuff and then classified into QRA for produce (11 models), seafood (10 models), composite (4 models), meat products (23 models), and dairy (18 models). As defined in the objectives of this study, this review focuses on dairy products only [7–23].

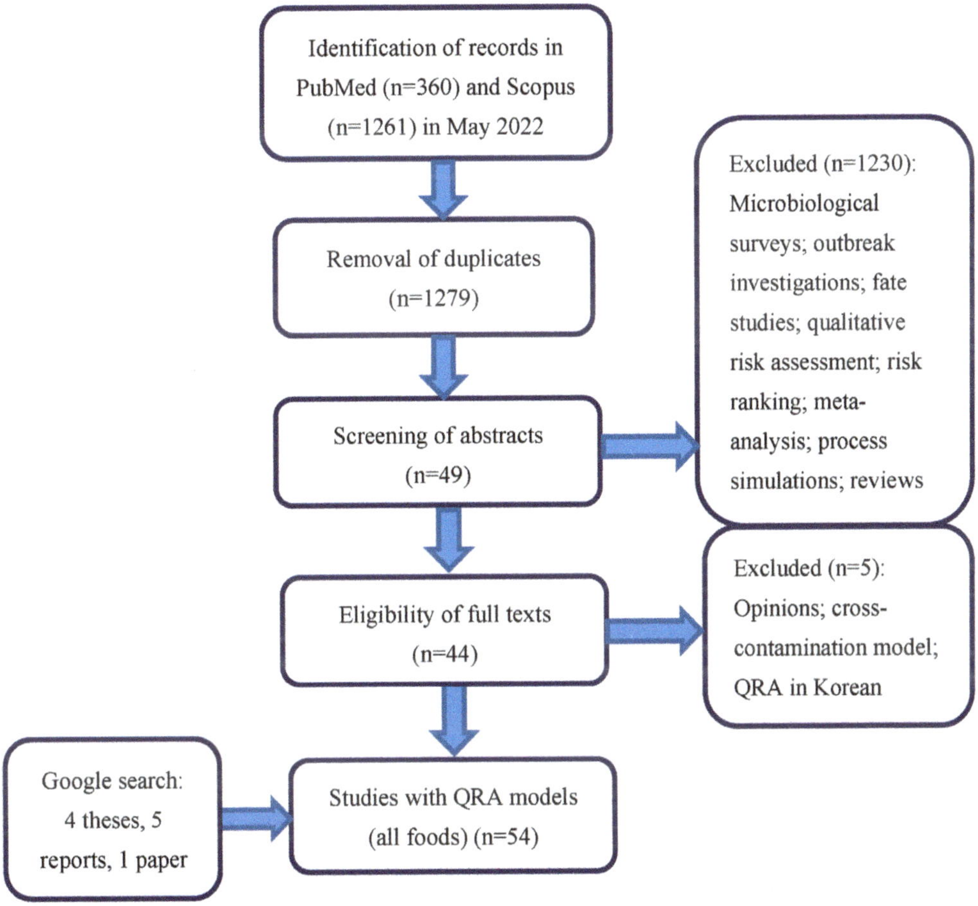

Figure 1. PRISMA chart of literature search of studies comprising quantitative risk assessment models on *L. monocytogenes* in foods published between January 1998 and May 2022.

3.2. Description of the QRA Models in Dairy Products

A total of 18 QRA models, published from 1998, investigating dairy products as sources of listeriosis were recovered (Table 1). Out of them, nine models represented the food production conditions of Europe, covering France (Bemrah et al. [13]; Sanaa et al. [16]; Tenenhaus-Aziza et al. [12]), Italy (Giacometti et al. [20]; Condoleo et al. [15]), Greece

(Koutsoumanis et al. [18]), Ireland (Tiwari et al. [14]), Denmark (Njage et al. [23]), and the EU (Pérez-Rodríguez et al. [10]). From the American continent (five QRA models), three of them pertained to the USA and Canada (FDA-FSIS [7]; Latorre et al. [19]; FDA-HealthCanada [9]), followed by Mexico (Soto-Beltrán et al. [22]) and Brazil (Campagnollo et al. [17]). Retrieved models also originated from Korea (Yang and Yoon [21]), whereas two from FAO/WHO [8] and one from EFSA BIOHAZ [11] were not particularly linked to any geographical location. Most QRA models addressed the risk of listeriosis associated with ready-to-eat (RTE) foods; only 2 out of the 18 models pertained to non-RTE food, specifically raw milk. A total of 10 out of the 18 models assessed the risk from different kinds of cheeses, and in most of the cases, they covered long supply chains; this is, either the whole supply chain (4 farm-to-table scope models) or the processing-to-table chain (3 QRA models). Six out of the nine QRA models focused on raw milk types, which, except for *queso fresco*, comprised simulations over long supply chains. This signifies that these long supply chains have been the preferred choice of the scope of QRA in cheeses as a consequence of the general understanding that *L. monocytogenes* contamination in cheeses may occur at multiple points along the food chain. The sole consumer's practices module was simulated in two models, whereas nine models comprised end-of-processing or retail-to-table. Apart from cheese and milk, QRA models also investigated ice cream [8], yogurt [21], and cultured milk [23]. FDA-FSIS [7] investigated several RTE dairy products.

In the risk or exposure estimation procedure, most models did not perform separation of uncertainty, although the distinction was often made between uncertainty and variability distributions. Second-order simulations were undertaken in three models (16.6%). Despite the many opportunities for cross-contamination, only three QRA models attempted to characterize it. Tenenhaus et al. [12] proposed more complex discrete event models for use during cheesemaking, ripening, and packaging.

Predictive microbiology models were employed in 15 out of 18 QRA models, covering different functions for growth and survival, although the simple log-linear model for growth was often employed. Despite its importance in obtaining more realistic exposure assessment or less conservative risk estimates, not all QRA models represented the lag phase duration. The lag phase was approached in only four models (22.2% of those taking advantage of predictive microbiology). Similarly, only three QRA models solved the growth of the pathogen for time–temperature trajectories [10,12,18].

Only 1 QRA model considered mortality as an endpoint for risk estimation, whereas 18 models considered illness as an endpoint. One model did not perform any risk estimation as it only targeted the exposure assessment component. The exponential "single-hit" dose–response equation was the most frequent choice for risk characterization, being utilized in 14 QRA models (77.8%), from which the preferred approach was that of FAO/WHO [8] (9 models). The Weibull-gamma dose–response model was employed in three QRA models.

Table 2 summarizes the predictive microbiology models and main outcomes related to what-if scenarios and sensitivity analysis from the listeriosis QRA models linked to dairy products. Most of the dairy QRA models undertook separate simulations evaluating the impact of what-if scenarios, which constituted either risk factors or intervention strategies (82.4% of the models), whereas sensitivity analysis on *L. monocytogenes* concentration, on dose per serving or on risk measures as response variables was undertaken in 35.3% of the models.

Table 1. Main features of quantitative risk assessment models of *Listeria monocytogenes* (LM) from consumption of dairy products by scope *.

Scope	Food	RTE	Cross-Contamination	DR—Endpoint	Type of DR Model	DR Sub-Populations	Strain Variability	Temp Profiles/Lag Time	Country	Source
Farm-to-table	Raw milk soft cheese	Yes	No	WG—I, D	Farber et al. [24]	High-risk/Low-risk	Proportion of virulent strains	No/No	France	Bemrah et al. [13]
	Soft-ripened cheese	Yes	No	Exp—I	FAO/WHO [8]	High-risk/Low-risk	Strain diversity implicit in r; T_{min} and EGR_{20} represent strain variability	No/Yes	North America	FDA-HealthCanada [9]
	Raw/pasteurized milk cheese	Yes	Yes: processing, cheese smearing stage	WG—I	Bemrah et al. [13]	High-risk/Low-risk	Proportion of virulent strains	No/No	Ireland	Tiwari et al. [14]
	Sheep's raw milk semi-soft cheese	Yes	No	Exp—I	FAO/WHO [8]	High-risk/Low-risk	Strain diversity implicit in r; challenge test data from a mixture of strains	No/No	Italy	Condoleo et al. [15]
Processing-to-table	Raw milk cheeses: Camembert of Normandy and Brie of Meau	Yes	No	Exp—I	FAO/WHO [25]	High-risk/Low-risk	T_{min} and pH_{min} represent strain variability	No/No	France	Sanaa et al. [16]
	Raw milk semi-hard cheese and pasteurized milk soft cheese	Yes	No	Exp—I	FAO/WHO [8]	High-risk/Low-risk	Strain diversity implicit in r	No/No	Brazil	Campagnollo et al. [17]
	Pasteurized milk soft cheese	Yes	Yes: cheese making (pasteurized milk, cheese surface); ripening (cross-contamination, cheese surface); packaging	Exp—I	FAO/WHO [8]	Generic	Strain diversity implicit in r; lag time distribution	Yes/Yes	France	Tenenhaus-Aziza et al. [12]
End Process-to-table	Pasteurized milk	Yes	No	.	.	.	T_{min} represents strain variability	Yes/Yes	Greece	Koutsoumanis et al. [18]
	Raw milk	No	No	Exp—I	FAO/WHO [8]	Multiple	Strain diversity implicit in r	No/No	USA	Latorre et al. [19]

Table 1. *Cont.*

Scope	Food	RTE	Cross-Contamination	DR—Endpoint	Type of DR Model	DR Sub-Populations	Strain Variability	Temp Profiles/ Lag Time	Country	Source
Retail-to-table	Soft/semi-soft cheese	Yes	No	Exp—I	EFSA BIOHAZ [11] based on Pouillot et al. [26]	Multiple (sex/age group)	Challenge test data from a mixture of strains; strain virulence and host susceptibility explicit in r distribution	No/No	Non-specific	EFSA BIOHAZ [11]
	Various dairy products	Yes	No	Mouse Epi—I	FDA-FSIS [7]	Multiple	Variability in the virulence of different strains represented in DR	No/No	USA	FDA-FSIS [7]
	Pasteurized milk	Yes	No	Exp—I	FAO/WHO [8]	High-risk/Low-risk	Strain diversity implicit in r	No/No	Non-specific	FAO-WHO [8]
	Ice cream	Yes	No	Exp—I	FAO/WHO [8]	High-risk/Low-risk	Strain diversity implicit in r	No/No	Non-specific	FAO-WHO [8]
	Raw milk	No	No	Exp—I	FAO/WHO [8]	High-risk/Low-risk	Strain diversity implicit in r	No/No	Italy	Giacometti et al. [20]
	Soft/semi-soft cheeses	Yes	No	Exp—I	Pouillot et al. [26]	Multiple	Challenge test data from a mixture of strains; strain virulence and host susceptibility explicit in r distribution	Yes/Yes	EU	Pérez-Rodríguez et al. [10]
	Yogurt	Yes	No	Exp—I	FAO/WHO [8]	High-risk/Low-risk	Challenge test data from a mixture of strains; strain diversity implicit in r	No/No	Korea	Yang and Yoon [21]
Consumption	Raw milk cheese (Queso fresco)	Yes	No	WG—I	Farber et al. [24]	High-risk/Low-risk	Proportion of virulent strains	No/No	Mexico	Soto-Beltrán et al. [22]
	Cultured milk	Yes	No	Exp—I	FAO/WHO [25]	Multiple	Strain variability modeled by class (cold, acid, salt, desiccation stressed) from WGS data	No/No	Denmark	Njage et al. [23]

* DR: dose–response; Exp: exponential; WG: Weibull gamma; I: illness; D: death: Mouse-Epi: mouse epidemiological model; EGR_x: exponential growth rate at x °C; r: parameter of the exponential dose–response model; T_{min}: minimum temperature for microbial growth; WGS: whole-genome sequencing.

Table 2. Predictive microbiology models and main outcomes related to what-if scenarios and sensitivity analysis from quantitative risk assessment models of *Listeria monocytogenes* (LM) from consumption of dairy products *.

Scope	Food	Predictive Microbiology Models	What-If Scenarios	Sensitivity	Complexity	Source
Farm-to-table	Raw milk soft cheese	-	(1) Excluding mastitis source of LM decreases median counts in cheese from 2.53 to 1.88 CFU/g; (2) reducing mean prevalence of contaminated farms from 3% to 2% decreases median counts in cheese from 2.53 to 0.024 CFU/g; (3) excluding mastitis source of LM and reducing the mean prevalence of contaminated farms from 3% to 2% reduces the mean incidence of listeriosis by 80%; (4) decreasing servings/person/year from 50 to 20 reduces the mean number of cases by 60%.	-	Low	Bemrah et al. [13]
	Soft-ripened cheese	Growth (linear EGR and square root, RLT); inactivation	(1) Consuming an artisanal raw milk soft-ripened cheese increases the mean risk per serving 157 times in comparison to the pasteurized one; (2) the mean risk per raw milk soft-ripened cheese serving is ~24–37 times smaller when every milk collection (bulk tank) is tested for LM, than when no testing is conducted; (3) reducing LM in raw milk by 3 log CFU/mL at the beginning of cheese manufacturing reduces the mean risk by a factor of 7–10 compared to baseline raw milk cheese; (4) testing batches of cheeses and removing non-compliant ones reduces the risk by 7–12 times that of the mean risk of non-tested pasteurized cheeses.	(1) Halving the EGR_{20} of LM reduces the mean risk per contaminated serving by a factor of ~8. Doubling the EGR_{20} multiplies the mean risk by a factor of ~4; (2) an increase of 1 °C in the home fridge temperature increases the mean risk per contaminated serving by a factor of 1.7; (3) shortening the maximum duration of the home storage from 56 to 28 days reduces the mean risk per contaminated serving by a factor of 1.4. **Outcome—risk per serving**: LM in contaminated cheese after home storage (r = 0.95); after retail storage (r = 0.83); after transport (r = 0.75); after aging (r = 0.64); LPD (r = −0.54); EGR_{20} (r = 0.45)	High: Meta-analysis, previous adjustment of mixed distributions, models for mixing and partition, lag phase modeled as "work to be done"	FDA-HealthCanada [9]
	Raw/pasteurized milk cheese	Growth (Gompertz)	(1) An increase in the initial LM in raw milk at farm level from 0.03 to 10 CFU/mL up to a maximum of 100 CFU/mL increases the final mean concentration by 35% in raw cheese and by 45% in pasteurized cheese; (2) when there is no further contamination during retail storage (only cross-contamination through smearing), the counts decrease by 24% in raw cheeses and 97% in pasteurized ones; (3) improper storage temperature above 4 °C at retail increases by 39% LM in raw cheeses and by 64% in pasteurized ones.	**Outcome—counts in raw/pasteurized milk cheese**: Temperature at retail (r = 0.65/0.75); cheese consumption (r = 0.28/0.48); storage time at retail (r = 0.15/0.20); fecal/silage/farm contamination factors (r = 0.15/0.27); cross-contamination from smearing machine (r = 0.05/0.12)	Medium: a separate Bayesian model to estimate sources of contamination on farms; discrete differential equation modeled transmission during smearing from (1) contaminated cheese to machine, (2) machine to cheese, and (3) machine to the surrounding environment and environment impact on the cheese	Tiwari et al. [14]
	Sheep's raw milk semi-soft cheese	Growth (linear and EGR square root)	(1) Cheese from mastitis-free flocks decreased concentration of LM in bulk tank milk by 24%, in comparison to contaminated random flocks; (2) flocks with a single mastitis case increase risk per contaminated serving seven times that of contaminated random flocks; (3) cheeses from mastitis free flocks present 0.07 times the risk per contaminated serving; (4) cheeses from family flocks (10 animals maximum) have a risk eight times higher.	-	Low	Condoleo et al. [15]

Table 2. *Cont.*

Scope	Food	Predictive Microbiology Models	What-If Scenarios	Sensitivity	Complexity	Source
Processing-to-table	Raw milk cheeses: Camembert of Normandy and Brie of Meau	Growth/no growth, Growth (modified logistic, cardinal models for temperature and pH)	-	-	Medium: growth was computed in the core and rind of cheeses considering modeled pH profiles	Sanaa et al. [16]
	Raw milk semi-hard cheese and pasteurized milk soft cheese	Growth (empirical functions)	(1) Anti-listerial LAB reduced from 7.7 log CFU LM/g of raw milk semi-hard cheese in the baseline scenario to 1.1 log CFU/g of cheese, reducing risk by >6 log RR; (2) addition of anti-listerial LAB to milk contaminated with LM at 1 log CFU/mL (the same concentration used in the baseline) reduced the risk 4.6-fold in pasteurized milk soft cheese in the general and vulnerable population.	-	Low	Campagnollo et al. [17]
	Pasteurized milk soft cheese	Growth (logistic growth model with delay and growth rate as secondary cardinal growth model with interactions)	(1) When the initial number of cells in the ripening room environment decreases from 2000 to 500 cells, the risk is divided by 3.7; (2) when the primo-contamination event occurs on the smearing machine, instead of during cheese-making, with 500 cells, the risk is divided by 350; (3) when the generation time of LM in the environment extends from 24 (base) to 48 h, the risk of listeriosis is divided by 546.	-	High: many recontamination and cross-contamination modules	Tenenhaus-Aziza et al. [12]
End Process-to-table	Pasteurized milk	Growth (linear model, polynomial functions for growth rate and lag phase duration)	(1) Changing domestic storage temperature from 5 to 4 °C increases the proportion of milk cartoons with no growth from 55 to 59%; (2) excluding the door shelf from the fridge increases the proportion of cartoons with no growth from 55% to 62%.	**Outcome—counts in milk at consumption**: Domestic storage time (r = 0.482); domestic temperature door-shelf (r = 0.288); retail storage temperature (r = 0.181); retail storage time (r = 0.174)	Medium: temperature profiles from the Greek chill chain of pasteurized milk, including transportation to retail, retail storage, and domestic storage; the lag time at a certain temperature was calculated based on the h_0 physiological state parameter	Koutsoumanis et al. [18]
	Raw milk	Growth (linear model, square root for growth rate)	(1) Increasing LM prevalence in bulk tank milk from 6% to 25% increases the mean risk four times; (2) a five-fold decrease in the median listeriosis cases per year is observed if a raw milk testing program was in place (i.e., conducting monthly testing of one sample of milk and recall of milk).	**Outcome—probability of illness**: temperature of the home refrigerator (r = 0.55–0.77); temperature of retail/farm fridge (r = 0.55); storage time in the home refrigerator (r = 0.27–0.36); serving size for raw milk purchased directly from milk tanks and milk consumed on farms (r = 0.19–0.30)	Low	Latorre et al. [19]

Table 2. *Cont.*

Scope	Food	Predictive Microbiology Models	What-If Scenarios	Sensitivity	Complexity	Source
Retail-to-table	Soft and semi-soft cheese	Growth (Rosso model, EGR 5 °C)	Probability of cheeses containing > 2.0 log CFU/g is 0.022 at retail and 0.024 at consumption (USA data)	Sensitivity analysis was conducted, taking together various RTE food classes.	Low: generic model; only demands some knowledge in R to utilize it	EFSA BIOHAZ [11]
	Various dairy products	Growth (linear model, square root model for EGR)	(1) If the maximum refrigerator temperature is set at 7 °C (instead of 16 °C in the baseline), the number of cases of listeriosis is reduced by 69%, and limiting the refrigerator temperature to 5 °C further reduces the number of cases to >98%; (2) reducing the maximum storage time from the 14-day baseline to 4 days reduced the annual incidence of listeriosis cases by 43.6%; (3) in *queso fresco*, the risk per serving is 43 times greater for the perinatal population and 36 times greater for the elderly population if cheeses were made from raw milk compared to pasteurized milk.	-	Medium: ten different dairy products considered; dose–response models developed for three subpopulations	FDA-FSIS [7]
	Pasteurized milk	Growth (linear model, square root growth model)	(1) If all milk were consumed immediately after purchase at retail, the number of cases in both susceptible and healthy populations would decrease 1000-fold; (2) if temperature distribution was shifted so the median increased from 3.4 to 6.2 °C, the mean number of illnesses increased > 10-fold for both populations; (3) when the storage time distribution was shifted from a median of 5.3 days to 6.7 days, the mean rate of illnesses increased 4.5-fold and 1.2-fold for the healthy and susceptible populations.	-	Medium: dose–response models for healthy and susceptible populations developed; a hierarchical beta-binomial model for the prevalence of LM in pasteurized milk	FAO-WHO [8]
	Ice cream	No growth/no death	-	-	Medium: D–R models for healthy and susceptible populations developed; a hierarchical beta-binomial model for the prevalence of LM in ice cream	FAO-WHO [8]
	Raw milk	Growth (linear model)	-	-	Low	Giacometti et al. [20]
	Soft/semi-soft cheeses	Growth (Baranyi model with Jameson effect, EGR 5 °C)	(1) Slicing the cheese increases the risk of infection by two times; (2) increasing storage temperature by 3–4 °C produces an increase of 530% cases per million servings; (3) decreasing storage temperature did not produce a substantial variation in the incidence of listeriosis (~4%) since temperature conditions in the baseline scenario did not allow for growth of LM in RTE cheese; (4) decreasing maximum mean initial LM counts values from 5 to 3 log CFU/g produces a decrease in up to 98% the cases; (5) decreasing time to consumption by 25% produced a decrease of 33% in the incidence of listeriosis cases per million servings; (6) adding the lag time effect produced a reduction of 30% in the number of cases per million servings.	-	Medium: time–temperature dynamic profiles from retail to consumption, and microbial competition models used solved with the RK4 algorithm	Pérez-Rodríguez et al. [10]

Table 2. *Cont.*

Scope	Food	Predictive Microbiology Models	What-If Scenarios	Sensitivity	Complexity	Source
	Drinking yogurt/regular yogurt	Survival (Weibull model, secondary polynomial model)	-	**Outcome—risk of illness from drinking and regular yogurt**: prevalence of LM (r = 0.67, 0.21), storage time at market (r = −0.31,−0.35), consumption (r = 0.08, 0.02)	Low	Yang and Yoon [21]
Consumption	Queso fresco	-	-	-	Low	Soto-Beltrán et al. [22]
	Cultured milk	Growth (linear model)	-	The increase in the proportion of tolerant LM resulted in an increased association between the estimated cases per million and an increase in the concentration of the pathogen during consumer storage. This is due to the increase in the concentration of the pathogen during storage for the scenarios involving 0%, 25%, and 75% tolerant proportion groups, which were 236 ± 139, 255 ± 150, and 293 ± 172 CFU/g, respectively, compared to 274 ± 161 for the 50% tolerant proportion	High: WGS data was used to model population heterogeneity in microbial phenotypic stress responses to integrate it into predictive models	Njage et al. [23]

* LPD: lag phase duration; RLT: relative lag time; EGRx: exponential growth rate at x °C; RR: risk reduction; r: coefficient of correlation in sensitivity analysis; CFU: colony forming units; '-': absence of information in the study.

4. Discussion

4.1. Risk Factors and Control Measures Assessed at Primary Production

Many sources of contamination exist in the farm environment, such as silage, soil, water, and inadequate sanitation and housing conditions, which prompt dissemination to and between animals. In addition, *L. monocytogenes* mastitis is an important source of contamination—that increases the risk associated with contamination of raw milk—which, although it has an extremely low between- and within-herd prevalence, when present, animals may have prolonged shedding of the bacteria in the milk. *L. monocytogenes* is transmitted from animal to animal through fecal–oral routes, usually via manure contamination of the pasture or silage with the microorganism. Next to this, bulk tank milk, milk filters, milking machines, milk handlers, and poor on-farm hygiene during milking are also considered sources of contamination. The QRA model of Bemrah et al. [13], reflecting French on farm conditions at the time, showed that the contamination load due to the environment was much stronger than that of animal mastitis in the presence of *L. monocytogenes* in raw milk soft cheese. In a scenario representing lower environmental contamination that reduced the mean prevalence of contaminated farms from 3% to 2%, the median concentration of *L. monocytogenes* in raw milk cheese decreased by 99% (from 2.53 to 0.024 CFU/g) (Table 2). However, in another scenario that assumed the absence of mastitis, the median concentration of *L. monocytogenes* in raw milk cheese was reduced by only 26%—from 2.54 CFU/g (baseline scenario assuming the probability of herds with *L. monocytogenes* to be 10%) to 1.87 CFU/g.

Comparable results concerning the relative importance of mastitis, yet in sheep, were obtained from a QRA model from Italy [15], whose scenario simulations showed that the median concentration of *L. monocytogenes* in bulk tank raw milk from mastitis-free flocks decreased in only 24% (from 0.56 CFU/mL to 0.43 CFU/mL) when compared to the baseline scenario of contaminated random flocks (Table 2).

In the listeriosis QRA models available for cheese, no sensitivity analysis comparing the contributions of the environment contamination and the mastitis animals has been conducted. Only the study of Tiwari et al. [14] estimated coefficients of correlation of 0.27 and 0.15 between fecal/silage/farm contamination factors with the *L. monocytogenes* counts in raw and pasteurized milk, respectively (Table 2). Nevertheless, despite their relative importance, on-farm contamination sources, either from shedding animals or from the broad environment, have been demonstrated by different QRA models to impact the exposure dose and the risk of listeriosis, in particular from raw milk cheeses. For instance, Condoleo et al. [15] estimated that sheep's raw milk cheeses from mastitis-free flocks presented 0.07 times the risk per contaminated serving of those from contaminated random flocks, whereas raw milk cheeses from family flocks consisting of a maximum of 10 animals each could present 8 times higher risk per contaminated serving. Similarly, increasing the initial *L. monocytogenes* population in raw milk at the farm level (between 0.03 and 10 CFU/mL for Ireland conditions) up to a maximum of 100 CFU/mL (worst-case scenario of contamination) would increase the final mean concentration of the pathogen by 35% for raw milk cheese and by 45% for pasteurized milk cheese [14]. Latorre et al. [19] tested a scenario whereby a four-fold increase in the risk per serving would occur if the prevalence of *L. monocytogenes* in bulk tank milk increased from 6% (baseline) to 25%. In the same line, according to the QRA model of FDA-Health Canada [9], a 3 log/mL reduction in *L. monocytogenes* concentration in raw milk at the beginning of cheese manufacturing—which can be interpreted as the result of the application of animal husbandry strategies for mitigating the contamination of bulk milk as raw material for cheese-making—can reduce the mean risk per serving by a factor of 7–10.

One such on-farm strategy to control the risk of listeriosis associated with raw milk is the bulk tank and tank truck milk testing in order to reduce the concentration of *L. monocytogenes* in dairy silo milk. Whereas Latorre et al. [19] estimated that a five-fold decrease in the median listeriosis annual cases for raw milk consumers would occur if a raw milk testing program were put in place (i.e., conducting monthly testing of one sample of milk and recall of milk), FDA-Health Canada [9] estimated that in raw milk soft-ripened cheeses, the milk collection testing would reduce the mean risk per serving by 24–37 times that of the risk when no testing at all is conducted (Table 2).

4.2. *Risk Factors and Control Measures Assessed at Processing*

It is widely known that pasteurization of milk is effective in destroying *L. monocytogenes*. The effectiveness of milk pasteurization as a key mitigation strategy to reduce the risk from the consumption of cheese was quantified by FDA-Health Canada [9] and FDA-FSIS [7]. The former estimated that consuming an artisanal raw milk soft-ripened cheese increased the mean risk per serving 157 times in comparison to consuming the pasteurized one in the general population. The latter estimated that the risk per serving of *queso fresco* is 43 times greater for the perinatal population and 36 times greater for the elderly population if cheeses were made from raw milk compared to pasteurized milk (Table 2).

Another strategy to control *L. monocytogenes* that can be applied during processing is the use of bacteriocinogenic lactic acid bacteria (LAB). Nevertheless, only one QRA model [17] investigated the effect of an anti-listerial cocktail from indigenous LAB on the risk of listeriosis from cheese. These authors estimated that the addition of 6 log CFU of such a LAB cocktail per ml of raw milk reduced the concentration of *L. monocytogenes* in raw milk semi-hard cheese ripened for 22 days from 7.7 log CFU/g (baseline scenario without added LABs) to 1.1 log CFU/g, which in turn reduced the risk by over 6 log. In the case of pasteurized milk soft cheeses, the addition of the same LAB cocktail to pasteurized milk inoculated at 1 log CFU/mL of *L. monocytogenes* decreased the risk 0.22-fold in both the general and vulnerable populations. Other current processing strategies, such as the smearing of cheeses with plant-based extracts having antimicrobial properties or the use of antimicrobial packaging, were not tested as what-if scenarios in any of the QRA models collected (Table 2).

4.3. *Cross-Contamination during Processing*

Despite the effectiveness of pasteurization in inactivating *L. monocytogenes*, post-pasteurization contamination and cross-contamination can occur within the processing plants and are exacerbated by the pathogen's capacity to grow at normal refrigeration temperatures, and its ability to find damp spots or niches where they can reside and proliferate. Furthermore, if mechanical cleaning, disinfection, and rinsing are not well executed, the bacteria can form a biofilm on surfaces in contact with food, which then becomes difficult to remove by standard sanitation protocols [27]. Nonetheless, despite the relevance of cross-contamination, only two QRA models for dairy products comprised cross-contamination modules: Tenenhaus-Aziza et al. [12] and Tiwari et al. [14]. The study of Tenenhaus-Aziza et al. (12] conducted on pasteurized milk soft cheese produced in France proposes new methods for modeling cross-contamination and recontamination events. They utilize six contamination event modules, listed as follows: (1) the primo-contamination event at the cheese-making phase, whereby milk or products can be contaminated, for example, by cells from the environment or by cells arising from pasteurization failure; (2) the primo-contamination event at the ripening phase, whereby the environment of the ripening room and the smearing machine can be initially contaminated; (3) the cross-contamination during smearing, whereby a whole colony from the surface of a cheese could be transferred at a given probability to the machine or to the immediate surroundings close to the machine, through the smearing solution and the cheese matter detached from the surface of the product; (4) cross-contamination during packaging, which was modeled using the same approach but in a simplified form, where the compartments were the cheeses and the packaging machine; (5) the transfer of colonies from the smearing room to the ripening room; whereby colonies located in the environment of the smearing room are not assumed to adhere, since they come from the smearing machine and the duration between contamination of the environment and transit of the batch is not enough long to allow adherence of the cells to the environment surfaces; and (6) the recontamination during ripening, whereby during the transit of a batch inside or outside the ripening room, colonies from the environment of the ripening room can be transferred to the surface of products present in the ripening room.

This model scientifically corroborated that frequent hygienic operation is necessary in the facilities by proving that the concentration of contaminated products correlates with the total number of cells in the ripening environment. To this respect, two of the what-if scenarios estimated that when the initial number of cells in the ripening room environment decreases from 2000 to 500 cells, the mean risk of listeriosis is divided by 3.7, whereas when the primo-contamination event occurs on the smearing machine, instead of during cheese-making, with 500 cells, the mean risk is divided by 350. In the listeriosis QRA model for Irish cheeses, Tiwari et al. [14] borrowed the cheese-smearing cross-contamination model from Tenenhaus-Aziza et al. [12] and found a low correlation between the cross-contamination from the smearing machine and the counts in raw and pasteurized milk cheeses (r = 0.05 and 0.12, respectively), which sustained the scenario that if no further contamination occurred during the retail phase, but only cross-contamination due to smearing, the *L. monocytogenes* counts would decrease by 24% in raw cheeses and by 97% in pasteurized cheeses (Table 2).

In the QRA models, there is no estimate of the contribution of cross-contamination in processing plants to the final listeriosis risk. Nonetheless, it is widely known that cross-contamination is an important factor, as suggested by the many surveys throughout the world, which have reported varying prevalence levels in the environments of dairy processing plants of up to 25.0% and, as implied by the listeriosis outbreaks due to contaminated dairy products, directly linked to cross-contamination from the processing facilities [28]. Floor drains, floors, coolers, and areas of pooled water, such as washing machine areas, are sites of frequent recovery of *L. monocytogenes*, which often contaminate food contact surfaces [29]. This reinforces the importance of including sound modules to represent cross-contamination prevalence, patterns, and events in QRA models for both raw and pasteurized milk dairy products.

4.4. Risk Factors and Control Measures at Retail and Home

Ten dairy QRA models (58.8%) covered the contamination of *L. monocytogenes* in shorter chains (two for end-of-processing-to-table, six for retail-to-table, and two for consumption only) of dairy products, namely, raw milk (x2), pasteurized milk (x2), ice cream, yogurt, soft and semi-soft cheeses, queso fresco cheese, cultured milk, and various dairy products (Table 1). Due to the fragmented scope of these QRA models, these models address the dairy foods entering the distribution chain or the retail establishment with a certain level of contamination with *L. monocytogenes*, which is prone to multiply with prolonged storage, even from low initial concentrations. Moreover, the dairy industry's trend toward the production of refrigerated products with longer shelf lives further aggravates this problem. Other equally important factors for increasing the risk can take place in products permitting the growth of *L. monocytogenes* during transport and distribution, retail, and home consumption, namely retail/home refrigerator temperature fluctuations and abuse, long-term storage, cross-contamination, and inadequate handling practices at retail and at home. The high frequency and amounts of dairy foods consumption also contribute to increasing the risk of listeriosis.

Many dairy QRA models compared the importance of retail/home storage temperature and retail/home storage time, and they unanimously found that the effect of higher temperature is stronger than longer time. FDA-FSIS QRA model in 10 different RTE dairy products [7] found out that if the maximum refrigerator temperature was set at 7 °C (instead of 16 °C in the baseline), the mean number of cases of listeriosis would be reduced by 69%, whereas further limiting the refrigerator temperature at a maximum of 5 °C would reduce the number of cases by >98%. On the other hand, if the maximum storage time was reduced from 14 days (baseline) to the (unrealistic) 4 days, the annual incidence of listeriosis cases would be decreased by 43.6%. In the same line of results, the listeriosis QRA model of pasteurized milk from FAO-WHO [8] undertook what-if scenarios that proved the greater relative importance of temperature over time: when the temperature distribution was shifted so that the median increased from 3.4 °C to 6.2 °C, the mean rate of illnesses increased over 10-fold for both the healthy and susceptible population; however, when the storage time distribution was extended from a median of 5.3 days to 6.7 days, the mean rate of illnesses increased 4.5-fold and 1.2-fold for the healthy and susceptible populations, respectively.

Similarly, FDA-Health Canada [9] predicted that an increase of only 1 °C in the home refrigerator temperature increases the mean risk per contaminated serving by a factor of 1.7, whereas halving the maximum duration of the home storage from 56 days to 28 days reduces the same risk by a factor of 1.4. In a more recent listeriosis QRA model from soft and semi-soft cheeses, Pérez-Rodríguez et al. [10] also compared the impact of increasing and decreasing storage temperature. When storage temperature was increased by 3–4 °C, the number of cases/million servings increased by 530%, while a decrease in the storage temperature by 3–4 °C produced only a 4% decrease in the number of cases since the baseline temperature conditions did not allow for growth *of L. monocytogenes* in cheese. By contrast, decreasing the time to consumption (storage time) by 25% produced a decrease of 33% in the incidence of listeriosis cases per million servings.

An interesting scenario was tested by Koutsoumanis et al. [18], which consisted of storing pasteurized milk cartons away from the door shelf of the fridge. According to their simulation, the proportion of cartons with no growth of *L. monocytogenes* increased from 55% to 62%. Outputs of sensitivity analysis of dairy QRA models also coincided with the relative importance of storage temperature versus storage time. Spearman rank sensitivity analysis on the probability of illness from the consumption of raw milk presented a higher correlation with the temperature of the home refrigerator (r = 0.55–0.77) than with the storage time in the home refrigerator (r = 0.27–0.36) [19]. Similarly, Tiwari et al. [14] found higher correlations of the counts of *L. monocytogenes* in raw and pasteurized milk cheeses with temperature at retail (r = 0.65 and 0.75, respectively) than with storage time at retail (r = 0.15 and 0.20, respectively) (Table 2). Nevertheless, despite the strong contribution of temperature to the risk as well as the well-known importance of maintaining the cold chain to control risks, most of the dairy QRA models (3/18) utilized variability distributions of average temperatures in the supply chain. To enable a better assessment of growth, time–temperature profiles with credible trajectories and oscillations should be used. Only three QRA models [10,12,18] solved *L. monocytogenes* growth for dynamic temperature profiles at every iteration (Table 1).

None of the retail-to-table QRA models included cross-contamination or poor handling modules, in spite of the potential of cross-contamination occurring during retail and at home. Only Pérez-Rodríguez et al. [10], when comparing the risk of listeriosis from non-sliced and sliced soft/semi-soft cheeses, indirectly determined that the processing step of slicing doubled the risk of infection, suggesting, therefore, that cross-contamination happens during slicing.

4.5. Contributions of Retail and Consumer Practices to the Final Risk of Listeriosis

Although not directly exposed, results from the QRA models, in perspective, pointed towards a higher contribution of the consumer module than the retail module to the risk of listeriosis from dairy foods. For instance, for the QRA model of Koutsoumanis et al. [18], the storage time at home (r = 0.482) had a stronger effect on the counts of *L. monocytogenes* in pasteurized milk at consumption than both the retail storage temperature (r = 0.181) and the retail storage time (0.174). Latorre et al. [19] also showed that the temperature of the home refrigerator (r = 0.55–0.77) can have a stronger effect than the temperature of the retail/farm refrigerator (r = 0.55) on the probability of illness from raw milk. In the case of soft-ripened cheeses [10], the risk per serving was more heavily driven by the *L. monocytogenes* counts in cheese after home storage (r = 0.95) than the counts after retail storage (r = 0.83), and in turn, than the counts after transport (r = 0.75). An interesting scenario performed in FAO/WHO [8] illustrated the strong contribution of the consumers' practices to the risk of listeriosis by estimating that if all milk were consumed immediately after purchase at retail, the number of cases in both susceptible and healthy populations would decrease 1000-fold. All of the QRA models above coincide in that the consumers' practices can be more determinant of the risk of listeriosis than the retail practices or conditions.

To a lesser extent, consumption as serving size or frequency has also been investigated in the dairy QRA models. Their impact on the risk is more variable, although in general, sensitivity analysis has ranked consumption-related variables lower than risk factors such as the prevalence of the pathogen, storage temperature, and time; therefore, it is less effective. According to Latorre et al. [19], the correlation between the probability of illness with serving size for raw milk purchased directly from milk tanks and milk consumed in farms was low, ranging between 0.19 and 0.30, whereas, in Yang and Yoon [21], the amount of consumption of yogurt had no effect on the risk of illness associated to drinking and regular yogurt (r = 0.08 and 0.02, respectively). In the early model of Bemrah et al. [13], it was shown that the strategy of reducing the servings per person per year of raw milk cheeses from 50 to 20 would reduce the incidence of listeriosis cases by 60%, less effective than other strategies such as excluding mastitis source or reducing the mean prevalence of *L. monocytogenes* of contaminated farms.

4.6. *L. monocytogenes Growth Kinetic Parameters as Drivers of the Final Risk*

Finally, some of the QRA models have shown that the kinetic parameters of the pathogen have a strong impact on the estimated risk. For instance, in their model for soft-ripened cheese, FDA-Health Canada [9] found moderate correlations between risk per serving and lag phase duration (r = −0.54) and exponential growth rate at 20 °C (EGR_{20}) (r = 0.45); furthermore, halving the EGR_{20} of *L. monocytogenes*, decreased the mean risk per contaminated serving by a factor of ~8, and doubling the EGR_{20} multiplied the mean risk by a factor of ~4. In Tenenhaus-Aziza et al. [12], when the generation time of *L. monocytogenes* in the environment was extended from 24 h (baseline) to 48 h, the risk of listeriosis from pasteurized milk soft cheeses was divided by ~550, whilst Pérez-Rodríguez et al. [10] showed that incorporating the lag time effect to the baseline model produced a reduction of 30% in the number of cases per million servings. These findings reinforce the importance of obtaining good estimates of the microbial kinetic parameters to model the changes in microbial concentration between the point of contamination and human exposure to the pathogens.

To avoid the assumption that the *L. monocytogenes* populations are homogeneous and that their kinetic parameters represent average population behavior, the common strategy in the QRA models was to represent strain variability in parameters such as growth rate, minimum temperature for growth, minimum pH for growth and lag phase duration from growth challenge data that utilized a cocktail of *L. monocytogenes* strains [10,11,15,16,18,21]. A different approach for modeling strain variability for increased precision exposure assessment was proposed by Njage et al. [23], consisting of using whole-genome sequencing (WGS) data to unravel the biological variability that induces the diverse response by microorganisms to the differing environmental conditions. These authors employed finite mixture models to distinguish the number of *L. monocytogenes* sub-populations for each of the stress phenotypes: acid, cold, salt, and desiccation. Based on the performance assessment of the machine learning methods, they selected the support vector machine approach for the prediction of acid stress and the random forest approach for cold, salt, and desiccation stress responses. They used WGS data from a collection of 166 *L. monocytogenes* strains from Canada and Switzerland, as well as associated data on growth phenotypes during the different stress conditions. Njage et al. [23] showed that none of the four stress response categories could be represented by a unique population, instead, maximum growth rates of *L. monocytogenes* were multimodal distributions, which implied the presence of different subgroups of strains: (1) The relative growth rate distribution for the cold stress response class was estimated as consisting of 3% of cold susceptible strains (μ_{max} = 0.76) and 97% of cold tolerant strains (μ_{max} = 1.01); (2) The relative growth rate distribution for the acid stress response class was estimated as consisting of 4% of highly susceptible strains (μ_{max} = 0.41), 44% of susceptible strains (μ_{max} = 0.85), 50% of tolerant strains (μ_{max} = 1.13) and 3% of highly tolerant strains (μ_{max} = 1.50); (3) The relative growth rate distribution for the salt stress response class was estimated as consisting of 16% of susceptible strains (μ_{max} = 0.83), 77% of tolerant strains (μ_{max} = 1.01) and 7% of highly tolerant strains (μ_{max} = 1.18); and (4) The relative growth rate distribution for the desiccation stress response class was estimated as consisting of 21% of susceptible strains (μ_{max} = 0.87), 74% of tolerant strains (μ_{max} = 1.02) and 4% of highly tolerant strains (μ_{max} = 1.26). This approach, however, demands the collection of a database of strains with available WGS and phenotypic data on microbial adaptation to various inherent food characteristics and conditions encountered during food processing and handling [30].

It is crucial to bear in mind that the outcomes of a risk assessment are context-specific and influenced by factors such as the country and population under consideration. Moreover, risk assessment is inherently linked to queries posed by a risk manager. Consequently, the presentation of assessment results should be tailored to the specific question at hand, whether it involves estimating risk at the population level to gauge the overall burden or assessing risk per portion to evaluate the impact of control measures. Furthermore, certain nuances are challenging to convey accurately. For instance, the term "cheese" encompasses a diverse array of processes and microflora, rendering the transfer of models from one country to another a complex task.

Traditionally, in risk assessment, it is assumed that the matrix's effect influences exposure, where growth or inactivation is linked to the properties of the matrix. Conversely, it is presumed that the food matrix does not influence the virulence of strains, although several studies propose that it might [31,32]. It is also generally assumed that it does not affect the variability of virulence profiles among strains. Consequently, the distribution of values characterizing intraspecific variability in the dose–response relationship is presumed to be identical regardless of the food under consideration. Nevertheless, available data demonstrate that the distribution of *L. monocytogenes* sequence types differs among various food categories [33,34]. Employing a dose–response approach that considers the diversity of virulence profiles would enable a more accurate assessment of the role of cheese and dairy products in the risk of listeriosis. A preliminary suggestion by Fritsch et al. [35] advocates for such an application.

4.7. Availability of Models

Sharing risk assessment models is essential to ensure the transparency of the approach and ease of re-use. This is particularly important in scientific research, where reproducibility and open science are increasingly valued [36]. By sharing models, researchers can allow others to scrutinize their work, identify any potential biases, and apply the models to their own data. This can help improve the accuracy and reliability of risk assessments and ultimately lead to better decision-making. In addition to ensuring transparency, sharing models also facilitates re-use. This is especially beneficial for researchers who may not have the resources to develop their own models [37].

Of all the studies analyzed, four provide access to the codes or spreadsheets [10,11,16,23], and one proposes to make the models used available on request [9] (details of models sharing characteristics are available in the Supplementary Material of this article). For the other models, no indication is given as to the availability of the models. One study dating from 2004 refers to a site that no longer exists [7]. This latter questioned the challenge of reproducibility. Indeed, after a few years, as software evolves and resources disappear (maintenance of websites, for example), it becomes difficult to reproduce the calculations made [38,39].

5. Conclusions

Cheese as a source of listeriosis tended to be studied in QRA models under the full farm-to-table approach because of the many factors and forces of contamination that can occur along the chain, namely, on-farm environmental contamination sources such as silage, soil, water, and inadequate sanitation and housing conditions; extensive manipulation after milk heat treatment (if heat treated); the potential for recontamination after pasteurization and cross-contamination events during processing; the possible presence of contaminating niches in processing and retail facilities; *L. monocytogenes'* ability to grow during refrigeration storage; long shelf-life in case of ripened cheeses; and wide consumption of cheese. QRA models pointed out that storage practices at home could be more determinant of the risk of listeriosis than those of retail. Furthermore, since storage temperature has a stronger effect on the risk of illness than storage time, *L. monocytogenes* growth should be more accurately estimated by using realistic time–temperature profiles as opposed to constant temperatures. Validated microbial growth kinetic parameters, as affected by lactic acid bacteria, should be employed in QRA models, given the impact of growth rate and lag phase duration on the final risk estimates. Sound cross-contamination modules that represent frequencies and contamination patterns should be placed along the food chain of dairy food. Finally, QRA models should be able to describe the effects of new non-thermal technologies, such as thermosonication and cold plasma, and bio-intervention strategies, such as ad hoc starter cultures and bioactive packaging.

Supplementary Materials: The following supporting information can be downloaded at: https://www.mdpi.com/article/10.3390/foods12244436/s1.

Author Contributions: Conceptualization, M.S. and U.G.-B.; methodology, M.S. and U.G.-B.; software, U.G.-B.; validation, V.C., L.G., M.S. and U.G.-B.; formal analysis, M.S. and U.G.-B.; investigation, V.C., L.G., M.S. and U.G.-B.; resources, M.S.; data curation, L.G., V.C. and U.G.-B.; writing—original draft preparation, M.S., L.G. and U.G.-B.; writing—review and editing, U.G.-B.; visualization, V.C., L.G., M.S. and U.G.-B.; supervision, M.S. and U.G.-B.; project administration, M.S.; funding acquisition, M.S. All authors have read and agreed to the published version of the manuscript.

Funding: This research received no external funding.

Data Availability Statement: Data is contained within the article. The BibTeX file containing records used in this systematic review is available.

Conflicts of Interest: The authors declare no conflict of interest.

References

1. EFSA. The European Union one health 2020 zoonoses report. *EFSA J.* **2021**, *19*, 6971. [CrossRef]
2. Leclercq, A.; Kooh, P.; Augustin, J.-C.; Guillier, L.; Thebault, A.; Cadavez, V.; Gonzales-Barron, U.; Sanaa, M. Risk factors for sporadic listeriosis: A systematic review and meta-analysis. *Microb. Risk Anal.* **2021**, *17*, 100128. [CrossRef]
3. Dass, S. Exposure Assessment of *Listeria monocytogenes* in Vacuum Packed Cold-Smoked Salmon in the Republic of Ireland. Ph.D. Thesis, Technological University Dublin, Dublin, Ireland, 2011.
4. Bomfeh, K. Risk Assessment for *Listeria monocytogenes* in Traditionally Processed Fish from Informal Markets in Accra and Tema. Master's Thesis, University of Ghana, Accra, Ghana, 2011.
5. Tang, J. Risk Assessment of *Listeria monocytogenes* in Ready-to-Eat Meat from Plants to Consumption. Ph.D. Thesis, Virginia Polytechnic Institute and State University, Blacksburg, VA, USA, 2013.
6. Guzel, M. Quantitative Risk Assessment for *Listeria* monocytogenes on Fresh-Cut Lettuce and Fresh-Cut Cantaloupe. Master's Thesis, Texas A&M University, College Station, TX, USA, 2015.
7. FDA-FSIS. *Quantitative Assessment of Relative Risk to Public Health from Foodborne Listeria monocytogenes among Selected Categories of Ready-to-Eat Foods;* Center for Food and Safety and Applied Nutrition, Food and Drug Administration, U.S. Department of Health and Human Services, and Food Safety and Inspection Service, U.S. Department of Agriculture: College Park, MD, USA, 2003; pp. 1–541.
8. FAO/WHO. *Risk Assessment of Listeria monocytogenes in Ready-to-Eat Foods: Technical Report;* World Health Organization and Food and Agriculture Organization of the United Nations: Geneve, Switzerland, 2004; pp. 1–269.

9. FDA-Health Canada. *Joint FDA/Health Canada Quantitative Assessment of the Risk of Listeriosis from Soft-Ripened Cheese Consumption in the United States and Canada: Report*; Food Directorate/Direction des Aliments, Health Canada, Santé Canada and Center for Food Safety and Applied Nutrition, Food and Drug Administration (US Department of Health and Human Services): Ottawa, ON, Canada, 2015; pp. 1–177.
10. Pérez-Rodríguez, F.; Carrasco, E.; Bover-Cid, S.; Joffré, A.; Valero, A. Closing gaps for performing a risk assessment on *Listeria monocytogenes* in ready-to-eat (RTE) foods: Activity 2, a quantitative risk characterization on *L. monocytogenes* in RTE foods; starting from the retail stage. *EFSA Support. Publ.* **2017**, *14*, 1252E. [CrossRef]
11. EFSA BIOHAZ Panel. Scientific Opinion on the *Listeria monocytogenes* contamination of ready-to-eat foods and the risk for human health in the EU. *EFSA J.* **2018**, *16*, 5134. [CrossRef]
12. Tenenhaus-Aziza, F.; Daudin, J.-J.; Maffre, A.; Sanaa, M. Risk-based approach for microbiological food safety in the dairy industry: The case of *Listeria monocytogenes* in soft cheese made from pasteurized milk. *Risk Anal.* **2014**, *34*, 56–74. [CrossRef] [PubMed]
13. Bemrah, N.; Sanaa, M.; Cassin, M.H.; Griffiths, M.W.; Cerf, O. Quantitative risk assessment of human listeriosis from consumption of soft cheese made from raw milk. *Prev. Vet. Med.* **1998**, *37*, 129–145. [CrossRef] [PubMed]
14. Tiwari, U.; Cummins, E.; Valero, A.; Walsh, D.; Dalmasso, M.; Jordan, K.; Duffy, G. Farm to fork quantitative risk assessment of *Listeria monocytogenes* contamination in raw and pasteurized milk cheese in Ireland. *Risk Anal.* **2015**, *35*, 1140–1153. [CrossRef]
15. Condoleo, R.; Mezher, Z.; Marozzi, S.; Guzzon, A.; Fischetti, R.; Senese, M.; Sette, S.; Bucchini, L. Risk assessment of human listeriosis from semisoft cheeses made from raw sheep's milk in Lazio and Tuscany (Italy). *Risk Anal.* **2017**, *37*, 661–676. [CrossRef]
16. Sanaa, M.; Coroller, L.; Cerf, O. Risk assessment of listeriosis linked to the consumption of two soft cheeses made from raw milk: Camembert of Normandy and Brie of Meaux. *Risk Anal.* **2004**, *24*, 389–399. [CrossRef]
17. Campagnollo, F.; Gonzales-Barron, U.; Pilão Cadavez, V.; Sant'Ana, A.; Schaffner, D. Quantitative risk assessment of *Listeria monocytogenes* in traditional Minas cheeses: The cases of artisanal semi-hard and fresh soft cheeses. *Food Control* **2018**, *92*, 370–379. [CrossRef]
18. Koutsoumanis, K.; Pavlis, A.; Nychas, G.-J.E.; Xanthiakos, K. Probabilistic model for *Listeria monocytogenes* growth during distribution, retail storage, and domestic storage of pasteurized milk. *Appl. Environ. Microbiol.* **2010**, *76*, 2181–2191. [CrossRef]
19. Latorre, A.A.; Pradhan, A.K.; Van Kessel, J.A.S.; Karns, J.S.; Boor, K.J.; Rice, D.H.; Mangione, K.J.; Gröhn, Y.T.; Schukken, Y.H. Quantitative risk assessment of listeriosis due to consumption of raw milk. *J. Food Prot.* **2011**, *74*, 1268–1281. [CrossRef] [PubMed]
20. Giacometti, F.; Bonilauri, P.; Albonetti, S.; Amatiste, S.; Arrigoni, N.; Bianchi, M.; Bertasi, B.; Bilei, S.; Bolzoni, G.; Cascone, G.; et al. Quantitative risk assessment of human salmonellosis and listeriosis related to the consumption of raw milk in Italy. *J Food Prot.* **2015**, *78*, 13–21. [CrossRef] [PubMed]
21. Yang, S.Y.; Yoon, K.S. Quantitative microbial risk assessment of *Listeria monocytogenes* and enterohemorrhagic Escherichia coli in yogurt. *Foods* **2022**, *11*, 971. [CrossRef] [PubMed]
22. Soto-Beltran, M.; Mena, K.; Gerba, C.; Tarwater, P.; Reynolds, K.; Chaidez, C. Risk assessment of *Listeria monocytogenes* in queso fresco in Culiacan, Mexico. *J. Microbiol. Res.* **2013**, *3*, 111–116.
23. Njage, P.M.K.; Leekitcharoenphon, P.; Hansen, L.T.; Hendriksen, R.S.; Faes, C.; Aerts, M.; Hald, T. Quantitative microbial risk assessment based on Whole Genome Sequencing Data: Case of *Listeria monocytogenes*. *Microorganisms* **2020**, *8*, 1772. [CrossRef]
24. Farber, J.M.; Ross, W.H.; Harwig, J. Health risk assessment of *Listeria monocytogenes* in Canada. *Int. J. Food Microbiol.* **1996**, *30*, 145–156. [CrossRef] [PubMed]
25. FAO/WHO. *Risk Characterization of Salmonella spp. in Eggs and Broiler Chicken and Listeria monocytogenes in Ready-to-Eat Foods*; Joint FAO/WHO Expert Consultation on Risk Assessment of Microbiological Hazards in Foods; FAO Headquarters: Rome, Italy, 2001; pp. 1–40.
26. Pouillot, R.; Hoelzer, K.; Chen, Y.; Dennis, S.B. *Listeria monocytogenes* dose response revisited—Incorporating adjustments for variability in strain virulence and host susceptibility. *Risk Anal.* **2015**, *35*, 90–108. [CrossRef]
27. Lakicevic, B.; Nastasijevic, I. *Listeria monocytogenes* in retail establishments: Contamination routes and control strategies. *Food Rev Int.* **2016**, *33*, 247–269. [CrossRef]
28. Beno, S.M.; Stasiewicz, M.J.; Andrus, A.D.; Ralyea, R.D.; Kent, D.J.; Martin, N.H.; Wiedmann, M.; Boor, K.J. Development and validation of pathogen environmental monitoring programs for small cheese processing facilities. *J. Food Prot.* **2016**, *79*, 2095–2106. [CrossRef]
29. D'amico, D.J. Microbiological quality and safety issues in cheesemaking. *Microbiol. Spectr.* **2014**, *2*, 251–309. [CrossRef] [PubMed]
30. Guillier, L.; Palma, F.; Fritsch, L. Taking account of genomics in quantitative microbial risk assessment: What methods? what issues? *Curr. Opin. Food Sci.* **2022**, *48*, 100922. [CrossRef]
31. Midelet-Bourdin, G.; Leleu, G.; Copin, S.; Roche, S.M.; Velge, P.; Malle, P. Modification of a virulence-associated phenotype after growth of *Listeria monocytogenes* on food. *J. Appl. Microbiol.* **2006**, *101*, 300–308. [CrossRef] [PubMed]
32. Pricope-Ciolacu, L.; Nicolau, A.I.; Wagner, M.; Rychli, K. The effect of milk components and storage conditions on the virulence of *Listeria monocytogenes* as determined by a Caco-2 cell assay. *Int. J. Food Microbiol.* **2013**, *166*, 59–64. [CrossRef] [PubMed]
33. Moller-Nielsen, E.M.; Björkman, J.T.; Kiil, K.; Grant, K.; Dallman, T.; Painset, A.; Amar, C.; Roussel, S.; Guillier, L.; Benjamin, F.; et al. Closing gaps for performing a risk assessment on *Listeria monocytogenes* in ready-to-eat (RTE) foods: Activity 3, the comparison of isolates from different compartments along the food chain, and from humans using whole genome sequencing (WGS) analysis. *EFSA Support. Publ.* **2017**, *14*, 1151E.

34. Painset, A.; Björkman, J.T.; Kiil, K.; Guillier, L.; Mariet, J.F.; Félix, B.; Amar, C.; Rotariu, O.; Roussel, S.; Perez-Reche, F.; et al. LiSEQ–whole-genome sequencing of a cross-sectional survey of *Listeria monocytogenes* in ready-to-eat foods and human clinical cases in Europe. *Microb. Genom.* **2019**, *5*, e000257. [CrossRef]
35. Fritsch, L.; Guillier, L.; Augustin, J.C. Next generation quantitative microbiological risk assessment: Refinement of the cold smoked salmon-related listeriosis risk model by integrating genomic data. *Microb. Risk Anal.* **2018**, *10*, 20–27. [CrossRef]
36. Plaza-Rodríguez, C.; Haberbeck, L.U.; Desvignes, V.; Dalgaard, P.; Sanaa, M.; Nauta, M.; Filter, M.; Guillier, L. Towards transparent and consistent exchange of knowledge for improved microbiological food safety. *Curr. Opin. Food Sci.* **2018**, *19*, 129–137. [CrossRef]
37. Filter, M.; Nauta, M.; Pires, S.M.; Guillier, L.; Buschhardt, T. Towards efficient use of data, models and tools in food microbiology. *Curr. Opin. Food Sci.* **2022**, *46*, 100834. [CrossRef]
38. Rougier, N.P.; Hinsen, K.; Alexandre, F.; Arildsen, T.; Barba, L.A.; Benureau, F.C.; Brown, C.T.; de Buyl, P.; Caglayan, O.; Davison, A.P.; et al. Sustainable computational science: The ReScience initiative. *PeerJ Comput. Sci.* **2017**, *3*, e142. [CrossRef]
39. Perkel, J.M. Challenge to scientists: Does your ten-year-old code still run? *Nature* **2020**, *584*, 656–659. [CrossRef] [PubMed]

MDPI AG
Grosspeteranlage 5
4052 Basel
Switzerland
Tel.: +41 61 683 77 34

Foods Editorial Office
E-mail: foods@mdpi.com
www.mdpi.com/journal/foods

www.ingramcontent.com/pod-product-compliance
Lightning Source LLC
Chambersburg PA
CBHW061449210926
51850CB00008B/1657